The Thermal Environment

Yutaka Tochihara
Editor

The Thermal Environment

From Viewpoints of Physiological Anthropology and Environmental Ergonomics

Springer

Editor
Yutaka Tochihara
Department of Human Life Design
and Science, Faculty of Design
Kyushu University
Fukuoka, Japan

ISBN 978-981-96-3008-0 ISBN 978-981-96-3009-7 (eBook)
https://doi.org/10.1007/978-981-96-3009-7

This Springer imprint is published by the registered company Springer Nature Singapore Pte Ltd.
The registered company address is: 152 Beach Road, #21-01/04 Gateway East, Singapore 189721, Singapore

Preface

Humans originated in Africa approximately seven million years ago and began walking on two legs in the jungle; approximately four million years ago, they expanded into the hot, dry savannahs and achieved stable bipedalism. At approximately two million years ago, *Homo erectus* emerged, ate meat, enlarged their brains, and began making tools. During these periods, they adapted to the hot, dry African environment by shedding their fur and increasing their eccrine glands to prevent hyperthermia during prolonged hunting. Approximately, 60,000 years ago, *Homo sapiens* dispersed from Africa and gradually expanded its habitant to areas with colder climates through biological (physiological and morphological) and cultural (clothing and housing) adaptations. They reached the southern tip of South America approximately 10,000 years ago. Humans have overcome harsh natural environments with the power of science and technology, gradually expanding their range of survival and creating more comfortable artificial environments. Cooling and heating systems are now widely used in many countries around the world. However, artificial thermal environments can have various adverse effects on health. In addition, recent global warming is predicted to make the Earth's thermal environment more severe. It has been pointed out that the current advanced artificial thermal environment may hinder the development of human potential and lead to the accumulation of unfavorable qualities. There may be a risk of such adverse effects on the adaptive capacity of humans to the environment in the future. Human thermal adaptation problems have been one of the main research topics in physiological anthropology.

Among the physical environments, the thermal environment is one of the most influential in terms of human comfort, performance, and health. The lowest temperature reported on Earth to date is −89.2 °C at Vostok Station, Antarctica, and the highest is 56.7 °C in Death Valley, the United States. The difference between these two extremes is 145 °C. Naked humans can only live comfortably in a temperature range of 10 to 35 °C, but with the invention of clothing and air conditioning, we are now living in a temperature range of over 100 °C. In addition to these natural thermal environments, humans are exposed to many artificial thermal environments. In low-temperature environments, such as cold stores, the average temperature can be as low as the −20 °C range, and there are examples of ultra-low temperature industrial environments, such as cold stores used to store tuna, where temperatures can reach as low as −50 °C. Similarly, there are a number of examples of artificial high-temperature environments, such as pre-furnace work areas in steel mills; although the automation of industrial processes and the widespread use of air conditioning have significantly reduced the number of such workplaces in which human workers are present. However, there is still the case of firefighting and work involving hazardous materials (nuclear, toxic gases, viruses, etc.), where full protective clothing must be worn regardless of the ambient environmental conditions; thus, a new artificial environment of extremely hot and humid conditions is becoming more prevalent. Environmental ergonomists have studied human comfort, physiological response, and performance in thermal environments, and have developed many international standards, including standards under the International Organization for Standardization (ISO).

The thermal environment was often reviewed by physiologists, occupational hygienists, architects, environmental engineers, and meteorologists. This is the first book-length work in which the thermal environment is considered from the viewpoints of physiological anthropology and environmental ergonomics, based mainly on the authors' original findings. The concepts and history of physiological anthropology and environmental ergonomics are described in Chap. 1, including an overview of human evolution as it relates to the thermal environment. Chapter 2 discusses the physical evaluation of the thermal environment, particularly in relation to clothing. Chapter 3 describes the evaluation methods for cutaneous thermal sensitivity and thermal sensation, and discusses them from the perspective of thermal adaptation. Chapter 4 introduces the physiological basis of thermoregulation. In Chapters 5 and 6 review in detail the characteristics of physiological responses in cold and hot environments, and the adaptation to thermal environments from a view point of physiological anthropology. Furthermore, from an environmental ergonomic perspective, the performance in cold and hot environments is discussed. Chapter 7 describes the methods for evaluating thermal stress when wearing protective clothing (a new artificial environment). Chapter 8 describes the workload in artificial severely cold environments. Residential thermal environments for the elderly are evaluated in Chap. 9, in particular suitable environments for sleeping and bathing. Lastly, international standards for thermal environment evaluation are introduced in Chap. 10, using ISO, the American Society of Heating, Refrigerating and Air-Conditioning Engineers and the Ameican Conference of Governmental Industrial Hygienists standards as examples.

The authors of this book were post-doctoral fellows of the Japan Society for the Promotion of Science and/or Ph.D. students at the Tochihara Laboratory (Environmental Ergonomics Laboratory), Department of Human Science, Faculty of Design, Kyushu University. Many of the authors are currently working as associate professors or professors at leading universities worldwide. Some of the studies presented in this book were conducted at the "Research Center for Human Environmental Adaptation" at Kyushu University.

The research results presented in this book would not have been possible without the cooperation of many coworkers. We would like to thank Prof. Masahiko Sato, a leader in physiological anthropology research; and Prof.

Ingvar Holmér, Prof. Mike Tipton, and Prof. Nigel Taylor, thermal physiologists, and authorities on protective clothing research, who guided us in these research topics. In completing this book, we would like to express our gratitude to Prof. Satoshi Muraki, Prof. Ping Yeap Loh, Mutsuhiro Fujiwara, Shizuka Umezaki, Sachi Takahashi, and Tomoko Yoshimura of Kyushu University for their academic advice and technical supports. For the editorial work, we would like to thank Chihiro Haraguchi, Stephanie Sibin, and Saki Kasai of Springer Nature for their assistance.

Fukuoka, Japan
October 2024

Yutaka Tochihara

Contents

Notes on Contributors

Takako Fukazawa is a Professor at Department of Education, Kyoto University of Education, Japan. Her field of research is clothing hygiene and environmental ergonomics. She works on healthy, safe, and comfortable clothing systems, focusing in particular on the relationship between heat and water vapor transfer through clothing and human thermal physiology and thermal perception. She received her PhD in Clothing Environment from Bunka Women's University. She was a post-doctoral fellow (PD) of the Japan Society for the Promotion of Science at Kyushu University.

Nobuko Hashiguchi is a Professor at Department of Health Science, Kyushu University, Japan. She is currently researching the effects of thermal environment on health of elderly and nurses, specifically in the field of ergonomics and nursing management. She received her PhD from Kyushu Institute of Design. Previously, she was a post-doctoral fellow (RPD) of the Japan Society for the Promotion of Science at Kyushu University. She was the Director of the Nursing course, and the Head of the Research Center for Nursing Innovation and Implementation, Kyushu University.

Joo-Young Lee is a Professor of the Department of Fashion and Textiles, College of Human Ecology, Seoul National University, Korea. She received a PhD from the Department of Clothing and Textiles at Seoul National University in 2005 and conducted research at the University of Minnesota (USA, 2006–2008), Kyushu University (Japan, 2008–2011), and the University of Ottawa (Canada, 2018) as a research fellow. Her research focuses on body temperature regulation of workers in hot and cold environments and evaluating the effectiveness of personal protective clothing for workers in extreme environments. She organized the 20th International Conference on Environmental Ergonomics in 2024.

Hajime Oi is an engineer and researcher at Nissan Motor, Japan. His field of study is thermal comfort in vehicle cabins and thermal energy management in vehicles. He received his PhD from the Graduate School of Design, Kyushu University. He is an expert in the ISO/TC159/SC5/WG1 (Ergonomics/Ergonomics of the physical environment/thermal environment) and vice chairperson of the national mirror committee for ISO/TC159/SC5 in Japan. He contributed as project leader to the development of ISO 14505-4 (Evaluation of thermal environments in vehicles, Part 4: Determination of the equivalent temperature by means of a numerical manikin).

Su-Young Son is an Associate Professor of the Department of Home Economics Education, Kyungpook National University, Korea. She is currently researching the effects of thermal environment on the health of elderly and women, specifically in the field of clothing ergonomics and occupational safety. She received her PhD in Ergonomics and Physiological Anthropology from the Graduate School of Design at Kyushu University. She was selected as a post-doctoral fellow of the Japan Society for the Promotion of Science at Kyushu University.

Yutaka Tochihara is an Emeritus Professor at Kyushu University, Japan. His field of study is physiological anthropology and environmental ergonomics, with a particular interest in human comfort, performance, and health in thermal environments. He received his PhD in Medical Science from Showa University. He was head of the Environmental Physiology Section of the Institute of Public Health, Japan. He worked as a visiting researcher at the National Institute of Occupational Health, Sweden. He was a Distinguished Professor in the Department of Human Science, Faculty of Design, Kyushu University. He served as Vice President of Kyushu University. He organized the 10th International Conference on Environmental Ergonomics in 2002.

Hitoshi Wakabayashi is an Associate Professor at Faculty of Engineering, Hokkaido University, Japan. He is currently working on environmental ergonomics and thermal physiology, with a particular interest in human thermoregulatory response and physical performance in cold environments. He received his PhD in Health and Sport Sciences from University of Tsukuba. Previously, he was a post-doctoral fellow of the Japan Society for the Promotion of Science (JSPS) at Kyushu University, during which he conducted research abroad at the University of Portsmouth with support from the JSPS Excellent Young Researcher Overseas Visit Program.

Titis Wijayanto is an Associate Professor of Mechanical and Industrial Engineering at Universitas Gadjah Mada, Indonesia. He is currently working on the impact of climate change on human health, specifically in the areas of thermal environment, work physiology, and cognitive ergonomics. Titis received his PhD in Ergonomics and Physiological Anthropology from the Graduate School of Design at Kyushu University. Previously, he was a post-doctoral fellow of the Japan Society for the Promotion of Science at Kyushu University.

Humans in the Thermal Environment

1

Yutaka Tochihara

1.1 Human History and the Thermal Environment

1.1.1 Human Evolution

The Earth was born approximately 4.6 billion years ago, and signs of life were recognized on Earth approximately 450 million years ago (Mya), with the earliest vertebrates appearing in the sea and land 360 Mya. Reptiles appeared on land, where cycads and ferns flourished, and dinosaurs ruled the Earth for a long period starting from 230 Mya. However, at 66 Mya, a giant meteorite crashed into Earth, causing major climate change and extinction of the dinosaurs within a short time. During the age of dinosaurs, primates evolved 25 Mya from small mammals that lived nocturnally [1]. They choose to live in trees with few competitors. The ancestors of humans arose approximately 7 Mya from apes, such as chimpanzees and bonobos. Humans have evolved over 7 million years [2], including early hominins, *Australopiths, Homo erectus,* premodern humans, and modern humans *(Homo sapiens)*, and adapted to their natural thermal environment (Fig. 1.1). The thermal environment of the open, hot-dry habitats of Africa has had a profound impact on the evolution of physiological and morphological aspects of humans [3].

Y. Tochihara (✉)
Department of Human Life Design and Science, Faculty of Design, Kyushu University, Fukuoka, Japan
e-mail: tochihara@kyudai.jp

1.1.1.1 Early Hominins

Recent molecular genetic studies (mitochondrial DNA analysis) have revealed that the earliest human ancestor, the ape-man, diverged from chimpanzees as early as 7 Mya [4]. Fujiyama et al. [5] reported that the difference in the genome between the two was only 1.23%. Brunet et al. [6] discovered the oldest human fossil (*Sahelanthropus tchadensis*) in Chad, Central Africa, and suggested that the fossils date between 6 and 7.2 Mya. The cranial capacity was approximately 350 cc, which is comparable to that of chimpanzees. The next species was *Ardipithecus ramidus*, discovered in Ethiopia and thought to be approximately 4.4 million years old [7]. They were as small as chimpanzees, lived in African forests and ate mainly fruit. They began to walk on both legs, and it is believed that bipedalism provided them the advantage of freeing their hands to carry food and objects.

1.1.1.2 Australopiths

The 3.2-million-year-old hominid fossil discovered in 1974 in the Afar Valley of East Africa contained a whole-body skeleton. This *Australopithecus afarensis*, named "Lucy," had a skeleton that was distinctly different from other

Y. Tochihara (ed.), *The Thermal Environment*, https://doi.org/10.1007/978-981-96-3009-7_1

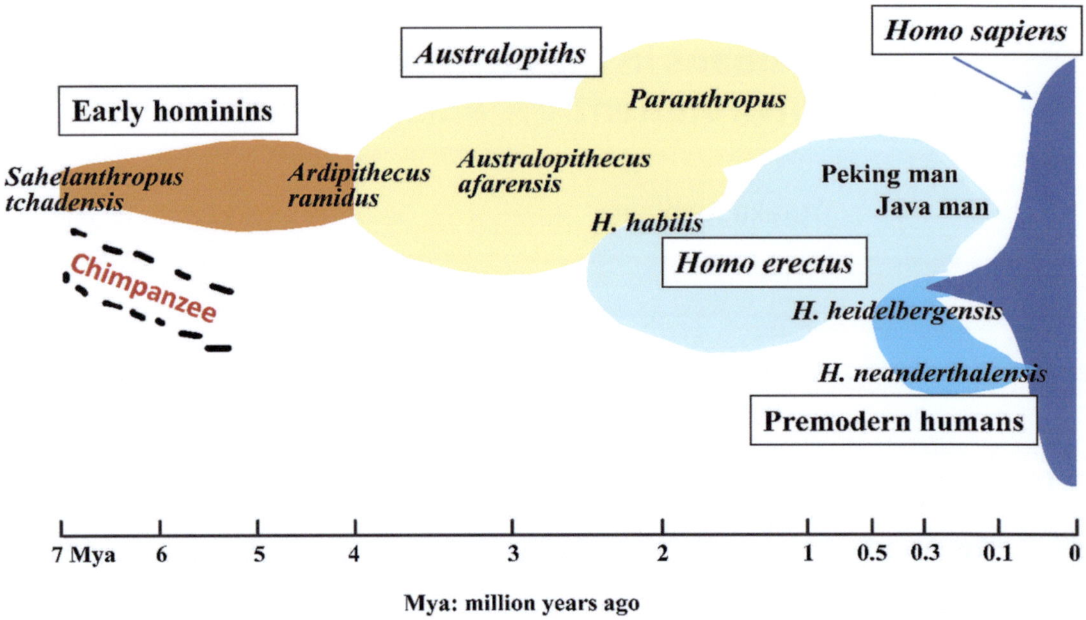

Fig. 1.1 Human genealogy. The ages on the horizontal axis are not evenly spaced and are approximate values. (Partially modified from Kono [2])

primates [8]. The broad pelvis, legs longer than the arms, and straight down point of the foramen magnum suggest that she was an upright biped. However, the cranial capacity was approximately 450 cc, which is considerably smaller than the 1450 cc of modern humans.

Although now denied, there was an intriguing hypothesis, named "East Side Story," regarding the origin of humans. Coppens [9] speculated that the background to the emergence of these *Australopiths* was a major crustal movement in East Africa (Great African Rift Valley) that occurred 10 Mya. Tropical rainforests have become savannahs of grasses and shrubs, with reduced rainfall and dry seasons in East Africa, due to the 3000-m mountain ranges created by the formation of the Great Rift Valley. Savannahs have one-fifth the number of edible plants compared to tropical rainforests. Primates grew to feed on the abundant fruits and leaves of rainforests and have faced great challenges. There were already diurnal and social groups of primates, but the group that left the rainforest was the ancestors of humans, the *Australopiths*, who walked upright on two legs.

There are many theories as to why humans evolved "upright bipedalism," but one interesting theory relates to the thermal environment. Wheeler [10, 11] speculated that humans began walking on two legs, a posture that receives less solar radiation than quadrupeds, to survive during activities in the scorching grasslands. When the sun is overhead, an upright hominid occupied only approximately 40% of the quadruped surface area. The thermoregulatory advantages of bipedalism in primates in the African savannah may have been a significant factor in the survival of hominins. Although the main advantage is a dramatic reduction in direct exposure to the sun, there are additional benefits from a higher distribution of body surfaces. Wind speeds are higher, and air temperatures are lower away from the ground, both of which increase the rate at which a biped dissipates heat via convection.

Loss of fur and an increase in eccrine sweat glands are thought to have occurred during this period, although some have placed it in *Homo erectus* [12, 13]. *Australopithecus* remained in Africa for a long time and went extinct 1 Mya.

1.1.1.3 Homo Erectus

In East Africa, savannahs have expanded as the climate has become drier. Approximately

2.5 Mya, *Homo habilis* (meaning man with good hands) also appeared in East Africa. They made simple stone tools, ate meat, used simple language, and gradually increased the cranial capacity (approximately 650 cc). It is believed that the human ancestors first walked on two legs and then used their free hands to make tools, stimulating the development of the cerebrum. Approximately 1.9 Mya, a tall primitive man called *Homo erectus* (meaning upright man) appeared. In 1984, an almost complete whole-body fossil of *Homo erectus*, named "Turkana Boy," was discovered. The limbs of this fossil are as long as those of modern humans, and the cranial capacity was approximately 900 cc.

Homo erectus was able to walk and run long distances to hunt large mammals [14]. Humans have many physical adaptations in our muscles and bones for long-distance running, traces of which first appear in *Homo erectus* fossils. They have short toes, a full arch of the foot, a long Achilles tendon, a narrower waist, and a large gluteus maximus [15]. These physical characteristics allow the energy cost of humans during walking to be a quarter of that of chimpanzees [16]. Similar to modern humans, they have evolved efficient locomotion, loss of fur, and increased eccrine glands. Consequently, they were capable of endurance running in the hot-dry savannahs of Africa [12].

Homo erectus made fine stone tools, used fire for the first time, and roasted and ate the meat. However, it is said that they did not start the fire themselves, but instead used a wildfire. Meat consumption is essential for brain enlargement, as the brain consumes 20–30% of energy, although it represents only 2% of the body weight [17]. The fossil of a primitive human (with a cranial capacity of approximately 900 cc) discovered in Kenya is called *Homo ergaster* (meaning a working human). Approximately 1.8 Mya, hominids first came out of Africa and then moved into Eurasia via the Middle East. This was the first African exodus (Fig. 1.2). By 1.5 Mya, they had migrated to Southeast Asia, where in 1891, Eugène Dubois discovered a 1.0-million-year-old fossil called the Java Man (*Homo erectus erectus;* cranial capacity of approximately 1000 cc). *Homo erectus* appeared in Europe 0.8 Mya and in East Asia 0.6 Mya. The winter around Zhoukoudian in Beijing (0.75 Mya), where the Peking Man (*Homo erectus pekinensis*; cranial capacity of approximately 1100 cc) was discovered in 1929, seems to have been cold even then, and traces of heating by fire have been found.

1.1.1.4 Premodern Humans

Approximately 0.5 Mya, *Homo heidelbergensis* appeared in Africa and Europe, and around 0.3 Mya in Asia. *Homo neanderthalensis* evolved from them at 0.4 Mya and lived in Europe and the Middle East. They had a cranial capacity of 1500 cc, equal to or greater than that of modern humans. However, development of the frontal lobe appears to be inferior to that in modern humans. During this period, Europe was subjected to an ice age of approximately every 100,000 years. *Homo neanderthalensis* therefore lived in a tundra-like climate but tolerated the cold well. They had no sewing skills; therefore, the only clothing they wore was fur. Their stockiness and muscular build were the result of adaptation to the cold. Because *Homo heidelbergensis* was born in Africa, it is thought that the skin color was dark to avoid intense tropical sunlight (ultraviolet rays). However, when they moved to Europe, they received less sunlight and could not synthesize vitamin D, which is necessary for calcium metabolism, if their skin remained dark. To prevent the development of rickets due to the lack of ultraviolet light, the melanin pigment in the skin was reduced, resulting in white skin similar to that of today's Europeans. They also bleached their hair and eyes, giving them blonde hair and blue eyes (Gloger's rule). *Homo neanderthalensis* also possesses these characteristics. Premodern humans are thought to have survived until 30,000 years ago.

1.1.1.5 Modern Humans (*Homo Sapiens*)

Genetic and fossil anthropological research suggests that modern humans (*Homo sapiens*: meaning intelligent man) appeared in Africa approximately 0.3 Mya. Cann et al. [18] measured mtDNA of 147 individuals from five geographical populations. They reported that all these mtDNAs originated from a woman who was

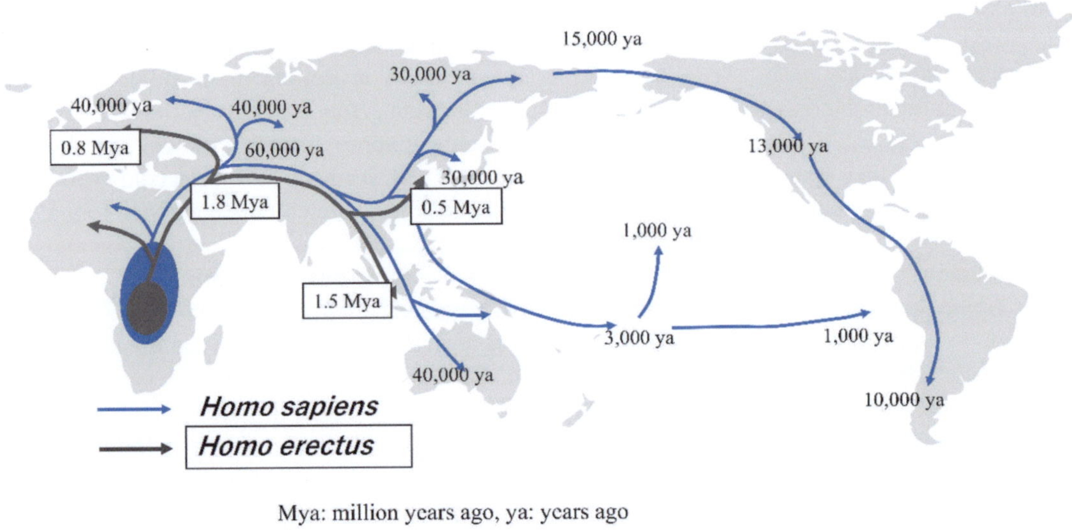

Fig. 1.2 History of human dispersal across the earth

postulated to have lived approximately 0.2 Mya, probably in Africa. Hublin et al. [19] identified a 315,000-year-old fossil hominin from Morocco. *Homo sapiens* experienced its second African exodus 60,000 years ago (Fig. 1.2). The Cro-Magnon man, a type of *Homo sapiens*, migrated to Western Europe 40,000 years ago and became the ancestor of modern Europeans. They used sophisticated stone tools, sewed their clothes with needles, and were able to fight colds. Since *Homo neanderthalensis* was thought to have died approximately 30,000 years ago, they have coexisted in Europe for almost 10,000 years.

Homo sapiens migrated to East Asia 45,000 years ago and advanced to Siberia 30,000 years ago. They had thicker subcutaneous fat and shorter limbs to better withstand cold conditions. The face was lowered along with the nose to prevent frostbite. A typical example is the cold-adapted Mongoloid, the Inuit. In addition, 15,000 years ago they crossed the Bering Sea, which was then frozen and connected to land, and moved into the Americas 13,000 years ago, reaching the southern tip of South America 10,000 years ago (Fig. 1.2). In addition, agriculture began in the Middle East approximately 11,000 years ago, and humans have developed civilization in the last 5000 years. *Homo sapiens* live on land worldwide because of their biological (physiological and morphological) and cultural adaptations (such as clothing, housing, and air-conditioning). It is highly unusual for a single species to be distributed in this way across all climates (temperatures and altitudes) of the Earth.

1.1.2 Human Morphological Thermal Adaptation

A morphological feature that, at first glance, distinguishes humans from other primates is bipedal locomotion. Bipedal locomotion began almost 7 Mya. In addition, ancient humans showed important morphological transitions in their skin as they migrated from the hot-humid jungles of Africa to the hot-dry savannahs, and in their body shape as they left Africa for colder regions.

1.1.2.1 Naked Skin and Increase in Eccrine Sweat Glands

Desmond Morris [20] states at the beginning of the introduction to his book "The naked ape," "*There are hundred and ninety-three living species of monkeys and ape. One hundred and ninety-two of them are covered with hair. The exception is a naked ape self-named Homo sapiens.*" Humans differ from other primates in many ways, but perhaps no derived human characteristic is more striking than the naked skin. Several hypotheses have been proposed to explain naked skin, including the alleviation of external parasite infestation, sexual selection, and even an aquatic lifestyle [20]. As the hunter ape abandoned its nomadic past and settled on fixed home bases, its burrows became heavily infested with skin parasites. The use of the same sleeping place night after night is thought to have provided an unusually rich breeding ground for a variety of ticks, mites, and bugs to the point where the situation presents a serious risk of disease. By shedding their hairy coats, they were better able to address this problem. Second, the human ancestors shed their fur on sexual attractiveness. It can be argued that by exposing their naked skin to each other during sexual encounters, both males and females become more sensitive to erotic stimuli. Another theory, that is not widely accepted today, is that before becoming a hunting ape, the primordial ape that left the forest went through a phase of becoming an aquatic ape. During this process, ancestors lost their hair, similar to other mammals that returned to the sea. Only the head, protruding from the surface of the water, retained a hairy coat to protect it from direct sunlight.

Probably the most widely accepted explanation by anthropologists is that human ancestors shed their fur to dissipate excess heat generated by hunting prey in hot-dry savannahs. Bipedalism is a critical adaptation in response to humans moving from a hot-humid jungle to a hot-dry savannah [15, 21]. In other words, long-distance running with both legs is closely associated with human epilation. To avoid falling prey on the predators, humans were forced to be active during the day when predators were resting. Moreover, lacking sharp claws and teeth, humans must chase their prey for long periods. While shedding fur increases the absorption of radiant heat from the sun during the day and increases the risk of injury, the release of heat from the body when running in pursuit of prey is more critical. Many mammalian quadrupeds can run faster than humans for short periods, however, as discussed below, their only means of heat dissipation is through panting, so they become hyperthermic on the run, allowing humans to acquire them. This method of prey acquisition is only possible in hot, dry, and open environments. Prey can also be found from a distance and their tracks are easier to locate [12].

In addition to losing the heavy coat of hair, humans increased the number of sweat glands over the entire body surface, providing considerable cooling—by producing a generous film of evaporating liquid over the air-exposed limbs and trunk [20]. Naked skin and a massive increase in the number of sweat glands are the two most important modifications to avoid hyperthermia during physical activity (hunting) in the savannah.

There are two types of sweat glands in the human skin, apocrine and eccrine (Fig. 1.3). The apocrine glands consist of a large, spongy secretory duct located deep in the dermis, combined with a long, more tubular excretory duct leading to a hair follicle, often in conjunction with a sebaceous gland. The apocrine glands are controlled by sympathetic adrenergic nerves and secrete viscous fluids containing lipids, proteins, and steroids in conjunction with the sebaceous glands. Secretion is mostly associated with the release of odor, probably by adapting the apocrine glands to functions related to territorial marking or as part of an alarm response. The axillae, nipples, pubic region, and ear canal are the only locations in the human body in which apocrine glands are distributed. In contrast, eccrine glands are smaller than apocrine glands, found only in the outer part of the dermis, and consist of a coiled secretory tube and a relatively straight excretory duct that extends to the surface of the skin. Eccrine glands are distributed over most of the surface of human body. The excretory duct of eccrine glands is formed by a double layer of epithelial cells capable of absorbing some of the components of sweat, particularly salt, so that eccrine sweat is on average approximately 99% water. Unlike apocrine glands, eccrine glands are not directly associated with hair follicles and are innervated by sympathetic cholinergic nerves [3]. Best and Kamilar [22] showed that increased eccrine gland glycogen content and skin capillarization led to an increase in sweating capacity in humans in hot-dry, open environments.

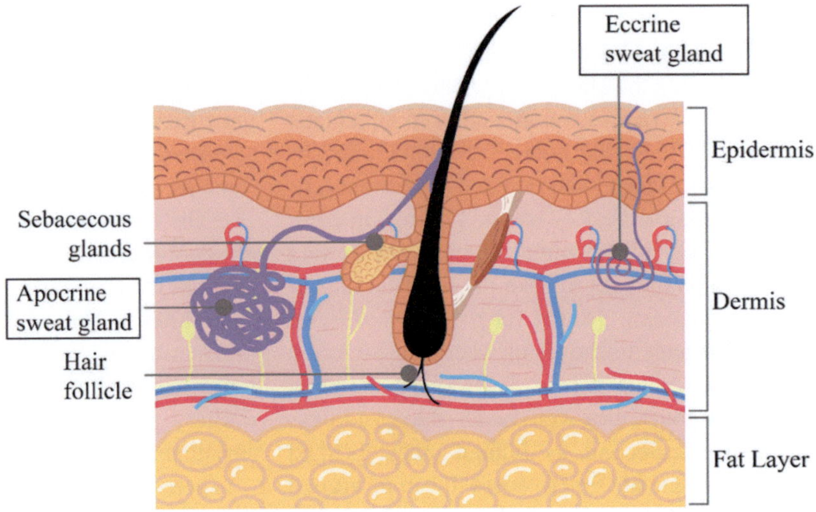

Fig. 1.3 Anatomical diagram of human apocrine and eccrine glands

In the human skin, the number of eccrine glands, which play an important role in thermoregulation, has increased significantly. Humans, like most mammals, have both apocrine and eccrine glands, but differ in the relative distribution and number of these gland types. The ratio of eccrine to apocrine glands varies among catarrhines: chimpanzees and gorillas have an eccrine/apocrine ratio of approximately 2:1; in macaques, mangabeys, baboons, and gibbons, the ratio is approximately 1:1 [23]; and in humans, almost all cutaneous glands on the general body surface are eccrine, numbering approximately 2–4 million [24].

Well-trained adult male humans can sweat up to 3 L/h or 11 L/day from their sweat glands [25], which is more than four times higher than that of chimpanzees [23]. If all 1000 cc of sweat evaporated, 585 kcal of heat would be lost from the body through evaporation. This is enough cooling power to lower the deep body temperature of a 70 kg person by approximately 10 °C, if heat production is ignored. Horses, like humans, can sweat profusely, but sweat from the apocrine glands is not as effective as humans in thermoregulation in hot environments. Therefore, trained humans can run for 2–3 h at 30 °C, but horses cannot. Mammals, except humans and horses, regulate their body temperature in hot environments by "panting." Panting is shallow, rapid breathing that significantly increases water evaporation due to increased ventilation and increases heat loss. Evaporation of water from the upper respiratory tract occurs mainly in the oral cavity, oropharynx, and upper part of the trachea. Because the respiratory epithelium of the pharynx is highly vascularized, evapotranspiration during panting efficiently cools the blood, thereby reducing the rise in core body temperature.

However, when humans perform such rapid breathing, the concentration of CO_2 in the blood decreases, which can lead to alkalosis and the risk of syncope. Using saliva is another technique used by many animals, particularly rodents and marsupials.

Humans are not hairless. This term, in relation to humans, implies the absence of a dense layer of thick fur. However, the number and density of hair follicles are similar to those of our closest primate relatives, and the uniqueness of human skin is that human body hair is miniaturized, very fine, and short, and the skin appears to be naked. In addition, human skin contains a special distribution of eccrine glands which, together with the reduction in body hair, creates a very specific thermoregulatory mechanism that allows us to cope very well with high temperatures, even in situations of strenuous physical activity [26]. Recently, Kamberov et al. [27] measured the density of hair follicles and eccrine sweat glands in several regions of the skin in three primate spe-

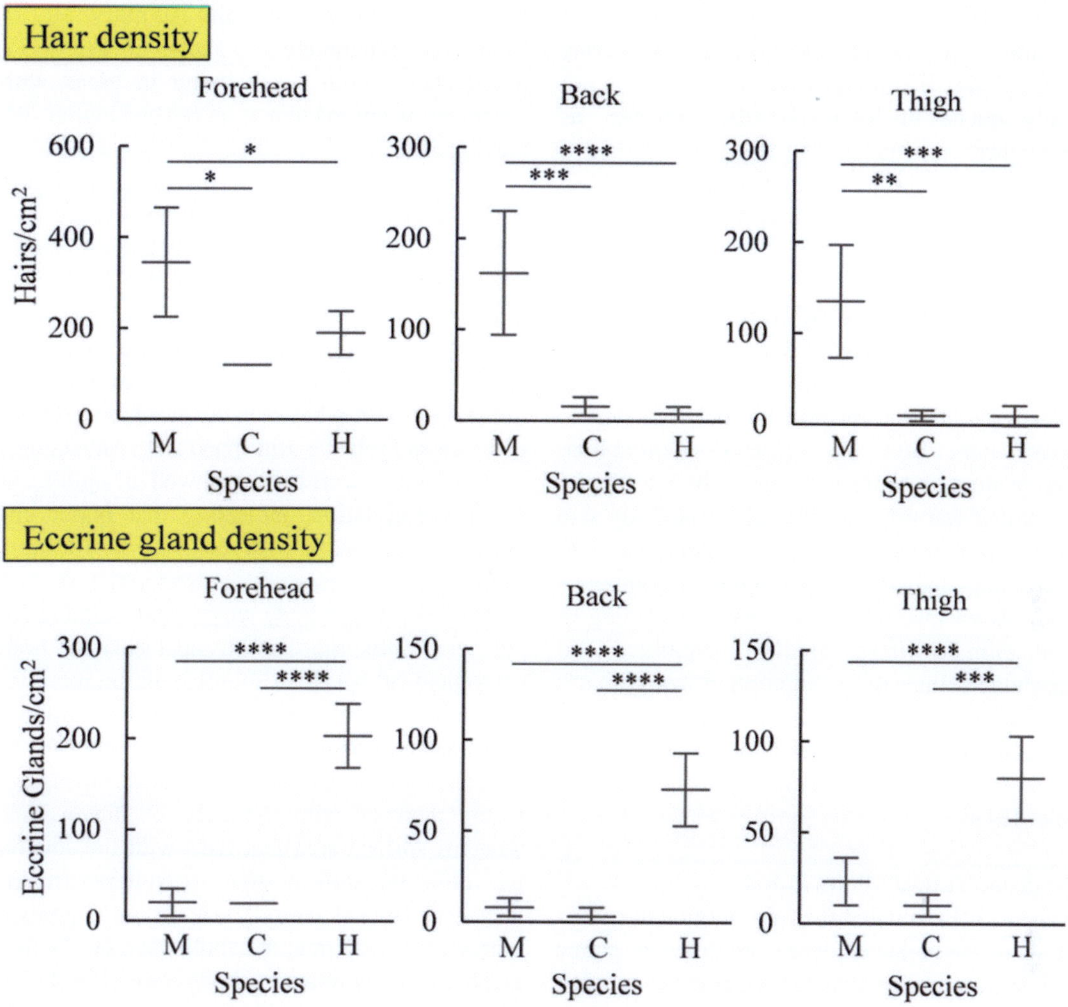

Fig. 1.4 Hair density (upper) and eccrine gland density (lower) in different body regions of three primates. The mean and standard deviation are shown for each species. M: Macaques, C: Chimpanzees, H: Humans. The standard deviation is not shown where the sample size was less than two individuals. (Partially modified from Kamberov et al. [27])

cies: macaques, chimpanzees, and humans (Fig. 1.4). Although human hair cover is greatly attenuated compared to that of close relatives, they found that humans have a chimpanzee-like hair density that is significantly lower than that of macaques. In contrast, the eccrine gland density in humans is on average 10 times higher than in chimpanzees and macaques, whose densities are strikingly similar. Their findings suggested that a decrease in hair density in the ancestors of humans and apes was followed by an increase in eccrine gland density and a reduction in fur cover in humans.

The loss of most of our body hair and the gain in the ability to dissipate excess body heat through eccrine sweating result in a dramatic increase in the size of our most temperature-sensitive organ, the brain. Humans first evolved through the initiation of bipedalism, loss of fur, and an increase in eccrine glands, followed by the development of the brain, in that order. The process of human evolution is closely linked to the thermoregulatory transition of our ancestors, who survived in the hot and dry African savannah. Thermoregulation may have been important in the evolutionary development of both hair loss and bipedality in humans [21]. Establishing when hominins lose their hair and increase their eccrine sweat glands is difficult because the skin is not easily traced in fossil records. However, as researchers have elucidated the genetic basis of the loss of fur and increase in sweat glands [28], they will be able to estimate when these DNA changes occur.

1.1.2.2 Bergmann's, Allen's, and Thomson's Rules

There are three well-known classical ecological rules regarding the relationship between human body shape and the thermal environment. Bergmann [29] observed that within a single, polytypic, warm-blooded species, the body size of different subspecies usually increased with decreasing habitat temperatures (Bergmann's rule). Allen [30] added that mammals living in cold climates have smaller extremities and appendages than those living in warm climates (Allen's rule). The human nose becomes relatively narrower as the mean annual temperature decreases; and this is known as the "Thomson's rule" [31].

Bergmann's and Allen's Rules

Robert [32] evaluated the first Bergmann rule in humans. He collected data from different ethnic groups with mean annual temperatures ranging from 10 to 80 °F and adult male body mass ranging from 40 to 77 kg. He found a negative relationship between body weight and mean annual temperature with a correlation coefficient $r = -0.600$ ($P < 0.001$). He showed that for every 1 °F decrease in annual temperature, the average weight of a human male increases by 0.305 kg. Newman and Munro [33] studied the relationship between the climatic environment and body size in 15,000 young White American males. The analyses showed a relationship that followed Bergmann's rule of greater weight and smaller body surface area (Ad) per unit body weight (mass) in colder climates [34]. They also showed that January temperatures correlated more closely with body size than heat stress data or the mean annual temperature. Sato [35] showed significant relationships between the mean annual temperature and Ad/mass ratio in Mongolian inhabitants, as shown in Fig. 1.5a. Residents of both sexes living in areas with higher mean annual temperatures had higher Ad/mass ratios.

As the length of the limbs increased in relation to the length of the trunk, the body surface area increased, indicating that the body had adapted to the high temperature environment. Allen's rule states that populations living in colder regions have shorter sitting-height to stature ratio than those living in warmer regions. An examination of the relationship between the mean annual temperature and relative arm span (arm span/stature) in 173 males worldwide showed a significant positive correlation, indicating that hotter climates are associated with relatively longer arms [36].

Why is a larger body shape more suitable for cold environments? The physical basis of both ecological rules is the difference in the relationship between body surface area (cm^2) and volume (cm^3), proportional to mass (kg) for organisms of different sizes. Because volumetric measurements increase with the cube of dimensions, whereas surface area increases with the square, the ratio of surface area to volume (mass) decreases as the overall size of an organism increases. In addition, metabolic heat production is most strongly related to body mass (Fig. 1.5b). Therefore, larger bodies are better suited to colder environments because they produce more heat and have a relatively smaller body surface area through which heat is dissipated [37].

Leonard and Katzmarzyk [37] compared their data with those of Robert's [32]. They also found a significant relationship between the mean annual temperature and Ad/mass, but the correlation coefficient decreased ($r = 0.59$ to $r = 0.29$), and the slope of the regression equation became gentle. The authors attributed this change to an improved global diet over the past 50 years. More recently, Bogin et al. [38] re-evaluated "Bergmann's rule." They suggested that Robert [32] actually described the effect of secular trends for increasing height in rich northern nations versus the stagnation or decline in the height of people in lower-income, colonial-era southern

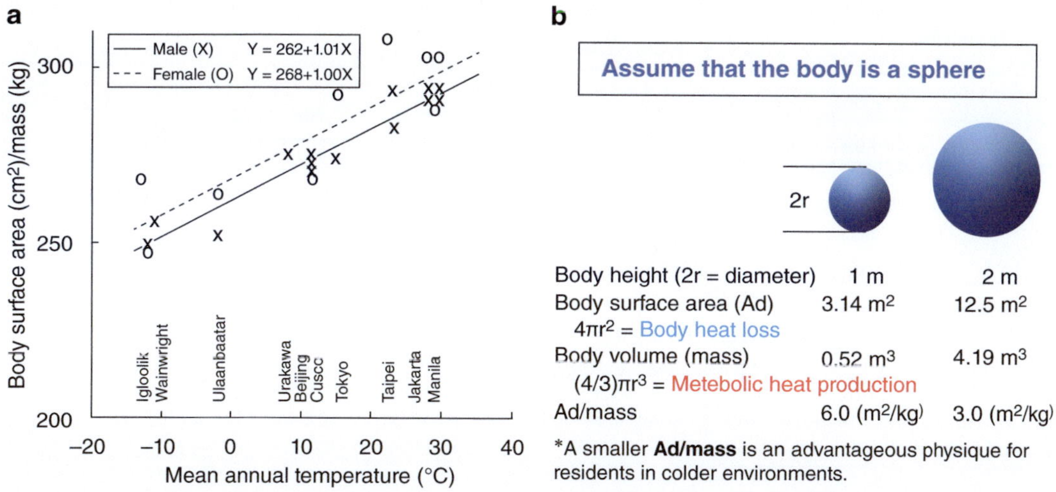

Fig. 1.5 (**a**) Relationship between the Ad/mass ratio of mongoloids and the annual mean temperature of their habitat. (Adapted from Sato [35]) (**b**) Why is a larger body shape more suitable for cold environment?

nations in the 1950s. They concluded that body shape is not regulated by climate, but rather by socio-economic-political factors and interactions between diet and infection. There is a considerable variability in human body proportions, and not all populations conform so easily to Bergmann's and Allen's rules.

Thomson's Rule

Perhaps due to its dramatically variable appearance, the human external nose has been the subject of several distribution studies. Thomson and Buxton [31] studied the relationships between the nasal index (nose breadth/nose height) of 15,000 living males from 150 different groups and climate. They found positive correlations among the nasal index, temperature, and relative humidity. They concluded that the platyrrhine nasal index is associated with a hot-humid climate and the leptorrhine nasal index with a cold-dry climate. Weiner [39] re-examined their data [31] and calculated the correlation coefficients between the nasal index and three thermal factors (temperature, relative humidity, and absolute humidity). The correlation coefficients were 0.63, 0.42, and 0.82, respectively, and the highest correlation coefficient was observed between the nasal index and absolute humidity. The nasal index is most closely related to the physical factors primarily involved in the exchange of water from the nasal epithelium to inhaled air. He reasoned that because the nose is partly a humidifying organ, and because humidity depends on temperature, absolute rather than relative humidity should be closely related to the nasal shape.

In addition, Maddux et al. [40] attempted to build on the earlier work of Weiner [39] by examining global distributions of temperature and humidity in more detail. The closest relationship between the nasal index and absolute humidity was again clearly confirmed, as shown in Fig. 1.6. Noback et al. [41], using computer analysis, measured the nasal cavities of 100 skulls from 10 groups living in five different climatic regions. They found that people from cold-dry regions such as Greenland and Siberia had higher and narrower nasal cavities than those from hot-humid regions such as Papua New Guinea and Gabon.

When cold-dry air is inhaled by a person, it passes through the high and narrow nasal cavity, where it is moistened by the nasal mucous membrane and warmed by the blood flowing through it. This prevents the cold air from entering the lungs directly. In contrast, people who live in hot-humid climates have wider noses (higher nasal indices), which allow more heat to escape from the body.

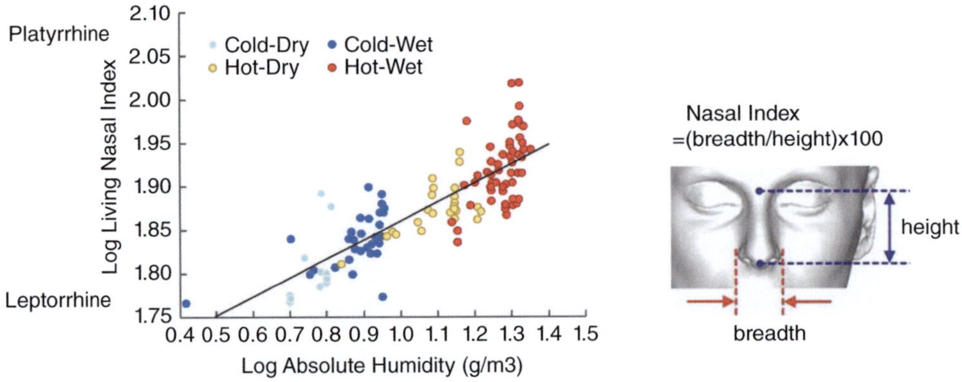

Fig. 1.6 Least squares regression of $\log_{10}$ nasal index on $\log_{10}$ mean annual absolute humidity for the 148 populations. (Adapted from Maddux et al. [40])

1.2 History of the Artificial Environment

Humans, who are believed to have emerged in the tropics of Africa approximately 7 Mya, and they gradually expanded their habitats into areas with cold climates through biological and cultural adaptation. Approximately 10,000 years ago, they had spread all over the world. Currently, humans have overcome harsh natural environments with the power of science and technology, are gradually expanding their range of survival, and are creating more comfortable artificial environments. The artificial environment created by highly developed science and technology provides a place where people can live comfortably by freely manipulating various factors of the living environment, such as thermal, light, air quality, and sound environments. Moreover, we are expanding our activity to oceans and outer space, places that humans could only dream of inhabiting 100 years ago [42].

Nevertheless, a significant problem is whether humans can physiologically adapt to such new artificial environments. In other words, the artificial environments in which modern humans spend most of their time may not be healthy. In fact, it goes against the biological characteristics of humans, and there is a risk that human health will be compromised [43]. Artificial environments created using technology often focus on the pursuit of convenience and an improved economy. They are not designed based on the physiological and psychological characteristics of humans. For example, bright lighting at night (artificial lighting) disturbs the biological rhythm (e.g., Eto and Higuchi [44]), and overcooling by the air-conditioning systems in the summer causes "cooling disorder," which will be discussed later. It cannot be denied that the artificial thermal environments have various adverse effects on health. Moreover, it has been pointed out that the current advanced artificial thermal environment may hinder the development of human potential and lead to the accumulation of unfavorable qualities. The thermal adaptation problems are one of the major research topics for physiological anthropologists [45–47].

Studies on adaptation to heat and cold have been conducted in Europe and the United States. Most often, these studies compared the physiological responses of men from European genetic backgrounds with those of men from other ethnically defined populations [48]. Many studies on the adaptation to heat and cold have also been conducted by Japanese physiologists [24, 49]. Internationally, the International Biological Program (IBP) was an advanced global research project on human-environmental adaptability. The IBP was planned in Europe in 1964 [50]. Physiologists, hygienists, geneticists, and anthropologists participated in the IBP under the theme "The biological basis of productivity and human welfare." One of the purposes of this program was to "examine the adaptability of human beings to various gradually changing environmental conditions." The research was based on "population growth," "food shortage," and "urbanization." It was predicted that as urbanization progressed, human physical strength and adaptability would change owing to changes in the living environment [51]. Currently (60 years since the IBP), artificial environments and global warming have become more serious problems. The use of air-conditioning has progressed rapidly. Most buildings are air-conditioned, and many houses have installed air-conditioning systems. Thus, people spend their whole year under comfortable thermal conditions. However, concerns about global warming have recently arisen. To save energy, temperature settings for cooling and heating must be increased and decreased, respectively, as much as possible. For example, the recommended temperature setting in Japan is 28 °C for cooling (Cool biz) and 20 °C for heating (Warm biz) [52]. "Cool biz" is a coined term combining "cool" and "biz" (the abbreviation of "business") in English, which signifies our campaigns to encourage people to dress more casually without a tie and a jacket so that they can work comfortably even in a room at 28 °C. "Warm biz" is the winter version of the aforementioned "Cool biz," which we have called for since 2005 with the aim of reducing the excessive use of heaters so that people can live comfortably at a room temperature of 20 °C. However, accurate physiological target values and their applicability remain unclear.

1.2.1 History of the Artificial Thermal Environment

Compared to natural environments, the history of artificial environments created by humans is extremely short. Much of the history is less than 140 years, beginning from when electricity became possible. In 1882, Thomas Edison began transmitting electricity from a thermal power plant he had built in New York City. He did this so that many people could use the incandescent light bulb that he invented in 1879. In 1887, Japan's first power plant was built, and electricity was sent to homes. There are many artificial environments, but here we briefly introduce the history of heating and cooling, mainly in Japan [43, 53].

1.2.1.1 History of Heating

Table 1.1 shows the history of the development of heating [43]. The heat sources for warming a room are roughly classified as "fire," "hot water," and "electricity." The first source of heat used by humans for heating was "fire" itself. Hominids began using wildfires for the first time 1.8 Mya. In the stratum of 800,000–600,000 years ago, burnt traces that are believed to be fireplaces have been discovered, and it is believed that primitive people who migrated to cold regions made fires to keep themselves warm. The fact that heating has existed since ancient Rome can be seen in the remains of a floor-heating system (hypocaust) in the famous baths of Emperor Caracalla. For the next 1000 years or more, wood-burning heating was mainstream, and it is believed that charcoal and coal were used in some areas. In the thirteenth century, coal began to be used at home in Europe, but its use was limited to the wealthy people. At the end of the thirteenth century, chimneys began to be used to increase heating efficiency [54]. For the next 500 years, people kept warm mainly by burning wood or coal in "fireplaces" and "stoves." Steam heating was first used around 1800, with the invention of the steam engine by James Watt in 1769. In 1836, large buildings in England and France began to adopt hot-air heating system, and in 1877, district heating began in 14 buildings in New York. Once electricity became available, the heating of buildings progressed rapidly, and from the 1950s onward, ducted central heating systems became widely used even in residential homes.

In Japan, in the famous book "*Essays in Idleness*" by Kenko Yoshida in the Kamakura period (1185–1333), the passage "*A house should be built with the summer in mind. In winter it is possible to live anywhere, but a badly built house is unbearable when it gets hot*," suggests that while the Japanese were quite concerned about how to cope with the heat of summer, they were less interested in heating houses. This is probably

Table 1.1 History of heating

Year	Events
1.8 Mya 0.7 Mya	Warming by wildfire Warming by bonfire
212	Heated floors (hypocaust) in baths of Caracalla in Rome Wood-burning heating
1200	Coal used in homes Use of chimneys begin
1300	Heating using "fireplace" or "stove"
1745	Development of steam heating
1769	Invention of the steam engine by James watt
1800	Development of steam heating systems in Spain Development of gas heating
1836	Development of warm air heating for large buildings in Europe
1877	World's first district heating for 14 buildings in NY
1877	Japan's first steam-heated building installed at a university
1879	Edison establishes the electric power supply company and popularizes steam heating in the USA
1929	Installation of radiant heating in classrooms in the Netherlands.
1930s	Popularization of electric heating systems
1950s	Ducted central heating begins in homes in the USA
1967	First district heating in Hokkaido, Japan
1970s	Oil heaters become popular in Japanese homes
1980s	Air-conditioned heating begins to spread in Japanese homes

Mya: million years ago

because Japan's winters are milder than those in Europe, and the winters are not cold enough to cause death. Therefore, the changes in heating methods in Japan are relatively simple. The first thing that appeared was the hearth. Hearths have been used since ancient times in Japan, not only for heating but also for cooking and lighting. Subsequently, wood- and charcoal-burning fires were used for heating for a long time; but in the Nara period (710–794), braziers began to be introduced into houses. The use of smokeless charcoal ensured a smokeless interior. Kotatsu (a traditional Japanese heating table covered with a blanket) first appeared during the Muromachi period (1334–1573) and became widespread during the Edo period (1600–1868). Subsequently, braziers and kotatsu continued to be used in many Japanese houses for a while after World War II. However, as the development of Hokkaido progressed during the Meiji period (1868–1912), conventional local heating became insufficient, and coal stoves such as those used in Europe and the United States were introduced. The first steam heating system in Japan was installed in a university building in 1877. In the 1960s, district heating systems were introduced, particularly in Hokkaido.

In Japanese houses, oil stoves were introduced after the war to heat entire rooms. The oil stove penetration rate, which was 15% in 1960, rapidly increased to 90% over 15 years. Since then, air-conditioning and heating, which do not pollute the air, have become popular. Subsequently, underfloor heating also spread to detached houses in Japan.

Worldwide, there are distinctive heating systems rooted in various regions, such as Russia's "pechka" that was developed from a fireplace, Korea's "ondol" floor heating system, and north-eastern China's "kan," all of which are famous.

1.2.1.2 History of Cooling

Table 1.2 shows the history of the development of air-conditioning [43]. It is very recent that air-conditioning has become possible. In the past, other than cooling rooms with ice or cold water, people had to rely on human-powered wind using folding fans for cooling. Subsequently, electric fans were introduced, but they could not cool the entire room. Full-scale air-conditioning had to

Table 1.2 History of cooling

Year	Events
1882	Automatic fans are commercialized in the USA
1902	First air-conditioning system installed in a printing plant in NY
1906	Carrier was granted a patent for the "apparatus for treating air"
1911	Carrier published a "psychrometric chart"
1914	Air-conditioning installed in commercial buildings in the USA
1920s	Air-conditioning installed in large buildings in the USA and Europe
1927	Japan's first fully air-conditioned building opens in Osaka
1928	Development of freon gas
1952	First air-conditioned oil tanker mess hall in Japan
1970s	Air-conditioned buildings begin to spread in Japan "Cooling disorder" becomes a social problem
1981	Development of inverter air conditioners
2000s	Household air conditioners become widely used

wait for Willis H. Carrier, known as the "father of air-conditioning" [55]. He installed the first air conditioner to regulate humidity in the New York Printing Plant in 1902. After obtaining his patent, he founded the Carrier Air Conditioner Company in 1908. In 1911, Carrier presented a "psychrometric chart" at the annual meeting of the American Society of Mechanical Engineers. In the 1920s, Europe and the United States introduced the use of many air conditioners in large buildings, such as department stores, theaters, and banks.

Japan's first cooling system, imported from Carrier, was installed in a spinning mill in 1907. In 1927, the first fully air-conditioned building was built in Osaka, Japan. After World War II, in 1952, air-conditioning was first installed as a workplace in the cafeteria of an oil tanker. After 1960, buildings with air-conditioning were rapidly built, and since then, people have benefited from "cool summers" due to air-conditioning. However, due to a lack of knowledge about the correct use of air-conditioning, many office workers in these buildings complained of "cooling disorders" in the 1970s. The most common complaints were coldness, fatigue, stiff hands and feet, sore throats, headaches, gastrointestinal disorders, susceptibility to cold, and skin irritation. Figure 1.7 shows the frequency of complaints of "cooling disorders" among workers in 256 offices in 1975. These syndromes are collectively known as "air-conditioning illnesses" and are thought to be caused by prolonged cooling of the body, as well as by the rapid and abrupt change in temperature from a hot environment outside to an air-conditioned room and from an air-conditioned room to a hot environment outside. Various studies have shown that those complaints are significantly more common in females than in males (Fig. 1.7). This is due to the following reasons [43]: (1) females tend to work indoors at a desk for a long time while males often work away from their desks; (2) females wear lighter clothes than males, at that time (1970s), women's miniskirts were in fashion; and (3) females are constitutionally more cold-sensitive than males. As the health problems caused by the indoor environment became a problem, the "Law for Maintenance of Sanitation in Buildings" was enacted ahead of the rest of the world in Japan in 1970. The purpose of this law is to specify matters necessary for sanitation for the maintenance and management of buildings used or utilized by numerous persons. Acceptable standards for indoor air quality are set for the thermal environment, carbon dioxide concentration, carbon monoxide concentration, amount of suspended particles, and concentration of formaldehyde. The thermal environmental conditions are currently as follows: 18 °C $\leqq$ temperature $\leqq$ 28 °C, 40% $\leqq$ relative humidity $\leqq$ 70%, airflow$\leqq$0.5 m/s.

Regarding room temperature, it has been shown that these "cooling disorders" are often a problem in offices where the average room temperature is less than 24 °C. For many mammals,

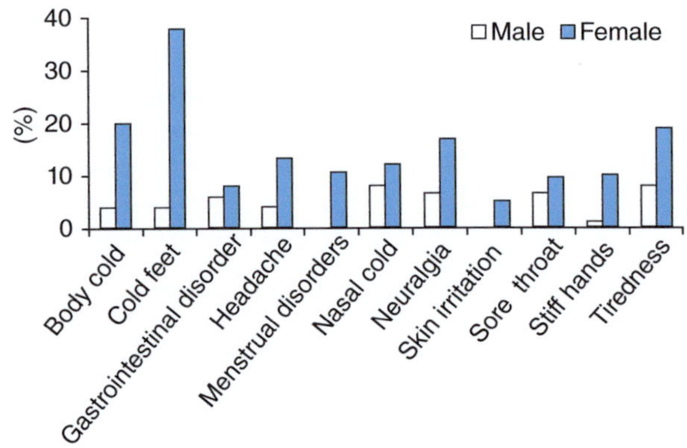

Fig. 1.7 Workers' complaints of "cooling disorder" in 256 offices and their gender differences

an environment with a temperature of 24 °C is not often considered a cold environment, but this is cold for humans when naked or lightly clothed. As a tropical species, humans have significantly lower cold tolerance than other mammals. Table 1.3 shows the lower critical temperatures (LCTs) for humans and mammals, which are often used as indices of cold tolerance. According to the Glossary of Terms for Thermal Physiology [56], LCT is defined as "the ambient temperature below which the rate of metabolic heat production of a resting thermoregulating tachymetabolic animal must be increased by shivering and/or non-shivering thermogenesis in order to maintain thermal balance." The LCT for humans is 22–27 °C [57–61], these values are considerably higher than those for other mammals [62]. The LCTs for humans are high and have a narrow range regardless of where they live, and as Scholander et al. [62] state in their paper, "Man is indeed a tropical animal carrying his tropical environment with him." In addition, Osada [63] found that the comfortable mean skin temperature, when naked and with various clothing conditions, is approximately 33 °C for both males and female in summer and winter. A mean skin temperature of 33 °C when naked is achieved at air temperatures of approximately 30 °C, which corresponds to the temperature of African savannah. This may also indicate that humans are animals of tropical origin.

The residential penetration rate of room coolers was only 5% in Japan in 1970, but it increased to 65% after 20 years, and the use of room coolers has spread rapidly among households. Today, one unit is installed in each room in every household.

1.2.2 History of Humans and the Artificial Environment

Table 1.4 describes the history of human evolution and the spread of artificial environments. In addition, for ease of understanding, the change over time from 7 Mya to the present is shown in terms of 1-year calendar. Humans were born on January 1, become *Homo erectus* in late September, Java men in mid-November, and *Homo sapiens* after December 15. Zero AD is equivalent to 21:30 on New Year's Eve (Fig. 1.8), and the Industrial Revolution began at approximately 23:40. The establishment of the power supply company by Edison in 1879 and the invention of the incandescent light bulb occurred 9 min after the Industrial Revolution. One hundred years ago, when artificial environments began to spread, it was only 8 minutes before the next year.

Humans have adapted to the natural environment for over 7 million years, from early hominins to *Homo sapiens*. Most of the time, their bodies adapted to withstand the climate of the

Table 1.3 Lower critical temperature (LCT) of humans and several mammals

Participants (*)	N	Age (range)	LCT (°C)	References	Mammals (**)	LCT (°C)
Caucasians	5	31–50	26	Erikson et al. [57]	Weasel	17
Lapps	9	26–57	27	Scholander et al. [58]	Lemming	15
Norwegians	3	20–22	27	Scholander et al. [58]	Squirrel	8
Japanese	5	25–44	24	Yoshimura and Yoshimura [59]	Polar bear	3
Caucasians	5	25–32	25.2	Wilkerson et al. [60]	Eskimo dog	−26
Japanese	5	22–26	21.7	Ishii [61]	White fox	−40

*All participants were almost naked males
**Adapted from Scholander et al. [62]

Table 1.4 History of human evolution and the spread of artificial environments. Seven million years of human history was converted to 1 year

Event	Years ago	Converted to 1 year
Emergence of humans	7.0 million	1 January
Ardipithecus ramidus	4.4 million	16 may
Australopithecus afarensis	3.2 million	15 July
Homo erectus	1.9 million	23 September
The first exodus from Africa	1.8 million	29 September
Migration to Southeast Asia	1.5 million	14 October
Peking man	0.6 million	30 November
Homo neanderthalensis	0.4 million	10 December
Homo sapiens	0.3 million	15 December
Migration to North America	60,000	31 December, 5:00
Beginning of agriculture	11,000	10:00
Beginning of ancient civilizations	4000	19:00
Industrial revolution	260	23:40
Electric power supply	144	23:49
Spread of the artificial environment	100	23:52

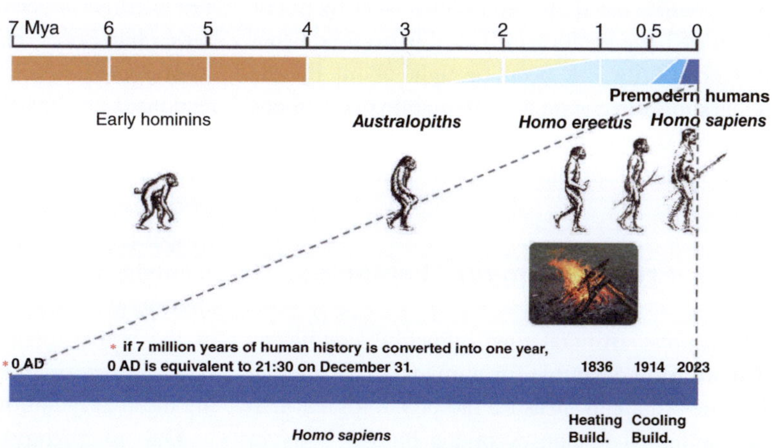

Fig. 1.8 Illustration of the history of humans and artificial environments

African savannah, and as recently as 200 years ago, they were active after sunrise and went to sleep after dark. Our bodies are designed to adapt to our natural environment. It is believed that from 20,000 years ago to the present day, there have been no biological changes in humans. However, over the last 100 years, the environment in which we live has rapidly changed due to the emergence of the artificial environment. We spend most of our time in artificial environments with lighting, air-conditioning, video, and sound. According to research, modern people spend 90% of their time (an average for 1812 people on weekdays) in artificial environments [64]. We have only had the experience of being cool during summer (air-conditioning) within the last 60 years since 1960, and the virtual reality environment has only been around for approximately 30 years.

Human ancestors, born 7 million years ago, adapted to the natural environment over a long period of time, and 300,000 years ago, *Homo sapiens*, who became modern humans, were born. Approximately 11,000 years ago, the hunter-gatherer lifestyle was replaced by farming and livestock farming, allowing humans to feed a large population. Moreover, the Industrial Revolution of 260 years ago significantly expanded the number of urban workers. Currently 8 billion *Homo sapiens* live on Earth, and this number is predicted to reach 10 billion by 2058. Never before has there been such a wealthy population of large mammals. There is no doubt that this prosperity has been brought about by the remarkable environmental adaptability of humans and the development of technology resulting from the development of the cerebrum.

1.2.3 Artificial Climatic Chambers

To propose artificial environmental conditions that are truly desirable for humans, experimental facilities that can measure the physiological and psychological responses of the human body in various artificial environments are necessary. Kyushu University's Faculty of Design, which serves as a research base for artificial environment design, has many facilities that can simulate artificial environments. A representative research facility is the Research Center for Human Environmental Adaptation, an experimental facility that aims to clarify the ability of humans to adapt to their environments [42]. This facility was originally established by Professor Masahiko Sato in 1971 and was completely reconstructed in 2001. It is a two-story building with a total floor area of 744 m^2, a machine room, an observation room, and a laboratory (thermal radiation chamber) on the first floor, and six laboratories (hyper and hypobaric, illumination, combined factor, thermal, water immersion, and living environment chambers) on the second floor. Subsequently, three sound-proof, high-shield chambers were set up in a separate building, making a total of 10 artificial climatic chambers. Table 1.5 shows the performance specifications of the Research Center for Human Environmental Adaptation. Except for research institutes such as NASA in the USA, no other educational research institution has such a large number of artificial climate chambers. Many of the studies presented in this book were conducted using these thermal chambers. In addition, many studies have been conducted using a hyper- and hypobaric chamber (e.g., Motoi et al. [65]), illumination chambers (e.g., Higuchi et al. [66]), and living environment chambers (e.g., Higuchi et al. [67]). Figure 1.9 shows the status of the experiment in each chamber; exercise experiments in the hyper- and hypobaric chamber, experiments on the effects of the color temperature of lighting on physiological responses in the illumination chamber, thermal environment evaluation experiments using thermal mannequins in the combined factor chamber, work evaluation experiments under ultra-low temperature in the thermal chamber, lighting environment and sleep experiments in the living environment chamber, scuba diving experiments in the water immersion chamber, and thermal comfort experiments in a two-wheeled vehicle in the thermal radiation chamber.

Due to changes in modern living environments and improvements in product performance demanded by people, the environmental conditions to be simulated are becoming more multi-

Table 1.5 Performance specifications of the Research Center for Human Environmental Adaptation at Kyushu University

	Air temperature	Relative humidity	Illumination	Remarks
No.1 hyper- and hypobaric chamber With a lavatory and shower room	−10–50 ± 0.5 °C	30–80 ± 5%	0–2000 lux	253–4052 ± 5 hPA, O_2 10–40 ± 0.1%
No.2 illumination chamber	0–50 ± 0.2 °C	30–90 ± 2%	0–20,000 lux Color temperature (3000–10,000 k)	
No.3 combined factor chamber	0–50 ± 0.2 °C Wall and floor temperature (10–45 ± 1 °C)	10–90 ± 2%	0–17,000 lux	Air velocity 0–1.5 m/s For each of the three layers
No.4 thermal chamber	−40–50 ± 0.2 °C	20–90 ± 3%	0–1500 lux	
No.5 living environment chamber With a bathroom, lavatory, and kitchen	0–40 ± 0.2 °C	30–80 ± 2%	0–2000 lux Color temperature of artificial window (3000–10,000 K)	
No.6 water immersion chamber	−10–50 ± 0.5 °C	30–80 ± 3%	1200 lux	Depth 0–5 m Temp. 5–45 °C
No.7 thermal radiation chamber With a bathroom and lavatory	−10–50 ± 0.5 °C	30–80 ± 3%	1500 lux	Thermal radiation 0–1.000 W/m^2
No.8[a] noise reduction and radio shield chamber	0–50 ± 0.2 °C	30–80 ± 3%	0–2000 lux	Electric shield
No.9[a] noise reduction and radio shield chamber	0–50 ± 0.2 °C	30–80 ± 3%	0–2000 lux	Electric shield
No.10[a] noise reduction and radio shield chamber	20–30 ± 1.0 °C	30–80 ± 15%	_	Electric shield

URL: http://www.id.design.kyushu-u.ac.jp/ninkou/english/homot.html
[a]Installed in a separate building

factorial and accurate. Therefore, part of this experimental facility has a bathroom, toilet, kitchen, bedroom, and living room so that real life can be simulated for a long time, and it is planned to allow study participants to freely change environments. The following experiments can be conducted at the experimental facility: physiological evaluation of new living spaces predicted for the next generation (underwater, deep underground, and ultra-high-rise buildings); studying the effects of living environments, including temperature, humidity, air pressure, lighting (illuminance and color temperature), oxygen concentration, odor, color, and sunlight on human physiological and psychological responses; studying on effects of the living environment and lifestyle (such as exercise and bathing) on sleep and biorhythms; suggestions based on physiological evaluation of air-conditioning, beds, bathrooms, toilets, and kitchen facilities for the elderly and disabled; and suggestions for safe and comfortable spaces based on three-dimensional motion analysis and electromyography analysis [43].

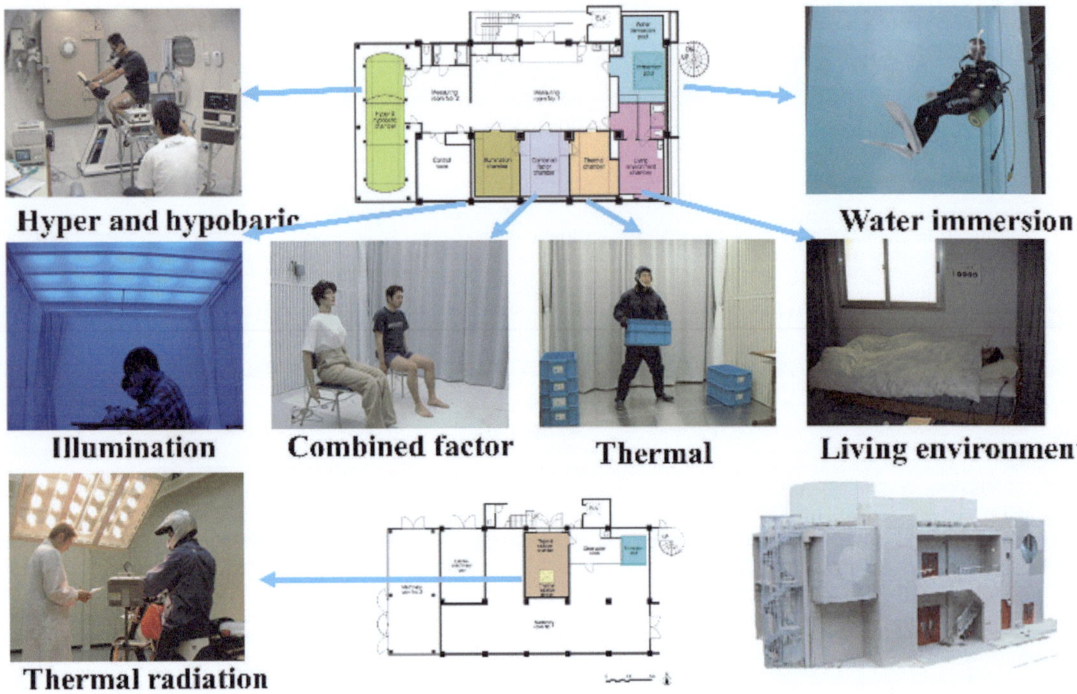

Fig. 1.9 Illustrations (photos) of the Research Center for Human Environmental Adaptation at Kyushu University

1.3 Physiological Anthropology

The first English book on "*Physiological Anthropology*" was published by Damon in 1975 [45]. This book consists of the following 13 chapters: the place of physiological studies in anthropology; the effects of light on man; the skin as an adaptive organ; the ecology of work; human adaptation to heat; environment and human adaptation in the Sahara; human adaptation to cold; human adaptation to high altitude; nutritional adaptation in man; anthropology and infectious disease; behavioral response and adaptation to environmental stimulation; effects of noise on human performance; and biological anthropology as an applied science. The authors were world authorities not only in anthropology but also in the following specialized fields: endocrinology and metabolism, periodontology, environmental medicine, cosmic medicine, clinical research, man-environment relations, psychology, and sociology.

The authors were staunchly anthropological in viewpoint, seeing things in terms of populations and evolutionary processes and demanding clear definitions of such concepts as "adaptation." Many chapters of the book described the effects of the physical environment on the adaptation of the human body.

Damon [45] noted that the present environment is dangerous in two ways: it contains harmful elements and changes so rapidly that humans cannot adapt quickly enough to maintain their health. Many man-created features of the human environment have important biological effects. However, there is currently no provision for fundamental human biology education for people entering these fields. He stated that the discipline that solves this problem is "physiological anthropology," which can be defined as applied biological anthropology.

Human biological adaptability was established during the hunter-gatherer era, which accounts for more than 99% of our existence; however, we now live in a very different environment. Nevertheless, our biological adaptability is fundamentally the same as it was in the past [68]. Therefore, physiological anthropologists must

Table 1.6 Location and organizer of the International Congress of Physiological Anthropology (ICPA)

	Year	Location	Country	Organizer		Year	Location	Country	Organizer
1	1991	Tokyo	Japan	Masaro Kaji	9	2008	Delft	The Netherlands	Johan F.M. Molenbroek
2	1994	Kiel	Germany	Hans W. Jürgens	10	2010	Fremantle	Australia	Alan Bittles
3	1996	Nara	Japan	Hiromi Tokura	11	2013	Banff	Canada	Douglas E. Crews
4	1998	Zagreb	Croatia	Pavao Rudan	12	2015	Chiba	Japan	Tetsuo Katsuura
5	2000	Seoul	Korea	Jang Yeul Sohn	13	2017	Loughborough	UK	Barry Bogin
6	2002	Cambridge	UK	N. Mascie-Taylor	14	2019	Singapore	Singapore	Fook Tim Chew
7	2004	Columbus	USA	Douglas E. Crews	15	2022	Eugene	USA	Josh Snodgrass
8	2006	Kamakura	Japan	Masahiko Sato	16	2023	Kota Kinabalu	Malaysia	Kamruddin Ahmed

answer the following questions when designing new environments. "What should we do? How much can people tolerate? What is optimal?" [45].

1.3.1 Organization of Physiological Anthropology in Japan

Organized activities in physiological anthropology were primarily conducted in Japan. The first association of anthropologists in Japan was founded in 1884, only 25 years after the founding of the Paris Anthropological Society organized by Pierre P. Broca. The Department of Anthropology of the Faculty of Science at the University of Tokyo was established in 1939 [69]. In 1948, Professor Toshihiko Tokizane of the Department of Physiology, Faculty of Medicine, the University of Tokyo began teaching physiological anthropology in Japan [70]. As a student in the Department of Anthropology, Masahiko Sato studied under Professor Tokizane and has been a leader in physiological anthropology research. Physiological anthropology activities were conducted as one of the research areas of the Anthropological Society of Nippon for some time. Organized activities in physiological anthropology began in 1978 as "The Conversation of Physiological Anthropology" with 51 researchers. This group grew into "The Research Society of Physiological Anthropology" in 1982 and developed into the "Japan Society of Physiological Anthropology" in 1993 [70]. Membership has markedly increased and reached 591 members in 2023. Furthermore, memorably, the first International Congress on Physiological Anthropology (ICPA) was held in Tokyo in 1991. The ICPA has been organized 16 times biennially by the International Association of Physiological Anthropology as shown in Table 1.6. The society began publishing a scientific journal entitled "*The Annals of Physiological Anthropology*" in 1983, which subsequently changed its name to the "*Journal of Physiological Anthropology*" in 2006. This journal is currently being published as an international open-access journal in BMC by Springer Nature.

1.3.2 Research Topics in Physiological Anthropology

Research topics in physiological anthropology in Japan can be divided into several periods [70]. The first period began with research activities around the 1950s and included studies that analyzed muscle activity and evaluated muscle fatigue using electromyography (EMG). In addition to EMG, measurements of the cardiac and respiratory cycles or energy metabolism gradually became more important as the research subject broadened. The second period of physiological anthropology research began in the 1970s. Using artificial climatic chambers (rooms with widely variable temperature, humidity, airflow, illumination, and air pressure) established by Professor Sato at Kyushu University in 1971, remarkable progress was made, especially in the study of environmental adaptability. The third period began in the late 1980s. Traditional studies on the effect of severe stress, such as heat, cold, and maximum work, were replaced by studies that assessed stress of a different quality than those of previous studies, such as mental work. Many work environments had to move from outdoors to indoors, including offices, to accommodate computer-based, creative work in comfortable, air-conditioned rooms. From this period, society began to focus on keywords such as "comfort" and "sensitivity," and scientific interpretations and objective evaluations of these keywords were pursued [71]. The period leading to the present began in the 1990s. In Japan, physiological anthropology has developed into a science that is responsible for social needs. Many applied studies, for example, have been conducted in collaboration with industries on thermal conditions [72] and lighting [73]. More recently, Yasukouchi [68], in his article "*The next stage of physiological anthropology*," defined physiological anthropology in comparison with human biology as follows: physiological anthropology is interested in physiological adaptability, especially in human-made environments, through individual thinking and variations in adaptive responses at both the individual and population levels. What Damon [45] defined and dreamed of 50 years ago, "applied biological anthropology" is now a reality in the Society of Physiological Anthropology.

1.3.3 Conceptual Framework in Physiological Anthropology

Sato [46] identified that the principal conditions necessary for physiological anthropology to claim to be a distinct science can be summarized by the following five keywords: "environmental adaptability," "functional potentiality," "physiological polytypism," "technological adaptability," and "whole body coordination." Currently, many studies based on these five keywords are being conducted. Because it is difficult to describe the meaning of each keyword in a short sentence, Table 1.7 shows the important terms in the text in which Sato [46] explained the five keywords of the conceptual framework in physiological anthropology. Yasukouchi [74] summarizes the relevance of these keywords as follows: among the five keywords, "physiological polytypism" is interrelated with "whole body coordination" and "functional potentiality" as a lower level, which can explain "environmental adaptability" in terms of its mechanism. Among them, "technological adaptability" is treated as adaptability to the environment of modern scientific and technological civilization. Moreover, further studies are now underway to address functional adjustment to the daily environment in association with genotype, such as the one by Nishimura et al. [75].

Table 1.7 Five keywords of the conceptual framework in physiological anthropology

Functional potentiality	Whole body coordination	Physiological polytypism	Environmental adaptability	Technological adaptability
Tolerance	Individual thinking	Homogeneous characteristic	Environmental stress	People living in cities
Acclimatization	Coordinative mechanism	Personality	High or low air temperature	Artificial life
Habituation	Sympathetic and parasympathetic	Individual within a population	Artificial environment	Modern civilization
Potential reserve	Automatic nervous system	Geographical and cultural variation	IBP project	Industry
Potential capacity	Nervous and endocrine systems	Genetic mechanism	Unimaginably dangerous environment	Adaptability to living environment

1.4 Environmental Ergonomics

1.4.1 Background of Environmental Ergonomics

Environmental ergonomics is a branch of ergonomics that focuses on the interaction between humans and their physical environments, particularly on how environmental factors affect human performance, comfort, and well-being. It involves studying the effects of various physical environmental conditions on human physiological and psychological responses, with the aim of optimizing the design of environments to enhance human performance and safety. Environmental ergonomics considers factors such as temperature, humidity, light, vibration, noise, and spatial layout in various settings, including workplaces, homes, vehicles, and outdoor environments [76]. Researchers in this field aim to understand how these environmental factors affect human behavior, cognition, and physiological functions, and how they can be modified to create more conducive and supportive environments. By considering the principles of environmental ergonomics, designers, architects, engineers, and other professionals can create environments that are better suited to human needs, thereby improving their overall comfort, safety, and performance.

The International Organization for Standardization (ISO) was established in 1947 to promote international standardization in various important technical fields [77]. The organization has a large number of Technical Committees (TCs). TC159 is the TC that deals with ergonomics. This TC has a subcommittee (SC), SC5, on the "Ergonomics of the Physical Environment (Environmental Ergonomics)," which includes WG1 (thermal environments), WG2 (light environments), WG3 (hazard signals and communication in noisy environments), WG4 (integrated environments), and WG5 (physical environments for people with special needs). Among the WGs of SC5, WG1 has been the most active, publishing 21 international standards and proposing many well-known thermal indices, such as predicted mean vote (PMV), wet bulb globe temperature (WBGT), and required clothing insulation (IREQ). See Chap. 10 for further information.

1.4.2 International Conference on Environmental Ergonomics

The first International Conference on Environmental Ergonomics (ICEE) was held in London and was attended by a few people. A major international conference held in Canada in 1986 was attended by many physiological anthropologists from Japan, and the program and abstracts were published in *The Annals of Physiological Anthropology* (Vol.5 No.3, 1986). Subsequently, biennial meetings were held in various countries, as shown in Table 1.8. ICEE participants included researchers, physicians, engineers, administrators, technicians, allied health professionals, and students from universities, private industries, and government research facilities around the world. Most research was focused on thermal environments, such as extreme heat, cold, and the role of (protective) clothing, as well as thermal comfort. Figure 1.10 shows the research presentations at the 10th ICEE (Fukuoka, Japan). Of the 206 presentations, 68% were related to the thermal environment. Within the physical environment, research presentations on the effects of the thermal environment on the human body dominated the ICEE presentations. The breakdown of research presentations related to the thermal environment was, in descending order, thermal comfort, cold stress, thermal physiology, heat stress, protective clothing, and thermal manikins and modelling [78]. Presentations on the following research topics were made at the 20th ICEE (Jeju, Korea): thermal stress and strain, heat and cold-related disorders, heat and cold adaptation, indoor and vehicle thermal environments, thermal physiology, hot and cold water immersion, personal protective clothing, manikins and modelling, international standards (ISO), and deep-sea and marine environments. The ICEE has become the leading international conference for research on human comfort, health, and performance in thermal environments.

Table 1.8 Location and organizers of the International Conference on Environmental Ergonomics (ICEE)

	Year	Location	Country	Organizers		Year	Location	Country	Organizers
1	1984	London	UK	Thomas P, Mekajavic I.	11	2005	Ystad	Sweden	Holmér I, Kuklane K, Gao C.
2	1986	Whistler	Canada	Mekajavic I, Kakitsuba N.	12	2007	Portoroz	Slovenia	Mekajavic I, Pisot R, Taylor N.
3	1988	Helsinki	Finland	Ilmarinen R, Pasche A.	13	2009	Boston	USA	Castellani JW, Endrusick TL
4	1990	Austin	USA	Wissler E, Nunneley S.	14	2011	Nafplio	Greece	Kounalakis S, Koskolou M
5	1992	Maastricht	The Netherlands	Lotens W, Havenith G.	15	2013	Queenstown	New Zealand	Cotter JD, Lucas SJE, Mündel T
6	1994	Montello	Canada	Frim J, Ducharme W, Tikuisis P.	16	2015	Portsmouth	UK	House JR, Tipton MJ
7	1996	Jerusalem	Israel	Shapiro Y, Moran D, Epstein Y.	17	2017	Kobe	Japan	Kondo N, Inoue Y, Nishiyasu T
8	1998	San Diego	USA	Hodgdon J, Heaney J.	18	2019	Amsterdam	The Netherlands	Gerrett NM, Daanen HAM, Teunissen LPJ
9	2000	Dortmund	Germany	Werner J, Hexamer M.	19	2022	Niagara Falls	Canada	Cheung S.
10	2002	Fukuoka	Japan	Tochihara Y, Ohnaka T.	20	2024	Jeju	Korea	Lee JY, Kim DH

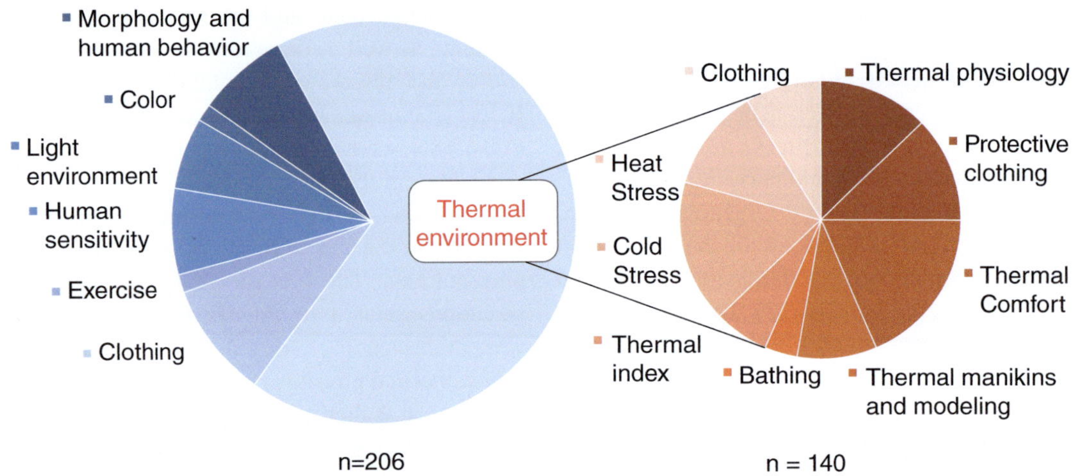

Fig. 1.10 Research presentations at the 10th International Conference on Environmental Ergonomics (Fukuoka, Japan)

References

1. Lloyd C. What on earth happened? The complete story of the planet, life, and people from the big bang to the present day. London: Bloomsbury Publishing Plc; 2008.
2. Kono R. What kind of humans were apes? (in Japanese). In: Ihara Y, Umezaki M, Yoneda M, editors. In pursuit of human nature: an introduction to physical anthropology. Tokyo: University of Tokyo Press; 2021. p. 23–42.
3. Lieberman DE. Human locomotion and heat loss: an evolutionary perspective. Compr Physiol. 2015;5:99–117.
4. Langergrabera KE, Prüferc K, Rowney C, Boesch C, Crockford C, Fawcette K, et al. Generation times in wild chimpanzees and gorillas suggest earlier diver-

gence times in great ape and human evolution. PNAS. 2012;109:15716–21.
5. Fujiyama A, Watanabe H, Toyoda A, Taylor TD, Itoh T, Tsai SF, et al. Construction and analysis of a human -Chimpanzee comparative clone map. Science. 2002;295:131–4.
6. Brunet M, Guy F, Pilbeam D, Mackaye HT, Likeius A, Ahounta D, et al. A new hominid from the upper miocene of Chad, Central Africa. Nature. 2002;418:145–51.
7. White TD, Suwa G, Asfaw B. *Australopithecus ramidus,* a new species of early hominid from Aramis, Ethiopia. Nature. 1994;371:306–12.
8. Johanson D, Edey M. Lucy: the beginnings of humankind. New York: Simon and Schuster; 1981.
9. Coppens Y. Human origins: the story of our species. London: Octopus Publishing Group; 2004.
10. Wheeler PE. The evaluation of bipedality and loss of functional body hair in hominids. J Human Evol. 1984;13:91–8.
11. Wheeler PE. The thermoregulatory advantages of hominid bipedalism in open equatorial environments: the contribution of increased convective heat loss and cutaneous evaporative cooling. J Human Evol. 1991;21:107–15.
12. Ruxton GD, Wilkinson DM. Thermoregulation and endurance running in extinct hominins: Wheeler's models revised. J Human Evol. 2011a;61:169–75.
13. David-Barrett T, Dunbar RIM. Bipedality and hair loss in human evolution revised: the impact of altitude and activity scheduling. J Human Evol. 2016;94:72–82.
14. Bramble DR, Liberman DM. Endurance running and the evolution of *Homo*. Nature. 2004;432:345–52.
15. Lieberman DE. The story of the human body: evolution, Health and Disease. New York: Pantheon; 2013.
16. Sockol MD, Raichlen DA, Pontzer H. Chimpanzee locomotor energetics and the origin of human bipedalism. PNAS. 2007;104:12265–9.
17. Leonard WR, Robertson ML, Snodgrass JJ, Kuzawa CW. Metabolic correlates of hominid brain evolution. Comp Biochem Phys A. 2003;136:5–15.
18. Cann R, Stoneking M, Wilson AC. Mitochondrial DNA and human evolution. Nature. 1987;325:31–6.
19. Hublin JJ, Ben-Ncer A, Bailey SE, Freidline SE, Neubauer S, Skinner MM, et al. New fossils from Jebel Irhoud, Morocco and the pan-African origin of *Homo sapiens*. Nature. 2017;546:289–92.
20. Morris D. The naked ape. London: Jonathan Cape; 1967.
21. Ruxton GD, Wilkinson DM. Avoidance of overheating and selection for both hair loss and bipedality in hominins. PNAC;2011b:108:20965–20969.
22. Best A, Kamilar JM. The evolution of eccrine sweat glands in human and nonhuman primates. J Human Evol. 2018;117:33–43.
23. Fork GE, Semken HA. The evolution of sweat glands. Int J Biometeorol. 1991;35:180–6.
24. Kuno Y. Human perspiration. Springfield: Charles C Thomas Publisher; 1956.
25. Torii M. Maximal sweating rate in humans. J Human Ergol. 1995;24:137–52.
26. Lupi O. Ancient adaptation of human skin: why do we retain sebaceous and apocrine glands? Int J Dermatology. 2008;47:65–654.
27. Kamberov YG, Guhan SM, DeMarchis A, Jiang J, Wright SS, Morgan BA, et al. Comparative evidence for the independent evolution of hair and sweat gland traits in primates. J Human Evol. 2018;125:99–105.
28. Aldea D, Atsuta Y, Kokalari B, Schaffner SF, Prasasya RD, Aharoni A, et al. Repeated mutation of a developmental enhancer contributed to human thermoregulatory evolution. PNAS. 2021;118(16)
29. Bergmann C. Über die verhältnisse der wärmoeökomie der thiere zu ihrer grösse. Göttinger Studien. 1847;3:595–708.
30. Allen JA. The influence of physical conditions in the genesis of species. Radical Rev. 1877;1:108–40.
31. Thomson A, Buxton HD. Man's nasal index in relation to certain climatic condition. J Roy Anth Inst. 1923;53:92–122.
32. Robert DF. Body weight, race, and climate. Amer J Phys Anthropol. 1953;11:533–58.
33. Newman RW, Munro EH. The relation of climate and body size in U.S. males. Amer J Phys Anthropol. 1955;13(1):1–17.
34. Newman RW. Human adaptation to heat. In: Damon A, editor. Physiological anthropology. New York: Oxford University Press; 1975. p. 80–92.
35. Sato M. Human adaptation to climate (in Japanese). In: The Anthropological Society of Nippon ed. Anthropology: Its diverse developments. Nikkei Science Inc; 1984. p. 158–69.
36. Steegmann AT. Human adaptation to cold. In: Damon A, editor. Physiological anthropology. New York: Oxford University Press; 1975. p. 130–66.
37. Leonard WR, Katzmarzyk PT. Body size and shape: climatic and nutritional influences on human body morphology. In: Muehlenbein MP, editor. Human evolutionary biology. Cambridge: Cambridge University Press; 2010. p. 157–69.
38. Bogin B, Hermanussen M, Scheffler C. Bergman's rule is a "just-so" story of human body size. J Physiol Anthropol. 2022;41:1–15.
39. Weiner JS. Nose shape and climate. Amer J Phys Anthropol. 1954;12:615–8.
40. Maddux SD, Yokley TR, Svoma BM, Franciscus RG. Absolute humidity and human nose: a reanalysis of climate zones and their influence on nasal form and function. Amer J Phys Anthropol. 2016;161:309–20.
41. Noback ML, Harvati K, Spoor F. Climate-related variation of the human nasal cavity. Amer J Phys Anthropol. 2011;145:599–614.
42. Tochihara Y, Wakabayashi H, Lee JY, Wijayanto T, Hashiguchi N, Saat M. How humans adapt to hot climates learned from the recent research on tropical indigenes. J Physiol Anthropol. 2022;41:27.
43. Tochihara Y. Creation of artificial environments (in Japanese). In: Tochihara Y, editor. Handbook of artificial environment design. Tokyo: Maruzen; 2007. p. 1–70.

44. Eto T, Higuchi S. Review on age-related differences in non-visual effects of light: melatonin suppression, circadian phase shift and pupillary light reflex in children to older adult. J Physiol Anthropol. 2023;42:11.
45. Damon A. Biological anthropology as an applied science. In: Damon A, editor. Physiological anthropology. New York: Oxford University Press; 1975. p. 360–7.
46. Sato M. The development of conceptual framework in physiological anthropology. J Physiol Anthropol. 2005;24:289–95.
47. Maeda T. Perspectives on environmental adaptability and physiological polymorphism in thermoregulation. J Physiol Anthropol Appl Hum Sci. 2005;24:237–40.
48. Baker PT. The physical environment. In: Harrison GA, Tanner JM, Pilbeam DR, Baker PT, editors. Human biology an introduction to human evolution, variation, growth, and adaptability. 3rd ed. New York: Oxford Univ Press; 1988. p. 450–78.
49. Yoshimura H. Acclimatization to heat and cold. In: Yoshimura H, Ogawa K, Itoh S, editors. Essential problem in climatic physiology. Tokyo: Nan- kodo; 1960. p. 61–106.
50. Worthington EB. The International Biological Programme. Nature. 1965;208:223–6.
51. Yoshimura H. Japanese adaptability (in Japanese). Tokyo: Kodansha; 1970.
52. Nakashima Y. Climate change policies in Japan/what are COO BIZ and WARM BIZ? Japan Environment Quarterly, Feature. 2013;3
53. SHASE (the Society of Heating, air-conditioning and sanitary engineers of Japan). History of air-conditioning and sanitary equipment technology (in Japanese). Tokyo: Maruzen; 1991. p.1–71.
54. Wright L. Home fires burning the history of domestic heating and cooking. London: Routledge & Kegan Paul; 1964. p. 1–219.
55. Ingels M. Father of air conditioning. Garden City: Country Life Press; 1952. p. 1–170.
56. IUPS Thermal Commission: Glossary of terms for thermal physiology. J Thermal Biology 2003;28:75–106.
57. Erikson H, Krog J, Anderson KL, Scholander PF. The critical temperature in naked man. Acta Physiol Scand. 1956;37:35–9.
58. Scholander PF, Andersen KL, Krog J, Lorentzen FV, Steen J. Critical temperature in Lapps. J Appl Physiol. 1957;10:2311–23.
59. Yoshimura M, Yoshimura H. Cold tolerance and critical temperature of the Japanese. Int J Biometeorol. 1969;13:163–72.
60. Wilkerson JE, Raven PB, Horvath SM. Critical temperature of unacclimatized male Caucasians. J Appl Physiol. 1972;33:451–5.
61. Ishii M. Cold tolerance of Japanese assessed by the lower critical temperature and so on (in Japanese). J Anthrop Soc Nippon. 1976;84:93–104.
62. Scholander PF, Hock R, Walters V, Johnson F, Irving L. Heat regulation in some arctic and tropical mammals and birds. Biol Bull. 1950;99:237–58.
63. Osada Y. Experimental studies on the sexual and seasonal differences of the optimal thermal conditions. J Human Ergol. 1978;7:145–55.
64. Shiotsu M, Yoshizawa S, Ikeda K, Nozaki A. Survey on human activity patterns according to time and place (in Japanese). Archit Plann Environ Eng AIJ. 1998;511:45–52.
65. Motoi M, Nishimura T, Egashira Y, Kishida F, Watanuki S. Relationship between mitochondrial haplogroup and physiological responses to hypobaric hypoxia. J Physiol Anthropol. 2016;35:12.
66. Higuchi S, Ishibashi K, Aritake S, Enomoto M, Hida A, Tamura M, et al. Inter-individual difference in pupil size correlates to suppression of melatonin by exposure to light. Neurosci Lett. 2008;440:23–6.
67. Higuchi S, Nagafuchi Y, Lee SI, Harada T. Influence of light at night on melatonin suppression in children. J Clin Endocrinol Metab. 2014;99:3298–303.
68. Yasukouchi A. The next stage of physiological anthropology. J Physiol Anthropol. 2023;42:3–7.
69. Sato M, Takasaki Y, Harada H, Yamasaki K, Watanuki S. Physiological anthropology in Japan between 1948 and 1982. Annals Physiol Anthrop. 1983;2:3–8.
70. Sato M. The progress of physiological anthropology in Japan. Appl Hum Sci. 1995;14:1–4.
71. Yasukouchi A. What is physiological anthropology? (in Japanese). In: Yasukouchi Y, Iwanaga K, editors. Physiological anthropology. Tokyo: Rikotosho; 2020. p. 3–10.
72. Tochihara Y. A review of Japanese-style bathing: its demerits and merits. J Physiol Anthropol. 2022;41:5.
73. Hayano J, Ueda N, Kisohara M, Yoshida Y, Yuda E. Ambient-task combined lighting to regulate autonomic and psychomotor arousal levels without compromising subjective comfort to lighting. J Physiol Anthropol. 2021;40:8.
74. Yasukouchi A. Perspective of keywords focused on physiological polytypism in physiological anthropology (in Japanese). Japanese J Physiol Anthropol. 2013;18:45–7.
75. Nishimura T, Katsumura T, Motoi M, Oota H, Watanuki S. Experimental evidence reveals the UCP1 genotype changes the oxygen consumption attributed to non-shivering thermogenesis in humans. Sci Rep. 2017;7(1):5570.
76. Parsons K. Environmental ergonomics: a review of principals, methods and models. Appl Ergon. 2000;31:581–94.
77. Parsons K. Human thermal environments the effects of hot, moderate and cold environments on human health, comfort and performance. London/New York: Taylor & Francis; 2003. p. 1–527.
78. Tochihara Y. Ohnaka. Environmental Ergonomics. The ergonomics of human comfort, health and performance in the thermal environment. Amsterdam: Elsevier; 2005. p. 1–522.

The Thermal Environment and Clothing Physical Properties

2

Takako Fukazawa

2.1 Properties of Heat Transfer from Human Body Surface

Human beings are a type of warm-blooded animal, whose body temperature remains almost constant at a preferable level regardless of the surrounding environment. In order to maintain a desired constant body temperature, thermoregulation is continuously carried out by balancing heat production and heat loss. The details of heat balance in humans have been described in Chap. 4. Therefore, in this section, the characteristics of heat transfer from the human body surface to the environment are described.

2.1.1 Sensible and Latent Heat Transfer Way

There are four physical pathways for heat transfer from the human body surface to the surrounding environment, as shown in Fig. 2.1: conduction, convection, radiation, and evaporation. The former three paths of conduction, convection, and radiation are also called sensible heat transfer, because they accompany temperature changes on the body surface, whereas the last path of evaporation is also called a latent heat transfer, because it changes the phase of water from liquid to vapour on the body surface.

Conduction is a phenomenon in which heat moves from a higher temperature to a lower temperature without mass transfer within or between objects in contact with each other. The amount of heat (K) transferred by heat conduction follows *Fourier's law*, as expressed in Eq. 2.1. The transferred heat flux is determined by the thermal conductivity (Table 2.1), which shows the heat transfer peculiar in a substance, and the temperature gradient within and/or between substances, and its thickness or travel distance.

For example, when a part of the human body touches two objects that indicate same temperature, a larger heat flux by conduction will be seen between the skin surface and the object with high thermal conductivity, while a smaller heat flux will be seen between the skin surface and the other object with low thermal conductivity. Therefore, when a part of the body surface touches an object with high thermal conductivity,

T. Fukazawa (✉)
Department of Home Economics, Faculty of Education, Kyoto University of Education, Kyoto, Japan
e-mail: fukazawa@kyokyo-u.ac.jp

Y. Tochihara (ed.), *The Thermal Environment*, https://doi.org/10.1007/978-981-96-3009-7_2

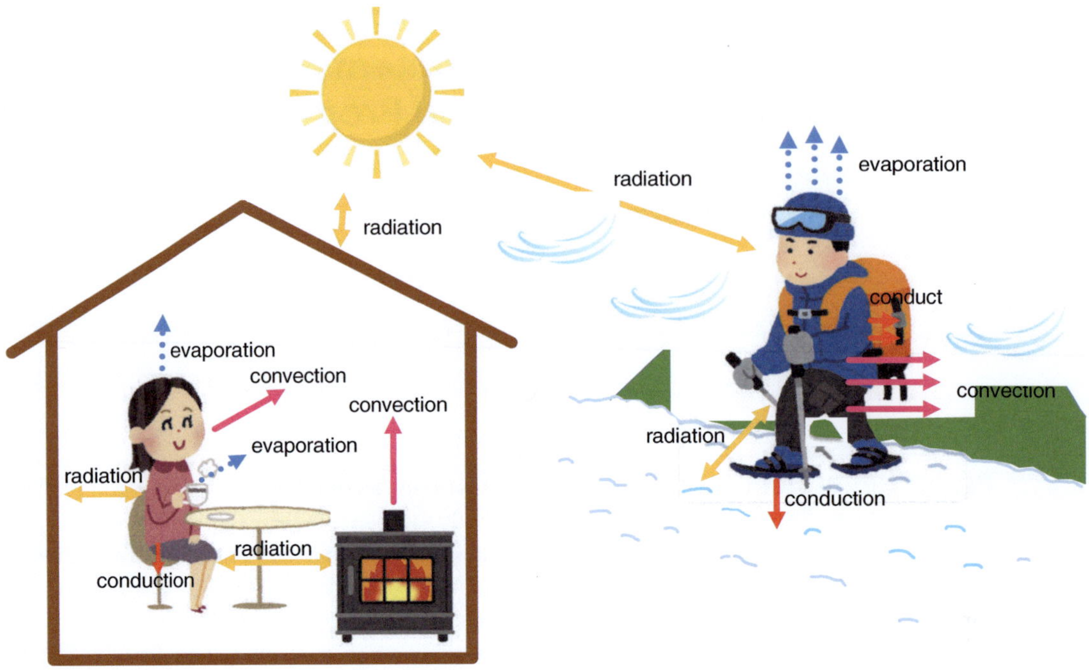

Fig. 2.1 A simple conceptual explanation of heat transfer pathways from the human body in thermal environments

Table 2.1 Thermal properties of fibres for substances and clothing materials

Materials			Thermal conductivity (W/mK)
Metals	SUS		15.080
	Gold		322.400
	Silver		431.600
	Copper		408.200
Fibres for clothing	Cotton	Longitudinal direction	2.879
		Transvers direction	0.243
	Linen	Longitudinal direction	2.831
		Transvers direction	0.344
	Wool	Longitudinal direction	0.480
		Transvers direction	0.165
	Silk	Longitudinal direction	1.492
		Transvers direction	0.118
	Rayon filament	Longitudinal direction	1.895
		Transvers direction	–
	Rayon staple	Longitudinal direction	1.414
		Transvers direction	0.237
	Nylon	Longitudinal direction	1.433
		Transvers direction	0.171
	Polyester filament	Longitudinal direction	1.257
		Transvers direction	0.157
	Polyester staple	Longitudinal direction	1.175
		Transvers direction	0.127
	Acrylic	Longitudinal direction	1.020
		Transvers direction	0.172

(continued)

Table 2.1 (continued)

Materials			Thermal conductivity (W/mK)
Otners	Air		0.026
	Water		0.608
	Ice		2.200
	Water vapour		0.024
	Paper		0.06–0.128
	Rubber		0.151
	Glass fibre	Longitudinal direction	2.250
		Transvers direction	0.509

These data are quoted and summarised from data in a book described by Fukusako [1] and reported by Kawabata [2]

if the temperature of the object is lower than that of the skin surface, then the human feels the object as cold.

Since the thermal conductivity of air, which human body is always contact with, is quite low (0.026 W/mK at 20 °C) [1], the amount of transferred heat flux from the human body to the environment through conduction becomes smaller if the body surface temperature is higher than that of the environment. On the other hand, if the human body is immersed in cold water, the amount of transferred heat flux by conduction becomes very large, because the thermal conductivity of water is quite high as 0.602 W/mK [1].

$$K = \frac{\lambda}{d}\left(T_1 - T_2\right) \tag{2.1}$$

where

K: heat flux by conduction (W/m^2)
λ: thermal conductivity (W/mK)
d: thickness or length between surfaces 1 and 2 (m)
T_1: temperature surface 1 (K)
T_2: temperature surface 2 (K)

Convection is a phenomenon in which heat is transferred by fluid flow, such as air. Heat flux from the human body by convection (C) is determined by the convective heat transfer coefficient, which indicates the degree of heat transfer by convection, and the temperature difference between the body surface temperature and the surroundings, as shown in Eq. 2.2. There are two types of convection: one is natural convection and the other is forced convection. Natural convection is a flow caused only by the buoyancy due to the temperature difference of the fluid. For example, the density of the air touching the body surface becomes smaller, in other words lighter, because it is warmed up by the human body, whereas the air, which is a bit far from the body surface, has a larger density due to its lower temperature. This density difference between warmer and cooler air causes natural fluid flow, with the less dense air rising and the denser air falling. Forced convection, on the other hand, is fluid flow caused by external factors such as movement or wind, e.g. airflow produced by fans is a typical form of forced convection.

$$C = h_c\left(T_1 - T_2\right) \tag{2.2}$$

where

C: heat flux by convection (W/m^2)
h_c: convective heat transfer coefficient (W/m^{2}K)
T_1: temperature surface 1 (K)
T_2: temperature surface 2 (K)

Radiation is a phenomenon in which an object emits and absorbs thermal energy in the form of electromagnetic waves depending on its surface temperature. Electromagnetic waves are also called infrared rays, which have shorter wavelengths than visible light and are invisible to humans. The thermal energy (R) emitted from an object obeys the *Stefan-Boltzmann law*, which means that it is proportional to the emissivity (ε) of the object and the fourth power of the absolute temperature indicated by the object, as expressed in Eq. 2.3. The emissivity of various materials is shown in Table 2.2 [3, 4].

Table 2.2 Emissivity of various materials for infrared radiation is referred from Holman [3] and Sugawara [4]

Sunstances and materials		Emissivity (−)
Black body		1.00
Metals	Aluminum with polished surface	0.04–0.08
	Aluminum with rough surface	0.06
	Copper with polished surface	0.07
	Copper with oxidized surface	0.50
Non metals	Skin surface of human	0.95–0.99
	Papers	0.70–0.93
	Glasses	0.95
	White potteries	0.86
	Red bricks	0.93
	Fire bricks	0.60
	Rubbers	0.86–0.94
	Fabrics	0.80–0.98
	Water	0.96
	Ice	0.96

Radiation between objects does not require a fluid such as air or liquid to transfer thermal energy. In the case of radiation between objects, heat release and absorption occur simultaneously regardless of temperature. If the emissivity of the objects is the same, the heat energy (R) by radiation among the objects is proportional to the fourth power of the temperature difference, as expressed in Eq. 2.4. For example, in the case of a human in a room, radiation heat transfer occurs between the human body and the wall. If the temperature of the wall is higher than that of human skin surface, the human body consequently receives heat energy by radiation from the wall.

$$R = \sigma\varepsilon\left(T\right)^4 \tag{2.3}$$

$$R = \sigma\varepsilon\left\{\left(T_1\right)^4 - \left(T_2\right)^4\right\} \tag{2.4}$$

where

R: heat flux by radiation (W/m^2)
σ: Stefan-Boltzmann constant (5.67×10^{-8} W/m^{2}K^4)
ε: emissivity (−)
T: surface temperature (K)
T_1: temperature surface 1 (K)
T_2: temperature surface 2 (K)

Evaporation is a phenomenon in which a liquid phase changes to a gas phase. Insensible evaporation from the body surface and respiratory tracts is related to the human body. In addition, during perspiration due to high body temperature, water in the sweat evaporates from the body surface. Heat flux by evaporation of water (E) is determined by the mass flux and the latent heat coefficient, which is defined as 2430 kJ/kg (Eq. 2.5). Evaporation is very important in thermoregulation. This is because, if a person weighing 60 kg evaporates 1000 g per day as insensible evaporation, the temperature rise in the body is suppressed by about 12 °C.

$$E = 2,430 \cdot \dot{m} \tag{2.5}$$

where

E: heat flux by evaporation (W/m^2)
$\dot{m}$: mass flux (kg/m^{2}s)

2.1.2 Heat and Water Vapour Transfer with Human Body Surface

Although we, as human bodies, are always exposed to various environments such as cool/warm climates or cold/hot environments, in terms of heat balance, the amount of heat release from the body is ideally equivalent to the amount of heat production of the body in order to keep core temperature at the appropriate level. The released heat flux from the body consists of four paths: three of which are conduction, convection, and radiation, called as sensible heat transfer, and the last of which is evaporation, called as latent heat transfer. Figure 2.2 [5, 6] shows the coefficients from several locations in the body and the whole body due to heat transfer by convection, including conduction (a), by radiation (b), and due to mass transfer (c). These coefficients have been obtained through measurements, in which wind

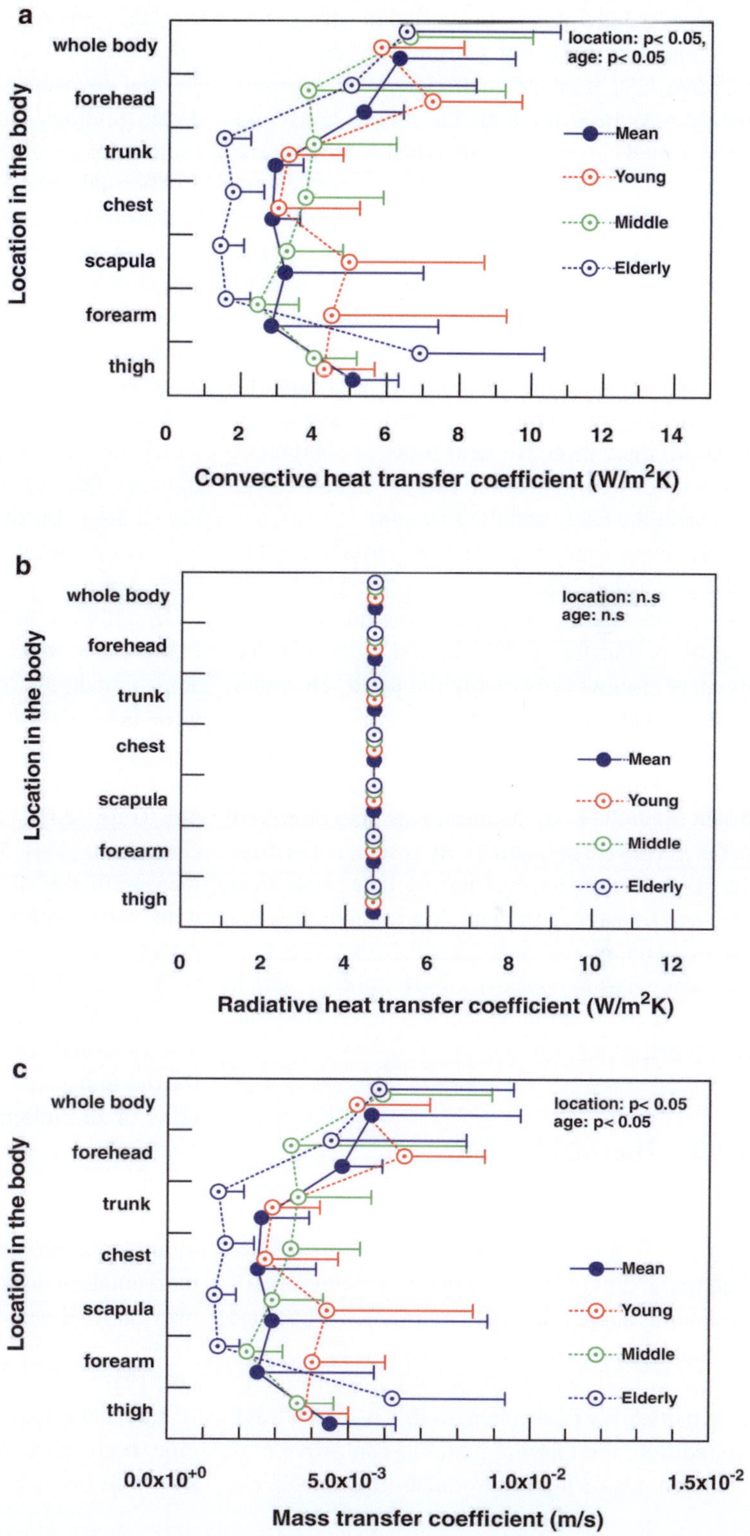

Fig. 2.2 Heat transfer coefficients convection including thermal conductivity (**a**), radiation (**b**), and mass transfer coefficients (**c**) for the whole body and for several regions in females. (These figures (**a**)–(**c**) were created originally by the author with reference data from Fukazawa et al. [5, 6])

speeds of 0.5–0.7 m/s were applied to sitting subjects under a thermally neutral condition. The radiative heat transfer coefficient was approximately 5 W/m^2K for both the whole body and each tested location. No differences were observed in the radiative heat transfer coefficients between the tested locations. On the other hand, the convective heat transfer coefficient of the whole body and each tested location of the body were significantly different depending on the location of the body. In detail, the convective heat transfer coefficient of the forehead was remarkably the largest among the tested locations. In addition, the convective heat transfer coefficient in the thigh was significantly larger than those in the chest, the back, and the forearm.

The mass transfer coefficient from the body surface showed different values throughout the whole body and in each part of the human body. The mass transfer coefficient was largest in the forehead, followed by that in the thigh. The mass transfer coefficients in the chest, the back, and the forearms were lower than those in the forehead and the thigh. A same tendency of dependence upon the body location was also observed in the mass transfer coefficient as seen in the convective heat transfer coefficient. This is because the mass transfer coefficient, in a manner, is similar to defining the heat transfer coefficient as well-known as *Reynolds-Colburn analogy* which expresses the relation between fluid friction and heat transfer (e.g. [3, 7]).

2.1.3 Thermal Manikin and Sweating Thermal Manikin

Thermal manikin is one of the useful devices for evaluating human thermal environments because it simulates precisely human heat production and release as well as body shape. So, it is like a reliable partner for researchers in this research field. In addition, the thermal manikin can provide us with benefits of high reproducibility, convenient use, time-saving, and so on. Therefore, many researchers working on the evaluation of human thermal environments deeply recognize that academic findings have been brought by research using a device of the thermal manikin.

Thermal manikin with various shapes and functions have been originally developed and installed in several laboratories in the world, although there are not so many facilities equipped with the thermal manikin. Figure 2.3a–f shows thermal manikins simulating heat production and morphology of adult male and female. In addition, thermal manikins shaped like adult male, which are movable, simulating human walking or running, have also been developed as shown in Fig. 2.3g–j. Furthermore, thermal manikin that represents sweating has also been developed (Fig. 2.3k).

In addition, not only thermal manikins simulating shape and heat production and release of adult female and male but also child and infant have been developed. Those thermal manikins are, specifically, simulating shapes of school child [8], infants aged 2 years [9] and 0.5 years old (Fig. 2.4a, [10]), and premature infant (Fig. 2.4b, [11]). While it is difficult to perform measurements of heat transfer from the body with participation of an infant or a premature infant due to ethical restriction reasons, these baby thermal manikins enable us to obtain the heat transfer properties from the body of infants. The obtained results and findings of evaluating the thermal environments using the thermal manikins of the infants would greatly contribute to discuss and make guidelines for designing the safety and health for infants and premature infants in daily life (e.g. [12–15]). This is because infant's and premature infant's behavioural thermoregulation does not entirely depend upon their own thermal sensations but depends upon their parents' and caregivers' thermal sensations (e.g. [16, 17]).

Local thermal manikins that reproduce specific body parts have also been developed as shown in Fig. 2.5.

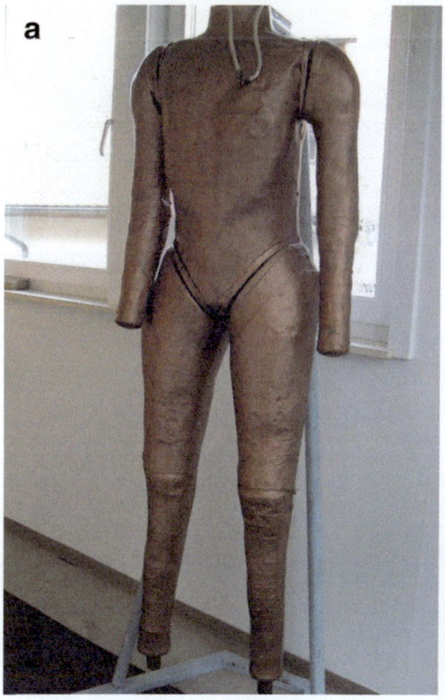

male (Hohenstein Institute, Germany)

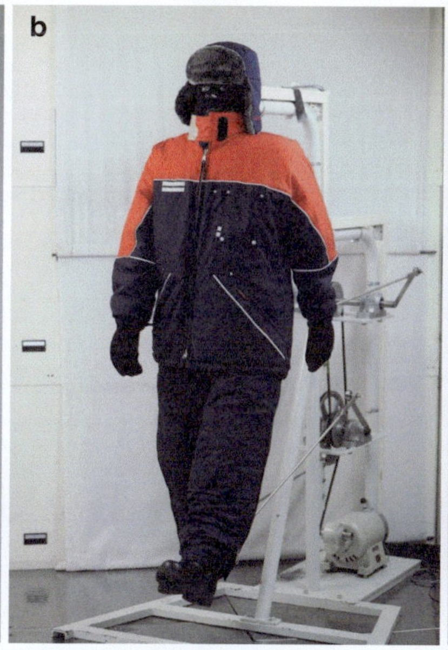

male named Charlie (Hohenstein Institute, Germany; the photo was provided by Hohenstein Institute)

male (Tampere University of Technology, Finland; the photo was taken at the Thermal Manikin Meeting in Tampere)

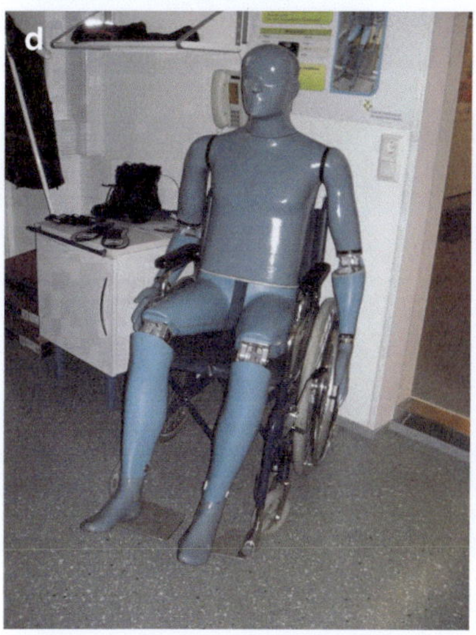

male (Finnish Institute of Occupational Health, Finland)

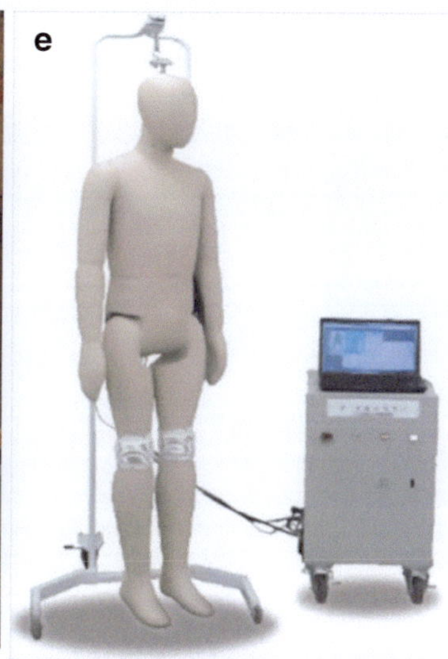

male (Kyoto Electronics Manufacturing Co., Ltd., Japan; the photo was provided by Kyoto Electronics Manufactureing Co., Ltd.)

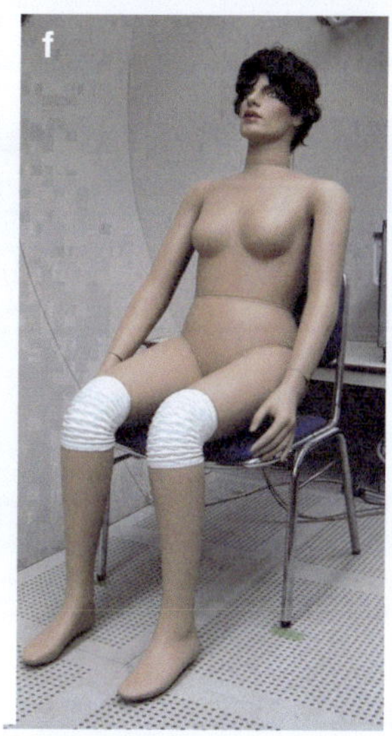

female (Kyushu University and Kyoto University of Education, Japan)

Fig. 2.3 Thermal manikins have been developed originally and/or installed in universities and institutes around the world. Some thermal manikins are able to represent human movement and sweating. They are actively employed in research on the human thermal environments. (**a** and **b**: Hohenstein Institute (Germany), **c**: Tampere University of Technology (Finland), **d**: Finnish Institute of Occupational Health (Finland), **e**: Kyoto Electronics Manufacturing, Co., Ltd. (Japan), **f**: Kyushu University and Kyoto University of Education (Japan), **g** and **h**: Lund University (Sweden), **i**: Seoul National University (Korea), **j**: Finnish Institute of Occupational Health (Finland), **k**: Kyoto Electronics Manufacturing, Co., Ltd. (Japan))

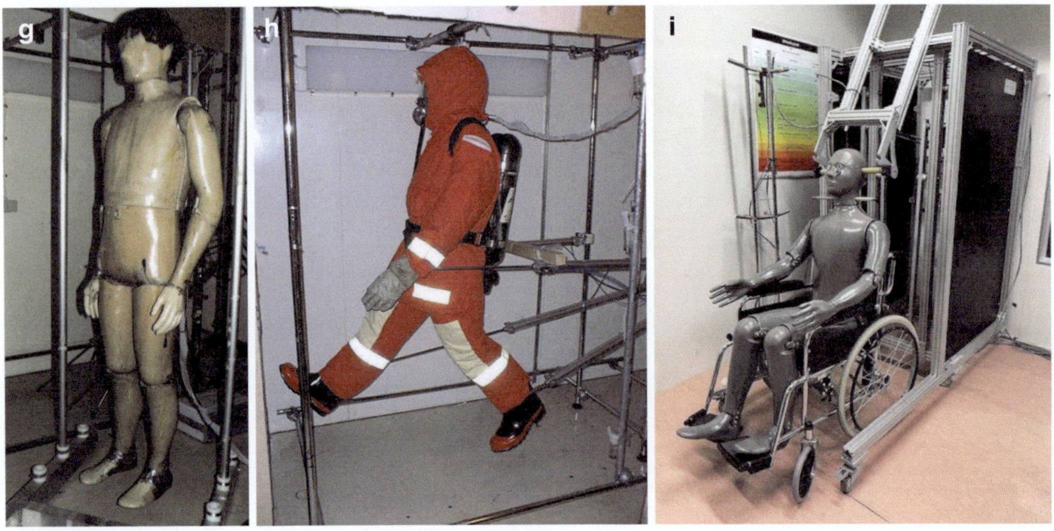

male, movable, named Tore (the Aerocol and Climate Laboratory, Lund University, Sweden)

g with an protective clothing system

male, movable (Seoul National University, Korea)

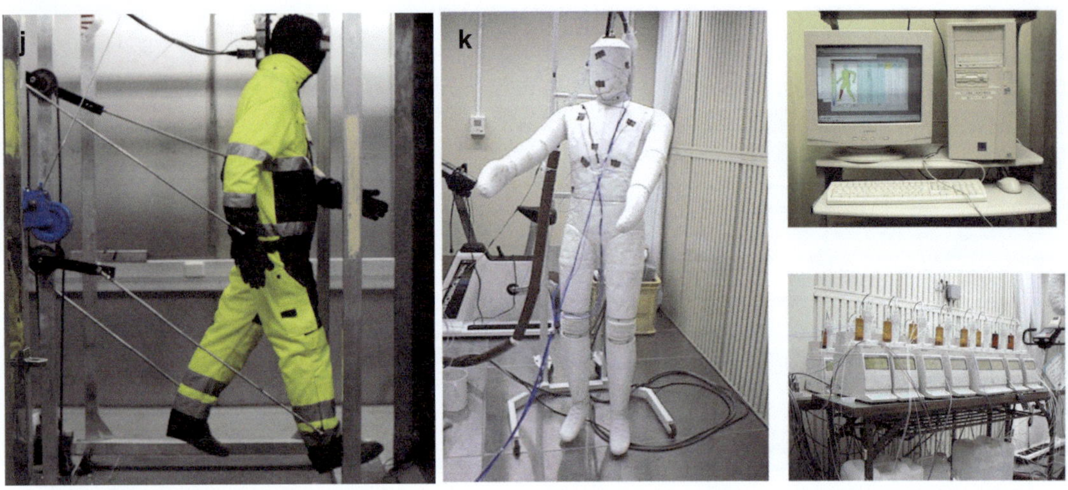

male, movable (Finnish Institute of Occupational Health, Finland; the photo was provided by Finnish Institute of Occupational Health)

male, movable, sweating (Kyoto Electronics Manufacturing Co., Ltd., Japan)

Fig. 2.3 (continued)

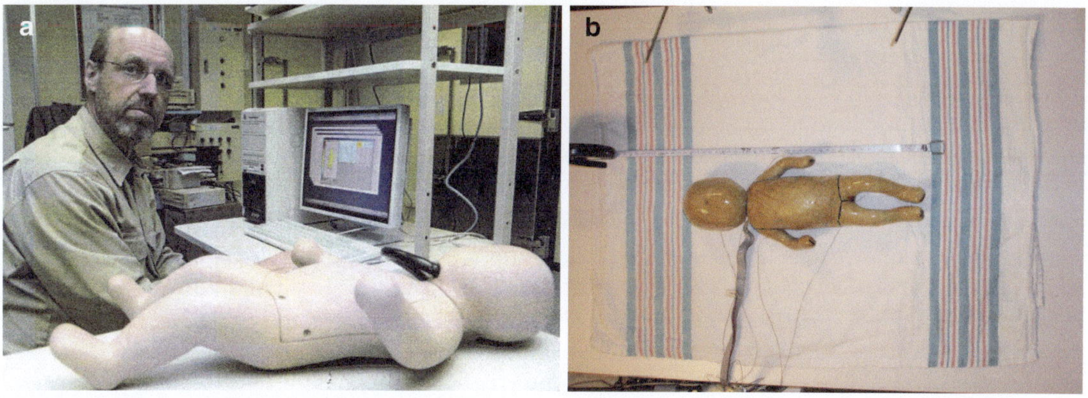

Prof. Dr. I Holmér (Lund University, Sweden) and a baby thermal manikin (aged 0.5 years old, Kyushu University and Kyoto University of Education, Japan)

prematurely born baby thermal manikin (Lund University, Sweden)

Fig. 2.4 Baby thermal manikins owned by (**a**) Kyushu University and Kyoto University of Education in Japan [10] and (**b**) Lund University in Sweden [11]. (The baby thermal manikin (**a**) has been developed through collaboration work with Kyushu University (Japan), Prof Dr. I Holmér and Prof. Dr. K Kuklane of Lund University (Sweden), and Kyoto Electronics Manufacturing, Co., Ltd. (Japan). The premature infant thermal manikin (**b**) has been developed originally by Lund University (Sweden))

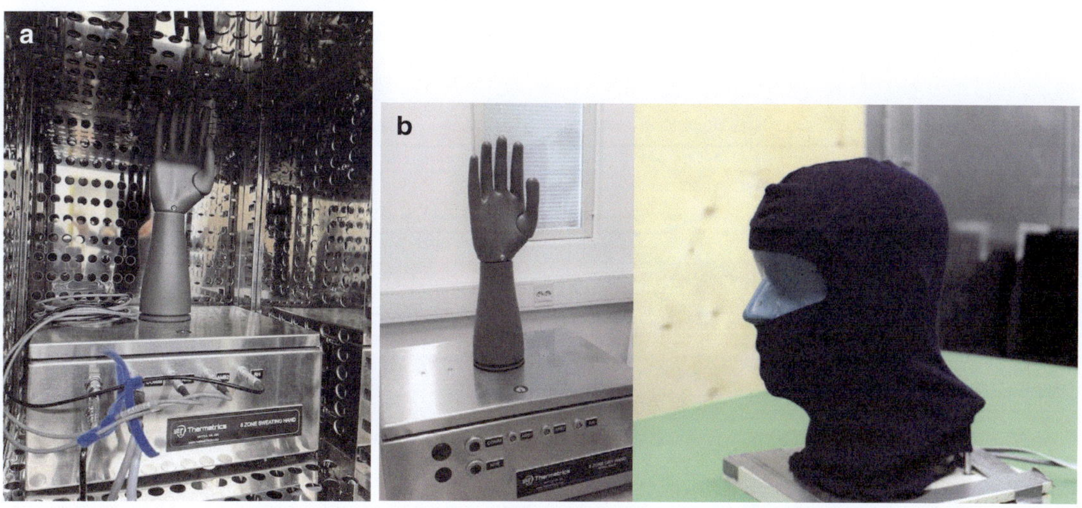

hand (Seoul National University, Korea)

hand and head (Finnish Institute of Occupational Health, Finland; these photos were provided by Finnish Institute of Occupational Health)

Fig. 2.5 Local thermal manikins simulating hand or head. (**a**: Seoul National University (Korea), **b**: Finnish Institute of Occupational Health (Finland))

2.1.3.1 Heat and Water Vapour Transfer with Thermal Manikins

Figure 2.6 shows that heat transfer coefficients by convection and radiation for the whole body and body regions measured using an adult female thermal manikin at 26 °C and 60–75%RH with two airflow conditions of 0.2 m/s and 1 m/s. In calculating the radiative heat transfer coefficient, a rate of 0.96 (−) was employed as emissivity of the thermal manikin's surface and a rate of 0.77 (−) was also employed for, taking into account, effective radiation area (e.g. [19]). For the whole body, the radiative transfer coefficients were approximately 5 W/m²K under the insensible airflow condition and about 4.7 W/m²K under the high airflow condition (Fig. 2.6b). Although those coefficients were quite close, a significant difference was observed between them. It might be due to the airflow condition difference. For the body locations, the radiative heat transfer

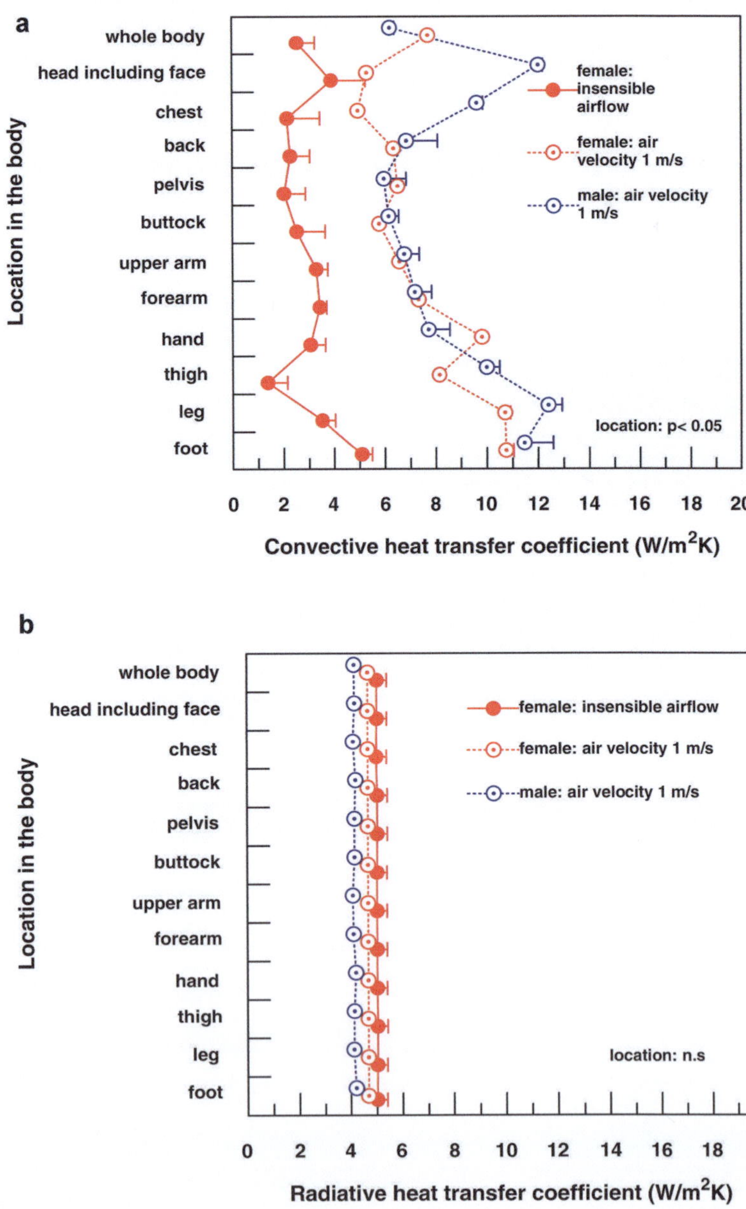

Fig. 2.6 Convective (**a**) and radiative (**b**) heat transfer coefficients of the whole body and locations measured using adult female [10] and male [18] thermal manikins

coefficients under both the insensible airflow and the high airflow conditions were almost the same as the whole body, respectively. It was, therefore, found to not have a significant difference among the locations in the body. As seen in Fig. 2.6a, the convective heat transfer coefficient for the whole body was remarkably larger under the high airflow condition than that under the insensible airflow due to being strongly affected by environmental airflow. The convective heat transfer coefficient in each location showed different values under the insensible airflow condition, with a significant dependence upon the body part. Remarkable differences between body locations were also observed in the convective heat transfer coefficient in each body region during exposure to the high airflow environment. The convective heat transfer coefficients in the front and back of the trunk were smaller than those in the extremities such as the leg and the arm. Tendencies were observed that a larger convective heat transfer coefficient was seen in the foot and in the hand among the lower limbs and upper limbs, respectively. These tendencies became more noticeable under the condition of the high airflow compared to that of the insensible airflow. Convective heat transfer across a cylindrical object under high airflow conditions is expressed by the semi-theoretical equation (Eq. 2.6) as follows [3, 20]:

$$h_c = Nu \frac{\lambda_{air}}{d} \tag{2.6}$$

$$Nu = 0.0266 \cdot Re^{0.805} \cdot Pr^{\frac{1}{3}} \tag{2.7}$$

$$Re = \frac{v \cdot d}{\nu_{air}}. \tag{2.8}$$

where

v: air velocity (m/s)
h_c: convective heat transfer coefficient (W/m²K)
d: diameter of cylinder (m)
Nu:Nusselt number (−)
Re: Reynolds number (−)
Pr: Prandtl number (−)
λ_{air}: thermal conductivity of air (W/mK)
ν_{air}: kinematic viscosity of air (m²/s)

As Eq. 2.6 indicates, the larger the diameter of the cylinder, the smaller the convective heat transfer coefficient, even when exposed to the same air velocity. This is one of the reasons why there is a difference in the convective heat transfer coefficient between locations in the body. The local convective heat transfer coefficient is plotted against the diameter of each body region in Fig. 2.7. It can be confirmed through the figure that the smaller the diameter of the body part, the larger the convective heat transfer coefficient [21, 22].

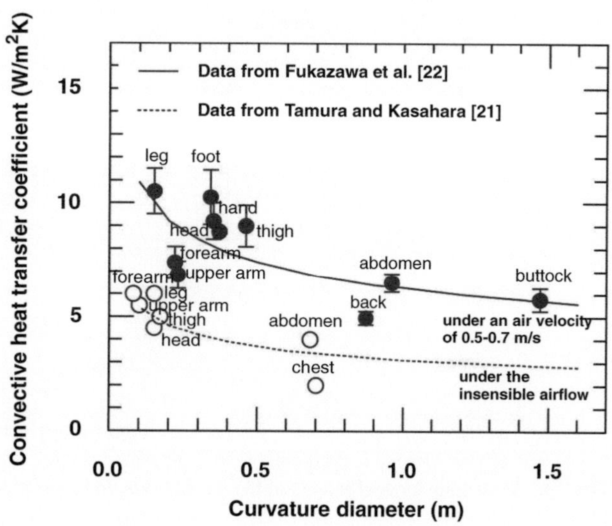

Fig. 2.7 Local convective heat transfer coefficient for the human body depends upon its morphological structure such as expressed as the local curvature diameter size [21, 22]

Heat transfer coefficients by convection and radiation for the whole body and locations measured using an adult male sweating thermal manikin at 5–6 °C and 70%RH with an air velocity of 1 m/s are additionally plotted in Fig. 2.6, expressed as blue symbols. The radiative heat transfer coefficients of whole body and each location obtained from the adult male sweating thermal manikin were lower than those obtained from the adult female thermal manikin. This might be due to the influence of the ambient temperature difference between the measurements. The convective heat transfer coefficients of the whole body and each location obtained from the adult male sweating thermal manikin were similar to those obtained using the adult female thermal manikin, except for the head, including the face, and the chest. The head, including the face, of the adult male sweating thermal manikin did not have eyes, nose, or mouth and was like a simple cylindrical shape with no irregularities. On the other hand, the head, including the face, of the adult female thermal manikin had a shape similar to that of human, with eyes, nose, ears, mouth, and the chest had two bulges that simulate breasts. Therefore, the difference in the convective heat transfer coefficient would be caused by the body shape difference between the two manikins [11]. This might be because, in the adult female thermal manikin, rounded elongated ridge such as the nose on the face and larger ridges like breasts on the chest act as stagnation points, which impede flow and reduce the convective heat transfer [20].

Mass transfer coefficient for the whole body and locations measured using the adult male sweating thermal manikin at 5–6 °C and 70%RH with an air velocity of 1 m/s is shown in Fig. 2.8. A significant local difference was also observed in the mass transfer coefficient. The mass transfer coefficients in both the upper and lower limbs were significantly larger than those in the front and back of the trunk, as same as the convective heat transfer coefficient. As mentioned above, this is because the similarity of heat transfer expressed as the *analogy of Reynolds-Colburn* can be applied for the mass transfer [3, 7].

It is very important to conduct experiments with human subjects to understand physiological and psychological responses in order to evaluate thermal environments. In the experiments with subjects, however, they might sometimes feel

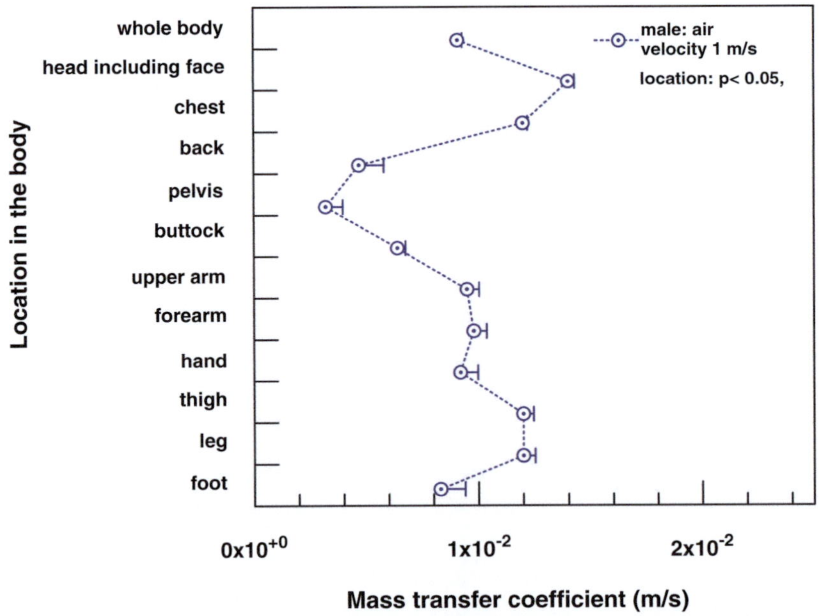

Fig. 2.8 Mass transfer coefficients of the whole body and locations measured by the sweating thermal manikin of male [18]

discomfort due to restrict of their posture and movements during the test because of various sensors attached to their bodies. It can be, therefore, noticed that employing thermal manikins and sweating thermal manikins instead of humans is much more effective for evaluating and understanding characteristics of the heat and water vapour transfer from the body surface as shown in Figs. 2.6 and 2.8.

There are two methods, which are expressed as serial method and parallel method, to calculate the whole body heat transfer coefficient (or thermal resistance) from measurements using a thermal manikin. In the former method, the heat transfer coefficient (or thermal resistance) of each location is calculated once, and next, in order to obtain the heat transfer coefficient of the whole body, the heat transfer coefficient (or thermal resistance) of each location is summed according to the ratio of the surface area of each location to the whole body. In the latter method, the heat transfer coefficient (or thermal resistance) of the whole body can be obtained by mean weighted surface temperature and mean weighted heat flux. Figure 2.9 shows thermal resistance of the whole body obtained through both serial and parallel methods [23, 24]. There was a difference in the resulting thermal resistance between the serial and the parallel methods as seen in the figure. It has been reported that a slight overestimation is produced by calculating thermal resistance using the serial method [26]. Therefore, as specified in ISO15831 [25], the parallel method should be employed for calculating the heat transfer coefficient (or thermal resistance) of the whole body measured with a thermal manikin. In the case of measurements using a sweating thermal manikin, the parallel method should also be employed to obtain the mass transfer coefficient (or water vapour resistance) of the whole body.

2.1.3.2 Reliability and Reproductivity

According to a report about experiments on the physiological and psychological responses during exposure to a cold of 17 °C at normobaric and hypobaric environments, which were respectively equivalent to sea level (0 m) and 5000 m elevation, the overall thermal resistances of the body surface were resulted in 0.14 ± 0.01 m^2K/W at sea level and 0.19 ± 0.02 m^2K/W at the hypobaric environment [23]. The overall thermal resistance of the body surface at sea level was significantly higher than that at the hypobaric environment. Thermal resistance measured by a thermal manikin under the same two environmental conditions was obtained as 0.13 ± 0.01 m^2K/W at sea level of 0 m and 0.19 ± 0.02 m^2K/W at the hypobaric environment of 5000 m altitude. It can be noticed that a good enough validity was obtained between

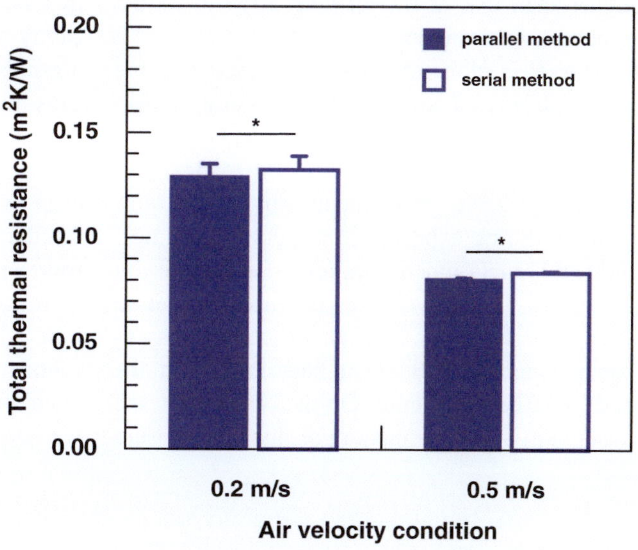

Fig. 2.9 Total thermal resistances of the whole body calculated [23, 24] by serial and parallel methods defined in ISO 15831 [25]

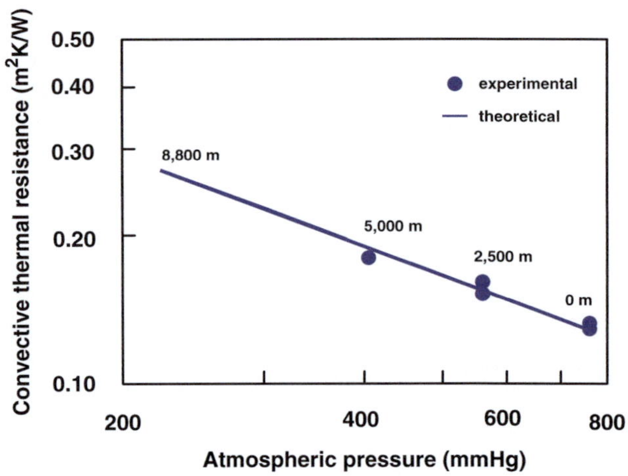

Fig. 2.10 Dependence of atmospheric pressure on the overall thermal resistance was measured experimentally using a thermal manikin and was also calculated theoretically [10, 23]

the overall thermal resistances obtained from measurements of the subject and thermal manikin because of an extremely high correlation with a correlation coefficient of 0.99 (−).

Fig. 2.10 shows the effect of atmospheric pressure upon the overall thermal resistance measured experimentally using a thermal manikin and calculated theoretically. The thermal resistance increases with the decreasing environmental pressure, in another words, the increasing altitude. This is because the kinematic viscosity of air increases due to the decrease in its density with the elevation. Thus, this is the reason why, in the referred report above, the overall thermal resistance from the body surface at sea level of the normobaric environment was significantly higher than that at the altitude 5000 m of the hypobaric environment.

The effect of increased thermal resistance due to the decreased atmospheric pressure is also reflected in human physiological and psychological responses. For example, referring to the report above, Fig. 2.11a shows mean skin surface temperature during exposure to the cold at the normobaric environment largely decreased by 27.4 ± 0.7 °C. This was because vasoconstriction took place due to the increase in activity of sympathetic nerve to diminish heat loss from the body surface in the cold. On the other hand, the mean skin surface temperature during exposure to the cold at the hypobaric environment decreased by 28.7 ± 0.5 °C but did not decrease as much as at the normobaric environment. This was because the activity of sympathetic nerve was depressed due to hypoxic exposure. The amount of heat dissipated from the body surface is approximately proportional to the temperature difference between the body surface and the environment; therefore, the larger the temperature difference between the body surface and the environment, the larger the released heat flux from the body surface. Consequently, a core temperature drop will be seen. Despite the greater temperature difference between the body surface and the ambient at hypobaric environment than the normobaric one, the core temperature at the hypobaric environment did not decrease significantly and remained at the same level as at the normobaric environment, as shown in Fig. 2.11b. This was because the thermal resistance was greater at the normobaric environment than at the normobaric one. In addition, it was found that even at the hypobaric environment, the subjects felt the same coldness as at the normobaric environment, as shown in Fig. 2.11c.

It can be noted that employing thermal manikins is indeed useful for evaluating thermal environments because they truly reproduce the phenomenon of heat transfer from the surface of the human body.

Fig. 2.11 Mean skin surface (**a**) and core (**b**) temperatures as physiological responses and thermal sensation of the whole body (**c**) as one of psychological responses during exposure to the cold of 17 °C at the normobaric environment of 1 atm (sea level) and at a hypobaric environment of 0.55 atm (equivalent to an elevation of 5000 m) [10, 23]

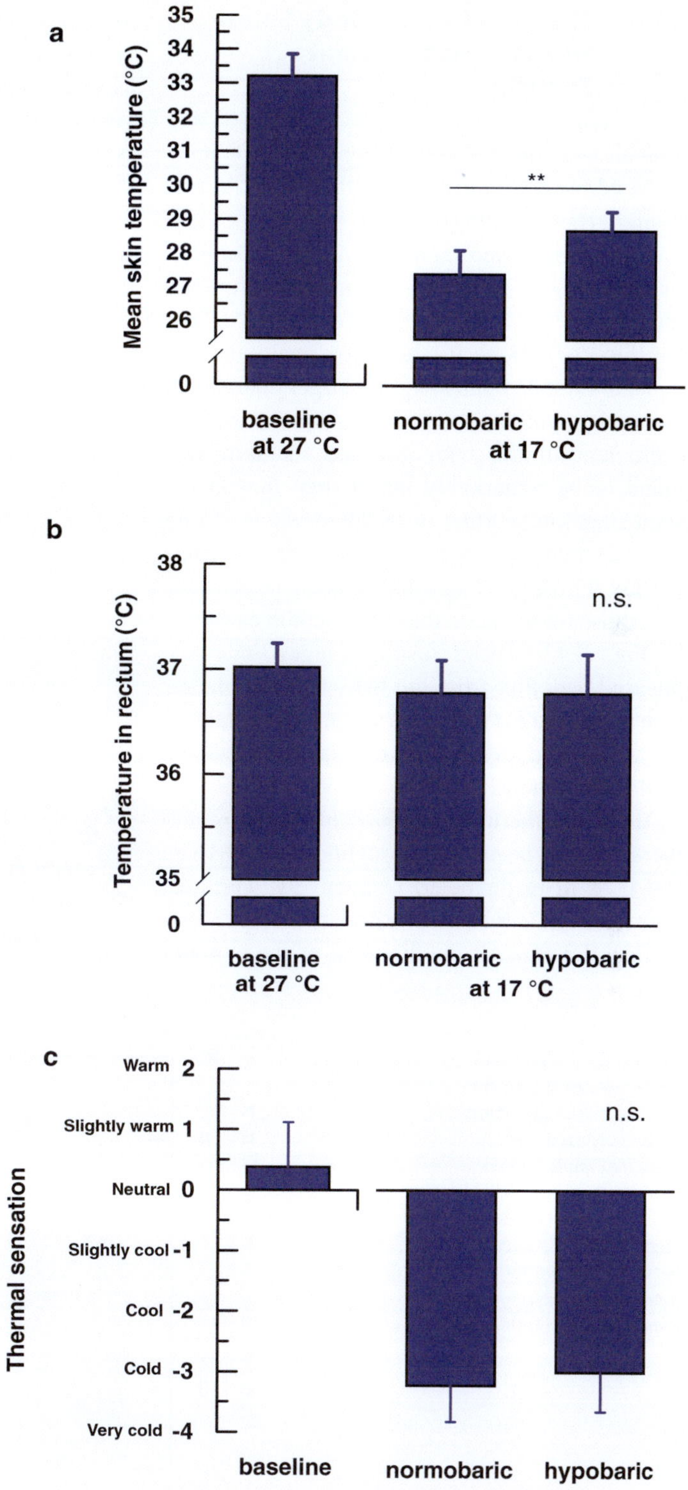

2.1.4 Effects of Human Body Shape and Structure on Heat Transfer

Radiant and convective heat transfer coefficients in the whole body measured with a baby thermal manikin (height 0.65 m, shown in Fig. 2.4a) and an adult female thermal manikin (height 1.68 m, shown in Fig. 2.3f) during a standing posture are shown in Fig. 2.12. While the radiative heat transfer coefficient in the baby manikin was found to be statistically slightly larger than that in the adult manikin, the convective heat transfer coefficient in the baby thermal manikin was found to be remarkably larger than that in the adult manikin. Furthermore, there was a strong tendency that local convective heat transfer coefficients for the baby manikin indicate larger rates than those for the adult thermal manikin as shown in Fig. 2.13. These results suggest that the released heat flux from the baby becomes larger than in the adult if their skin surface temperatures show the same level during exposure to the same cold environment.

Here, the thermal manikins of the baby and the adult can be treated as cylinders of 0.16 m and 0.28 m in diameter, respectively. The heat transfer from a cylinder is similar to that from a vertical flat plate, if the diameter of the cylinder is exceedingly larger than the thickness of the boundary air layer formed over the cylinder itself, as expressed in Eq. 2.9 [3]. The theoretical convective heat transfer coefficients for the whole body can be obtained for both thermal manikins through the relation between Nusselt (Nu (−)) and Rayleigh (Ra (−)) numbers, as expressed in Eqs. 2.10 and 2.11. The Ra is calculated as 2.9×10^3 for the baby manikin and as 6.4×10^4 for the adult manikin according to Eq. 2.11.

$$h_c = \frac{Nu \cdot \lambda_{air}}{L} \tag{2.9}$$

$$Nu = 0.59 Ra^{\frac{1}{4}} \tag{2.10}$$

$$Ra = \frac{g\beta L^3\left(\overline{T}_{sk} - T_a\right)}{\nu_{air}^2} \cdot Pr \tag{2.11}$$

where

h_c: convective heat transfer coefficient (W/m²K)
L: height of the thermal manikin (m)
g: acceleration of gravity (m/s²)
$\overline{T}_{sk}$: mean weighted surface temperature of the thermal manikin (K)
T_a: ambient temperature (K)
β: coefficient of volume expansion (1/K)
ν_{air}: kinematic viscosity of air (m²/s)

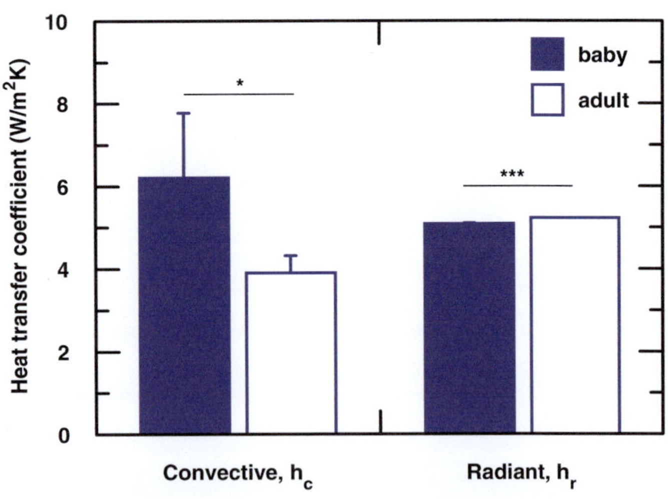

Fig. 2.12 Radiant and convective heat transfer coefficients in the whole body measured with baby and adult female thermal manikins [10, 12, 27]. Asterisk means significant difference between the two manikins in each location. *: $p < 0.05$, and ***: $p < 0.001$

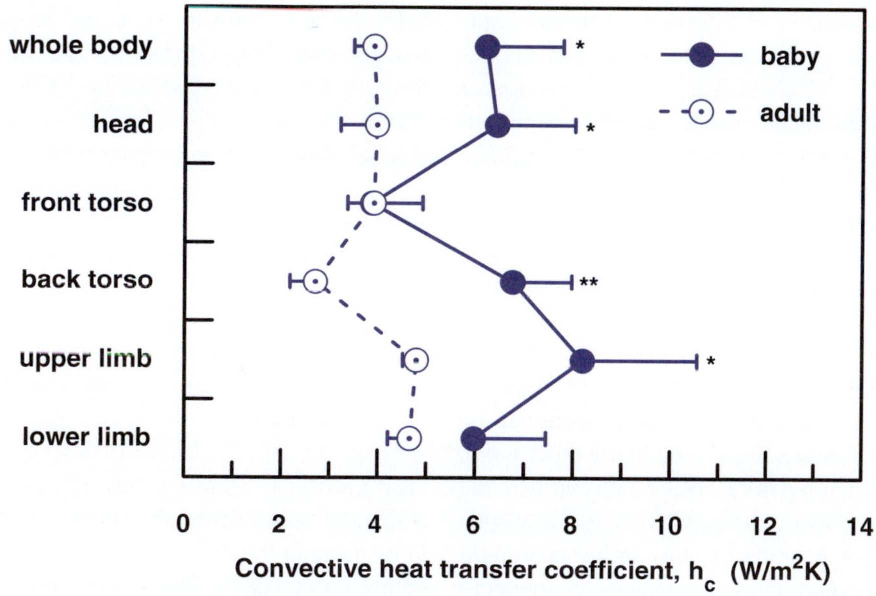

Fig. 2.13 Convective heat transfer coefficients in locations and in the whole body for the baby and the adult thermal manikins [10, 12, 27]. Asterisk means significant difference between the two manikins in each location. *: $p < 0.05$, and **: $p < 0.01$

Pr: Prandtl number (−)
Nu: Nusselt number (−)
Ra: Rayleigh number (−)

The *Nu* increases with the height to the power of 0.75, because the *Ra* increases the third power of the manikin's height as seen in Eq. 2.11. Therefore, the connective heat transfer coefficient increases with the increasing height to the power of −0.25. That is, the convective heat transfer coefficient becomes larger for the smaller manikin than for the taller one. This is the reason why the experimentally obtained convective heat transfer coefficient showed a larger rate in the baby thermal manikin than in the adult thermal manikin.

Figure 2.13 indicates local convective heat transfer coefficient of each location in the body. Since there is indeed the effect of height upon the convection mentioned above, local differences in the heat transfer coefficient are found to be significant in the whole body.

Thermal resistance from the body surface to the environment has been measured by Kuklane et al. [11] using two thermal manikins simulating an adult male (height 1.71 m) and an adult female (height 1.68 m). As a result, no statistical difference was found to be significant in the total thermal resistance, because the obtained thermal resistances for both the male and female manikins were approximately 0.10 m^2K/W, respectively. This was probably because the heights of both the thermal manikins were so similar, around 1.7 m, that the *Ra* numbers for both the manikins, mentioned above, were approximately 6×10^4, respectively. As strongly emphasized in ISO 15831 [25], it must be, however, noted that the size, such as height of the thermal manikin, has a significant impact on the evaluation of heat transfer from the body surface as well as the thermal resistance of clothing.

2.2 Heat and Water Vapour Transfer from Human Through Clothing

Human has a thermoregulatory function that tries to balance the produced and released heat amounts, because human is a homeotherm whose core temperature should be constantly regulated at a level around 37 °C. The human thermoregulation consists of both autonomic and behavioural mechanisms. Behavioural thermoregulation includes the act of dressing on/putting off clothes according to the exposed environment. If a human puts some clothes on, thin air layers are formed between the body surface and clothes and between the clothes. The innermost air layer among these formed air layers is called microclimate. If the microclimate is kept within the range of 32 ± 1 °C and 50 ± 10%RH with an air velocity of 0.15–0.25 m/s [28], then a thermally comfortable state can be achieved, as shown Fig. 2.14. On the other hand, if, e.g. the temperature and/or humidity of the microclimate is/are deviated from the appropriate range, a state of thermal discomfort is brought about. This thermal comfort range is independent of the environment. Behavioural thermoregulation employing clothes on/off is often selected to maintain the microclimate within this specific range, because it enables us to have easier fine control and can be done anywhere without using electric power.

Humans produce heat inside their bodies and dissipate the produced heat at the same time. In addition, insensible evaporation always occurs from the body surface, and sweating sometimes occurs. Therefore, in order to be in a thermally comfortable state for clothed humans, it should be better to understand the impacts of clothes on both the simultaneous heat and water vapour transfer from body surface to the environment through the clothes themselves. The effects of clothes on heat and water vapour transfer in clothed humans are divided into three main points: the first is the physical properties inherent to the materials of the clothes, the second is how the clothes are worn, and the third is movement.

2.2.1 Heat Transfer from the Body to the Environment Through Clothing

Heat transfer in clothing generally refers to sensible heat transfer from the surface of the human body through the clothing to the environment by pathways of conduction, convection, and radiation (Eq. 2.12). The total amount of sensible heat flux through these three pathways (q_s) is roughly proportional to the temperature difference between the body surface ($\bar{T}_{sk}$) and the environment (T_a) (Eq. 2.13). The total amount of sensible heat flux through these three pathways is determined by the thermal resistance (R_{s_t}) and the temperature difference between the human body surface and the environment. If it is necessary to prevent hypothermia in a cold environment, wearing clothing with a high thermal resistance can reduce the sensible heat flux transmitted from the body through the clothing. On the other hand, if it

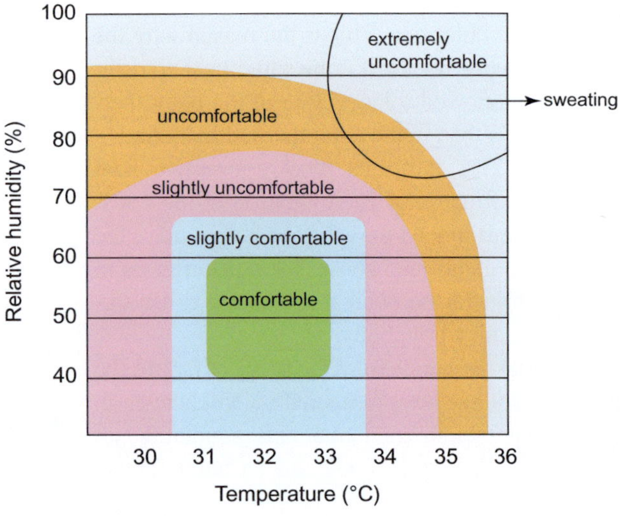

Fig. 2.14 Diagram of thermal comfort range expressed in terms of temperature and relative humidity in the microclimate, which is the innermost air gap between the body surface and the clothing [28]

is necessary to prevent a rise in core temperature in a hot environment, wearing clothing with a low thermal resistance does not interrupt the sensible heat transfer from the surface of the human body.

For a clothed human, the thermal resistance from the body surface to the outside surface of outer clothes is called the thermal resistance of clothing ($R_{s_{cl}}$) and it is also called the thermal resistance of the clothing ensemble in the case of layered clothing. The $R_{s_{cl}}$ is in series with the thermal resistance of outside-to-environment (R_{s_a}). Therefore, the R_{s_t} mentioned above consists of these resistances of clothing and outside-to-environment as expressed in Eq. 2.14.

$$q_s = K + C + R \quad (2.12)$$

$$q_s = \frac{\overline{T}_{sk} - T_a}{R_{s_t}} \quad (2.13)$$

$$R_{s_t} = R_{s_{cl}} + R_{s_a} \quad (2.14)$$

where

K: heat flux from the body surface by conductance (W/m^2)
C: heat flux from the body surface by convection (W/m^2)
R: heat flux from the body surface by radiation (W/m^2)
q_s: sensible heat flux from the body surface (W/m^2)
$\overline{T}_{sk}$: mean weighted body surface temperature (K)
T_a: ambient temperature (K)
R_{s_t}: thermal resistance (m^{2}K/W)
$R_{s_{cl}}$: thermal resistance of clothing (m^{2}K/W)
R_{s_a}: thermal resistance of outside-to-environment (m^{2}K/W)

2.2.1.1 Clo—One of the Units Expressing Thermal Resistance

Resistance of clothing is often expressed in units of clo, which was proposed by Gagge et al. in 1941 [29]. This unit of clo is defined as follows; clothing that has thermal insulation of 1 clo makes an adult male feel neither hot nor cold and stay thermally comfortable if he stays in a room in which the temperature and relative humidity are maintained at 21.2 °C and 50% with an airflow of 10 cm/s. According to this definition, his average skin surface temperature shows approximately 33 °C, and heat production is 50 kcal/m^{2}h, of which 76% is dissipated through clothing. In addition, the thermal resistance from the outside of clothing to the environment is also defined as 0.14 m^{2}hK/kcal. Therefore, 1 clo corresponds to 0.18 m^{2}hK/kcal, which can also be expressed 0.155 m^{2}K/W in the general unit of thermal resistance.

2.2.1.2 Obtaining Method of the Thermal Resistance of Clothing

A conceptual diagram of heat transfer in layered clothing is simplified in Fig. 2.15. In this concept, the thermal resistance of clothing includes total thermal resistance of clothing (I_t) as heat transfer from body surface to the environment though the clothing, effective thermal resistance of clothing (I_{cle}), and basic thermal resistance of clothing (I_{cl}), in which increase in heat transfer area because of wearing clothes is included, as heat transfer from the body surface to the outside of clothing. These

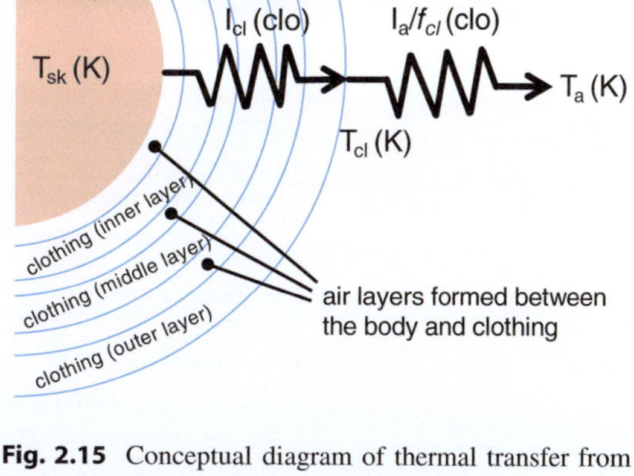

Fig. 2.15 Conceptual diagram of thermal transfer from the body to the environment through clothing

thermal resistances of clothing have their own definitions. The measurement method for the ratio of increase in heat transfer area due to wearing clothes, which is called as clothing area factor (f_{cl}), has not been, however, standardised yet. Therefore, it is often obtained by various methods, mainly the photographic method as reported by, e.g. Ueno and Sawada [30].

The I_t consists of the I_{cl} and the outer surface of the garment to the ambient environment (I_a) with consideration of the f_{cl}, as shown in Eq. 2.15. Therefore, the I_{cl} can be obtained by subtracting the I_a, considering the f_{cl}, from the I_t.

$$I_t = \frac{\overline{T}_{sk} - T_a}{0.155\ q_s} = I_{cl} + \frac{I_a}{f_{cl}} \qquad (2.15)$$

where

f_{cl}: clothing area factor which indicates ratio of increase in surface area due to wearing clothes to body surface area (−)

I_t: total thermal resistance of clothing from body surface to the environment though the clothing (clo)

I_a: thermal resistance of air layer which is from the outer surface of the clothes to the ambient environment (clo)

I_{cl}: basic (intrinsic) thermal resistance of clothing (clo)

q_s: sensible heat flux (W/m^2)

$\overline{T}_{sk}$: mean weighted skin surface temperature (K)

T_a: environmental temperature (K)

Thermal resistance is also expressed as effective thermal resistance of clothing (I_{cle}); it can be simply obtained by subtracting I_a from I_t (Eq. 2.16). For a single piece of clothing that is not layered, the effective thermal resistance of the clothing is expressed as I_{clu}. Table 2.3 shows values of the I_{clu}, which are often employed as daily wear [31].

Table 2.3 Thermal resistance of general clothing for daily use

Type of clothing		Thermal resistance (m^{2}K/W)	I_{cl} (clo)
Underwears for lower body	Panties	0.005	0.03
	Briefs	0.003–0.006	0.02–0.04
	Long pants	0.012–0.034	0.08–0.22
	a combination of panty and bra	0.0047–0.006	0.03–0.04
Underwears for upper body	Bras	0.002	0.01
	Waist length slips	0.022	0.14
	Knee length slip	0.025	0.16
	T-shirts	0.008–0.019	0.05–0.12
	Long sleeve shirts	0.017–0.039	0.11–0.25
Shirts, blouses	Collared short sleeve shirts made with thin fabric	0.029	0.19
	Collared short sleeve shirts made with plain fabric	0.037–0.039	0.24–0.25
	Collared short sleeve shirts made with thick fabric	0.039	0.25
	Collared long sleeve shirts made with thin fabric	0.023–0.028	0.15–0.18
	Collared long sleeve shirts made with plain fabric	0.026–0.051	0.17–0.33
	Collared long sleeve shirts made with thick fabric	0.039–0.053	0.25–0.34
	Collarless sleeveless shirts	0.020–0.033	0.13–0.21
	Collared long sleeve blouses made with thin fabric	0.023	0.15
	Long sleeved blouses with collar made of thick fabric	0.052	0.33

(continued)

Table 2.3 (continued)

Type of clothing		Thermal resistance (m^2K/W)	I_{cl} (clo)
Pants, trousers	Short pants, half pants	0.009–0.017	0.06–0.11
	Tight fitting trousers	0.028–0.037	0.18–0.24
	Loose fit trousers	0.033–0.039	0.21–0.28
	Work pants, overalls	0.037–0.039	0.24–0.28
Sweaters	Round neck/V neck sweaters made with thin fabric	0.020–0.031	0.13–0.2
	Round neck/V neck sweaters made with plain fabric	0.036–0.039	0.23–0.25
	Round neck/V neck sweaters made with thick fabric	0.048–0.056	0.31–0.36
	Turtleneck sweaters using with thin fabric	0.040	0.26
	Turtleneck sweaters using with plain fabric	0.057	0.37
	Turtleneck sweaters using with thick fabric	0.084	0.54
Jackets	Single jackets made with plain fabric	0.056–0.070	0.36–0.45
	Single jackets made with thick fabric	0.062–0.070	0.4–0.45
	Double jackets made of plain fabric	0.065	0.42
	Double jackets made of thick fabric	0.074	0.48
Skirts	Maxi length skirts	0.036–0.043	0.23–0.28
	Midi length skirts	0.026–0.030	0.17–0.19
	Knee length skirts	0.022–0.036	0.14–0.23
	Mini length shkirts	0.016	0.10
	Knee length flared skirts	0.022	0.14
	Knee length gathered skirts	0.019–0.028	0.13–0.18
	Knee length pleated skirts	0.025–0.040	0.16–0.26
Dresses	Knee length long sleeve dresses	0.050–0.073	0.32–0.47
	Knee length short sleeve dresses	0.045	0.29
	Knee length sleeveless dresses	0.036–0.042	0.23–0.27
Outdoor clothing	Coats	0.093	0.60
	Down jackets	0.085	0.55
	Hoodies	0.109	0.70
Nightdress	Long sleeve pajamas	0.074–0.088	0.48–0.57
	Short sleeve pajamas	0.065	0.42
	Long sleeve nightgowns	0.037–0.071	0.24–0.46
	Short sleeve nightgowns	0.028–0.039	0.18–0.25
Others	Ankle length socks	0.003	0.02
	Mid-calf length socks	0.005	0.03
	Over the calf length socks	0.004–0.009	0.03–0.06
	Thin insoles	0.003	0.02
	General thickness to thick insoles	0.006–0.008	0.04–0.05
	Gloves	0.008–0.012	0.05–0.08

These listed data are quoted from ISO 9920: Ergonomics of the thermal environment—Estimation of thermal insulation and water vapour resistance of a clothing ensemble [31]

Fig. 2.16 Relation between the thermal resistances of clothing ensembles, I_{cl}, and the sum of individual thermal resistance of clothing, $\sum I_{clu}$ (e.g. [12, 27])

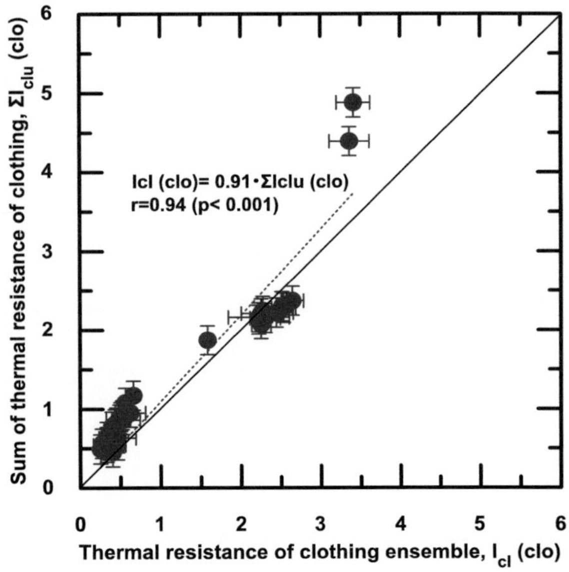

$$I_{cle} = I_t - I_a = I_{cl} + I_a\left(\frac{1}{f_{cl}} - 1\right) \quad (2.16)$$

where

I_{cle}: effective thermal resistance of clothing (clo)

It has been experimentally verified that the basic thermal resistance, I_{cl}, of layered clothing is approximately equivalent to the sum of the thermal resistances of the individual clothing that makes up the layered clothing, as shown in Eq. 2.17 and in Fig. 2.16 (e.g. [12, 27]).

$$I_{cl} = \sum_{i=1}^{n} I_{clu_i} \quad (2.17)$$

where

i: number of clothing layers

I_{clu_i} : effective thermal resistance of the clothing in the case of a single layered (clo)

In order to obtain the thermal resistance of clothing, thermal manikins, as described in the previous section, are often employed (Fig. 2.17). This is because not only the body shape but also the surface temperature or the heat flux can simulate almost exactly the same as human, and their representativity is also quite good.

2.2.1.3 Estimating Method of the Thermal Resistance of Clothing

Estimation methods for both the thermal resistance of the clothing, I_{clu}, and the clothing area factor, f_{cl}, have been given in ISO 9920 [31] as the following Eqs. 2.18–2.20.

For a single layered clothing:

$$I_{clu} = 0.043 \cdot A_{\text{cov}} + 1.4 \cdot d_{fab} \cdot A_{\text{cov}} \quad (2.18)$$

For layered clothing:

$$I_{cl} = 0.835 \sum_{i=1}^{n} I_{clu_i} + 0.161 \quad (2.19)$$

$$f_{cl} = 1 + 0.28 \cdot I_{cl} \quad (2.20)$$

where

A_{cov}: percentage of surface area covered with clothing in the whole body (%)

D_{fab}: thickness of clothing (m)

2.2.1.4 Major Factors Affecting Heat Transfer in Clothing

Heat transfer in clothing is influenced not only by the characteristics of clothing materials but also by the way the clothing is worn and by body

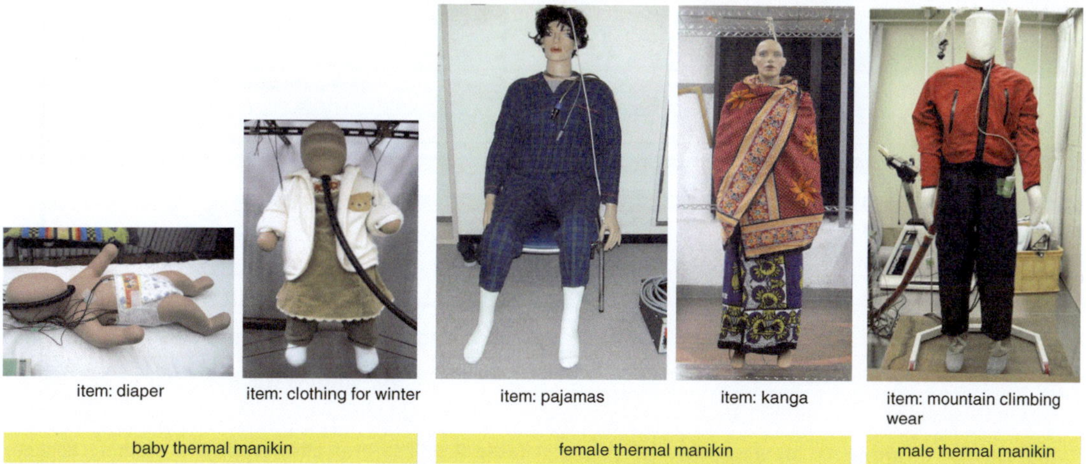

Fig. 2.17 Pictures of measuring thermal resistance of clothing using thermal manikins

movements; in detail, thermal resistance of clothing is affected by the thermal properties of the clothing materials themselves, clothing conditions such as the covered regions in the body by the clothing and their area, openings, thickness of the air layer under the clothing, and the number of layered clothing. In addition, bellows action caused by body movements accelerates pumping effects and ventilation in the clothing, and consequently the thermal resistance decreases.

Thermal Properties of Material for Clothing

As for the thermal properties, the thermal conductivity of fibres for clothing materials [2] is shown in Table 2.1, along with that of some substances (e.g. [1]). Although the thermal conductivities of fibres for clothing material are smaller than that of water, they are slightly larger than that of air. Table 2.4 shows the percentage of fibre in the volume of fabric for clothing [32, 33]. As you can see from this table, the ratio of fibres in the fabric volume is around 1% for down jackets, 10–20% for general clothing, and less than 30% for very thin and dense windproof jackets. Therefore, it can be noticed that the clothing contains both fibre and air in its volume. Furthermore, the clothing contains a lot of air, and its ratio of air to volume is extremely high.

Table 2.4 Thermal properties of fabrics for clothing are quoted from a report by Fujimoto and Seki [32] and by Fujimoto and Niwa [33]

Fabrics for clothing	Thermal conductivity/ Apparent thermal conductivity (W/mK)	Packing factor (%)
Furs	0.099–0.145	4.41–5.79
Faux furs	0.103–0.121	1.47–1.87
Down jackets	0.149	1.36
Newly develpoed material for cold climates	0.034	7.69
Natural leathers	0.048–0.056	23.81–30.19
Synthetic leathers	0.051–0.063	16.14–22.60
Worsted fabrics	0.029–0.053	10.19–18.20
Woven woolen fabrics	0.032–0.042	9.02–14.78
Woven cotton fabrics	0.036–0.054	12.62–22.98
Woven synthetic fabrics	0.027–0.043	16.84–30.96
Wool knit fabrics	0.038–0.050	10.27–21.53
Cotton knit fabrics	0.059	12.60
Synthetic knit fabrics	0.039–0.049	12.26–24.14
Blended fabrics	0.041	15.20
Woven silk fabrics	0.028	27.20

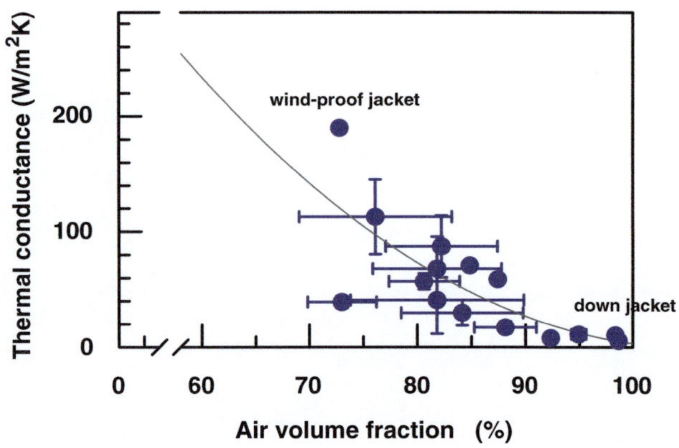

Fig. 2.18 Relation between the air volume fraction and its thermal conductance [32]. The air volume fraction of a down jacket is around 99%, while that of an air impermeable jacket such as a wind proof jacket is around 70%

Fabric for clothing is an assembly of fibres and air, so its thermal conductivity is expressed as apparent thermal conductivity (e.g. [32]). Thermal conductance of a fabric can be obtained by dividing the apparent thermal conductivity by the thickness, and its reciprocal is the thermal resistance of the fabric. As can be seen in Fig. 2.18, clothing containing a larger amount of air in the fabric has higher thermal insulation.

In radiation heat transfer, the emissivity of an object is one of the dominant factors in determining the amount of heat transfer by radiation. Table 2.2 [3, 4], as previously shown, indicates the emissivity of various materials for infrared radiation. The emissivity of fabrics for clothing ranges from approximately 0.8 (−) to 1.0 (−), because their surface structures are so uneven and complicated that there is not a large difference in the emissivity against infrared rays (0.8 μm–0.8 mm electromagnetic waves). On the other hand, under sunlight, the reflectivity and absorptivity of fabrics for clothing against the visible light with a wavelength of 0.4–0.8 μm are strongly influenced by their surfaces' colour. Clothing coloured in black or dark has high absorptivity, while that in white or pale has low absorptivity against the visible light. The surface temperature of clothing is, therefore, largely influenced by the colour of the clothing itself under the visible light. In Table 2.5 [34], the heat absorbability in clothing is expressed as a ratio of the amount of heat absorbed in white clothing to that absorbed in the other coloured clothing when exposed to the visible light.

Table 2.5 The heat absorbability in clothing depends upon the difference in colour

Colour of clothing	Heat absorbability (−)
Green	1.1–3.2
Black	1.4–5.5
Blue	2.0–4.7
Navy	1.9–5.4
White	1.0
Light bule	1.7–3.6
Pink	1.0–3.8
Light gray	1.5–2.1
Yellow	1.5–2.4

Listed data are quoted from Mitsui [34] and from the author's original unpublished experiment

Location and Its Area Covered with the Clothing in the Whole Body

A relatively still air layer is formed on the body surface and underneath the clothing when a human wears clothing. The thermal insulation effect of this static air layer suppresses heat dissipation from the body surface. Therefore, as shown in Fig. 2.19 [22, 35], the thermal resistance of the clothing increases with an increase in the covered area by clothing on the body.

However, even if the body surface area ratio covered by clothing is the same, the thermal resistance of clothing varies slightly depending upon the body parts. For example, the thermal resistance of clothing that covers the limbs is larger than that of clothing that covers the trunk. This is because the convective heat transfer coefficient from thinner parts of the body, such as arms or legs, indicates higher value compared to

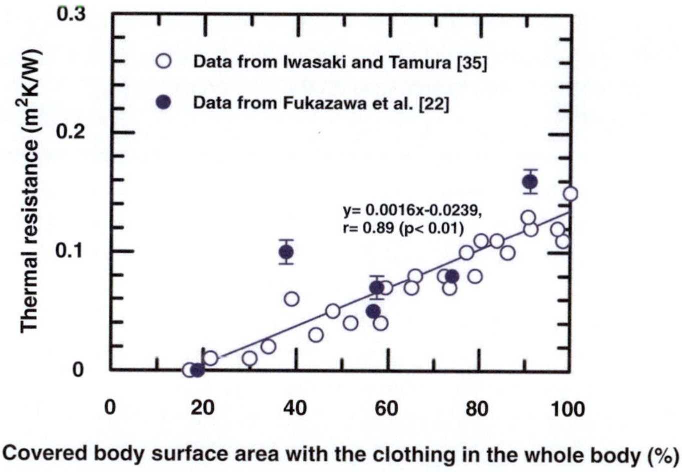

Fig. 2.19 The thermal resistance is plotted against the ratio of the covered body surface area with clothing in the whole body [22, 35]

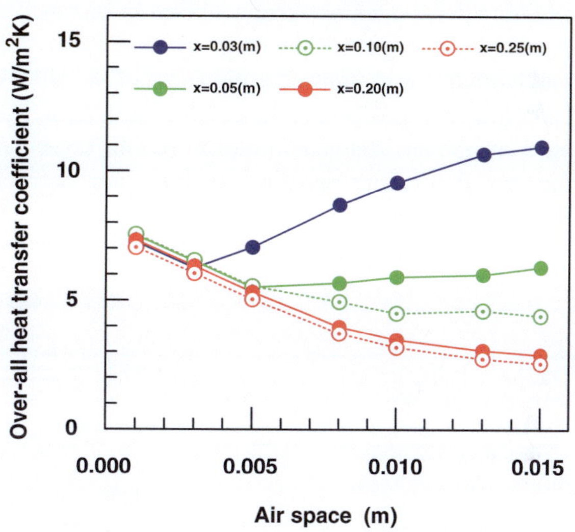

Fig. 2.20 Effects of the size of the air space on the overall heat transfer coefficient [36]. In the figure, *x* represents the height above the hemline

thicker parts of the body, such as the chest or back, as shown in Fig. 2.7 [21, 22].

Air Gap Thickness Under Clothing

Air layers are formed between the body surface and the innermost clothing, as well as between clothing when a human wears clothing. Although air is the best thermal insulator, the thermal resistance of each formed air layer in the clothing is not exactly proportional to its thickness. Figure 2.20 [36] shows the effects of air layer thickness and height on the heat transfer coefficient in clothing. At higher positions such as around the collar of clothing, e.g. $x = 0.25$ m or $x = 0.20$ m in the figure, the heat transfer coefficient decreases gradually as the air layer thickness increases. At lower positions such as at the hem of clothing, the heat transfer coefficient decreases in inverse proportion to the thickness of the air layer up to 3–5 mm due to viscosity of air; in other words, the thermal resistance increases in proportion to the thickness of the air layer up to around 3–5 mm. However, the heat transfer coefficient increases if the thickness of the air layer in the clothing exceeds about 5 mm, because air comes in from the openings of the clothing and/or permeates through the fabrics of the clothing (e.g. [36, 37]). This is because the incoming air creates a flow in the gap and promotes convection. Therefore, high thermal resistance would be obtained by clothing in which the thickness of the air layer could be maintained at about 3–5 mm. On the other hand, in the case where low thermal resistance is desirable, e.g. in summer, it would be better to keep the air layer in clothing larger.

Openings in Clothing

Clothes consist of several parts, some of which, such as collars, cuffs, and hem, are called as openings in clothing, because air from the environment comes in and air in the clothing comes out through them. Not only the size and shape of the openings but also their direction affect the flow of air in the clothing. Since the structure of the human body is long in the vertical direction, conditions of upward openings such as collars and downward openings such as hems have significant impacts on the flow of air layers in clothing. Even if, for example, the thickness of the air layer in clothing is kept at around 3–5 mm, fresh air from the environment comes in the air layer through downward openings if they are not closed due to being negative pressure inside. The inlet flow velocity of air from downward openings in clothing is affected by both negative pressure and air gap size in the clothing as shown in Figs. 2.21 and 2.22 [36, 38]. When airflow is generated and developed by inlet air through downward openings in clothing, heat dissipation will be promoted due to the low thermal resistance of the clothing if upward openings are not closed [39, 40], as shown in Fig. 2.23. On the other hand, even if inlet air through downward openings in clothing accelerates airflow inside, heat dissipation will be suppressed due to the higher thermal resistance of the clothing if upward openings are closed, as shown in Fig. 2.23. If both upward and downward openings in clothing are not closed, the thermal resistance of the clothing becomes much lower because fresh air comes quickly inside and warmed air comes out smoothly to the environment through the openings. On the other hand, if both upward and downward openings in clothing are closed, the thermal resistance of the clothing is higher because neither inlet nor outlet airflow occurs through the openings.

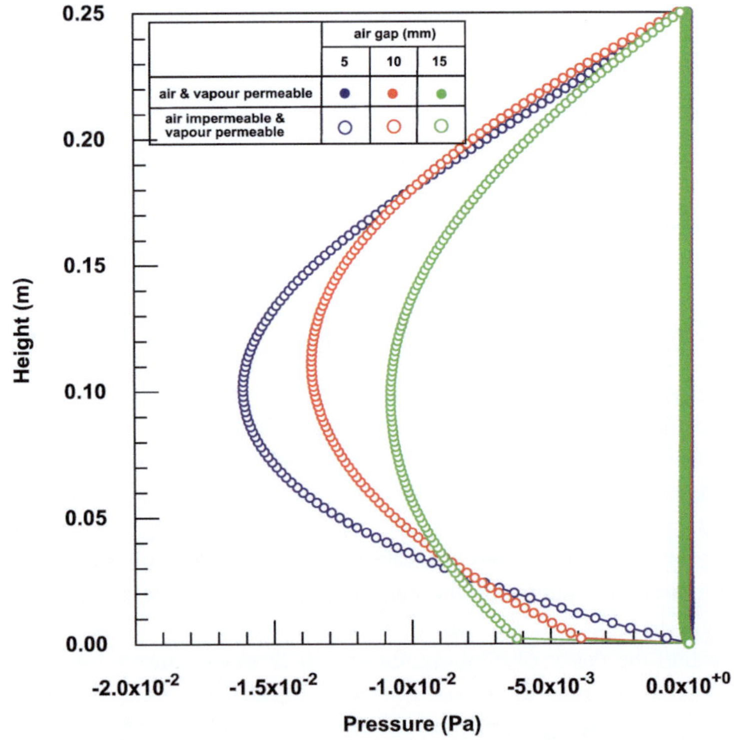

Fig. 2.21 Differences in negative pressure distribution in the vertical direction of the microclimate due to air and/or water vapour permeabilities of clothing [36, 38]

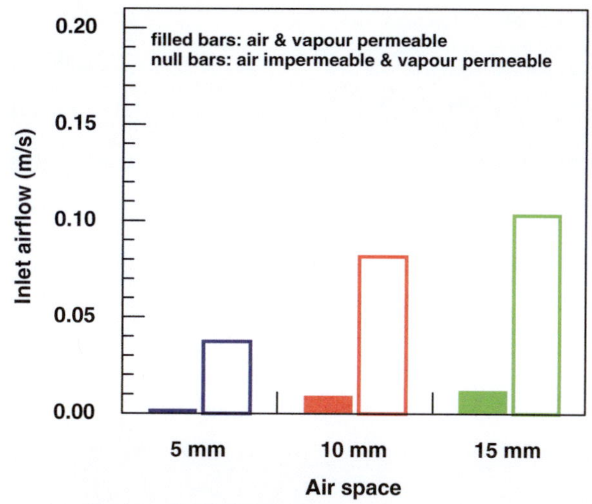

Fig. 2.22 Effects of air and water vapour permeabilities of clothing upon the inlet flow rate from the bottom opening [36, 38]

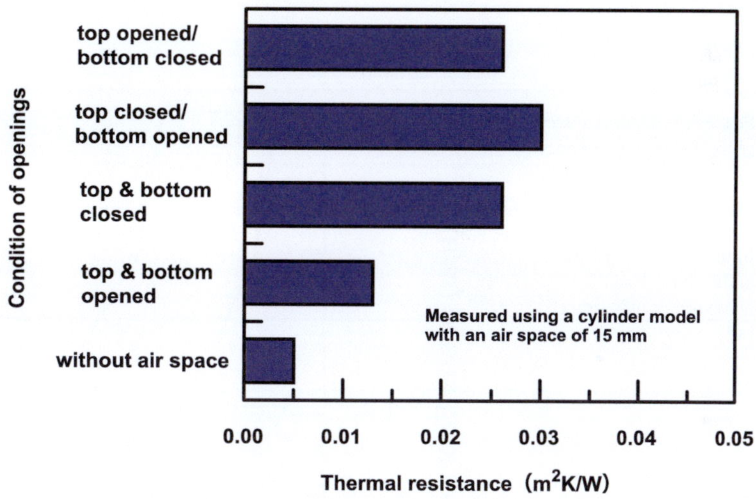

Fig. 2.23 Thermal resistance in clothing is greatly dependent upon its condition of openings [39, 40]

Ventilation and Pumping Effects by Movement

Thermal resistance of clothing is often measured with thermal manikins in static postures such as standing or sitting under controlled environmental conditions. The thermal resistance of clothing is, however, remarkably influenced by ambient airflow and body movement. For example, total thermal resistance of clothing, I_t, decreases with an increase in ambient airflow and also decreases with an increase in walking speed as shown in Fig. 2.24 [41]. In addition, the effect of walking speed on the decrease in the total thermal resistance is much larger than that of ambient air velocity [42] as seen in Fig. 2.25. Since the total thermal resistance of clothing is strongly affected by both air velocity and movement, ISO 9920 [31] provides an equation for correcting the total thermal resistance of clothing by using air velocity and walking speed as shown in Eq. 2.21. An applicable range of the equation is, however, 0.6–1.4 clo for the basic (intrinsic) thermal resistance of clothing, I_{cl}, during rest, or 1.2–2.0 clo for the total thermal resistance of clothing, I_t. In addition, note that the equation is available to modify under conditions in which ambient airflow velocity is from 0 m/s to 3.5 m/s and walking speed is up to 4.3 km/h.

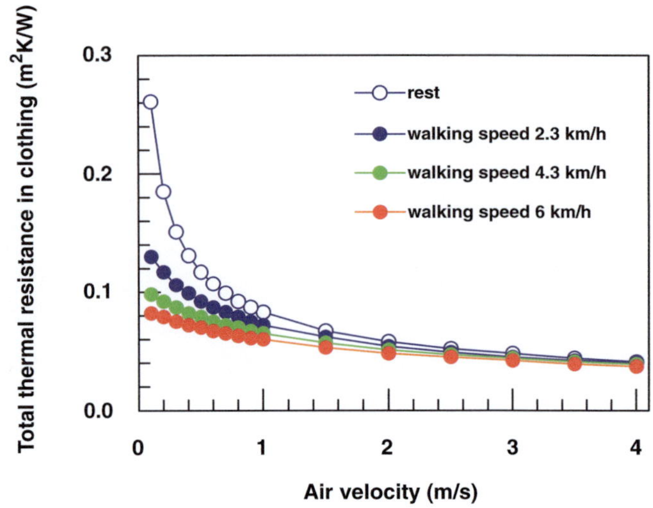

Fig. 2.24 Differences in the thermal resistance of clothing during several walking speeds under various air velocities [41]

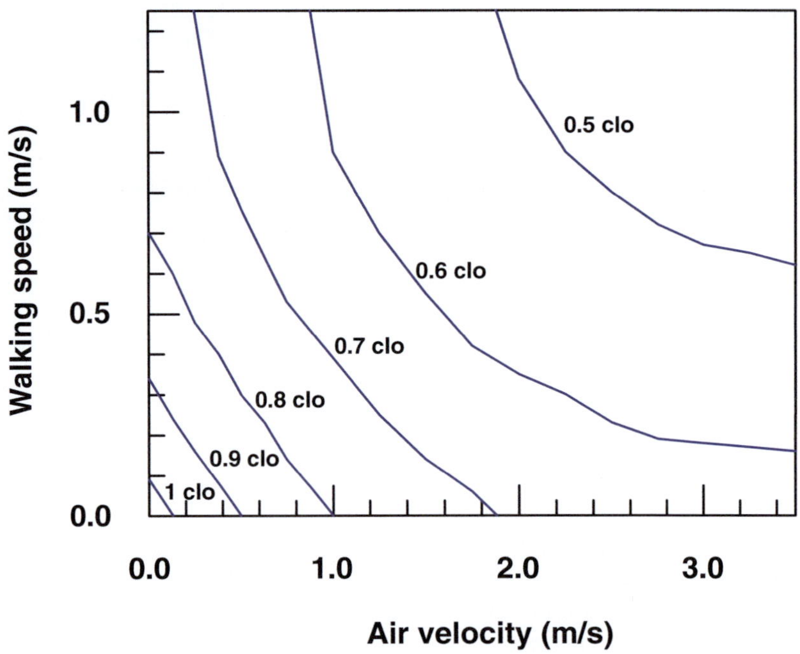

Fig. 2.25 Dependence of the thermal resistance of clothing upon both air velocity around the body and walking speed. The figure was created by the author based on Havenith et al.'s report [42]

$$I_{t,corr} = e^{\left(-0.281(v_{air} - 0.15) + 0.044(v_{air} - 0.15)^2 - 0.492 \cdot v_{walk} + 0.176 \cdot v_{walk}^2\right)} \cdot I_t \tag{2.21}$$

where

v_{air}: air velocity around the body (m/s)
v_{walk}: walking speed (m/s)
I_t: total thermal resistance of clothing during rest (clo)
$I_{t,corr}$: corrected total thermal resistance of clothing considering air velocity and walking speed (clo)

2.2.2 Water Vapour Transfer from the Body to the Environment Through Clothing

Water vapour transfer in clothing is the transfer of water vapour which evaporates from the body surface of a clothed human and transfers to the environment through clothing. A part of the produced heat in the body is consumed as latent heat energy when evaporation occurs over the body surface. The degree of water vapour transfer by insensible perspiration and sweat from the body surface through clothing can be expressed by an index as water vapour permeability resistance or evaporative resistance. For example, latent heat transfer from the body surface is suppressed during the wearing of clothing with a high water vapour permeability resistance, resulting in thermal discomfort being recognised during exercise or in a hot environment as a consequence.

2.2.2.1 Method of Obtaining Water Vapour Permeability Resistance of Clothing

The same concept as the thermal resistance of clothing is basically employed for the evaporative heat resistance of clothing; according to the concept, there are three expressions for the evaporative resistance of clothing: total water vapour permeability resistance of clothing, basic (intrinsic) water vapour permeability resistance of clothing, and effective water vapour permeability resistance of clothing. These water vapour permeability resistances of clothing are evaluated using water vapour concentration/pressure and mass flux as follows.

Total water vapour permeability resistance of clothing (R_{e_t}) indicates evaporative heat resistance from the body surface of a clothed human to the environment through the clothing as shown in Eq. 2.22. It can, therefore, be obtained and expressed by the basic (intrinsic) evaporative heat resistance of the clothing ($R_{e_{cl}}$) and evaporative heat resistance from the outer surface of the clothing to the environment (R_{e_a}) with consideration of the clothing area factor, f_{cl}, as shown in Eq. 2.23.

$$R_{e_t} = \frac{w\left(p_s^* - p_a\right)}{q_e} \tag{2.22}$$

$$R_{e_t} = R_{e_{cl}} + \frac{R_{e_a}}{f_{cl}} \tag{2.23}$$

where

p_a: water vapour pressure at the environment (kPa)
p_s^* : saturated water vapour pressure at the mean weighted skin surface (kPa)
q_e:latent heat flux from the body surface (W/m^2)
w:skin wettedness at the body surface (−)
R_{e_t} : total water vapour permeability resistance of clothing (kPa·m^2/W)
R_{e_a} : water vapour permeability resistance of air layer between surface of outer clothing to the environment (kPa·m^2/W)
$R_{e_{cl}}$: basic water vapour permeability resistance of clothing (kPa·m^2/W)

Water vapour permeability resistance is also expressed as effective water vapour permeability resistance of clothing ($R_{e_{cle}}$); it can be simply obtained by subtracting R_{e_a} from R_{e_t} as shown in the following equation.

$$R_{e_{cle}} = R_{e_t} - R_{e_a} \tag{2.24}$$

where

$R_{e_{cle}}$: effective water vapour permeability resistance of clothing (kPa·m^2/W)

As seen in the above equations for the water vapour permeability resistance of clothing, its physical value is obtained by measuring the water vapour transfer rate from a clothed human body or from a thermal manikin in the same way as the evaluation method of the thermal resistance. For obtaining water vapour permeability resistance of clothing, it is common to employ a sweating thermal manikin such as seen in Fig. 2.3k. In most cases, when measuring the water vapour permeability resistance of clothing, the surface of the sweating thermal manikin is uniformly maintained at a skin wettedness of 1 (−) by means of wearing a wetted knitted garment.

2.2.2.2 Method of Estimating the Water Vapour Resistance of Clothing

Although water vapour permeability resistance of clothing is affected by its fundamental physical properties of both air and water vapour permeabilities and is also affected by the thickness of the air layer underneath the clothing, there is a good correlation between the resistances of heat transfer and water vapour permeability in general clothing, except for water vapour impermeable clothing.

Specifically, water vapour transfer can be represented by the Lewis relation because of the analogy to heat transfer. Therefore, ISO 9920 [31] has proposed a method for estimating water vapour permeability resistance of clothing by employing the Lewis relation from the thermal resistance of clothing. In the estimation method, a water vapour permeability index of *im* (−), which expresses the level of water vapour permeability of clothing ranging from 0 to 1, has also been proposed [43, 44]. This index of *im* (−) is a ratio of clothing's thermal resistance to water vapour permeability resistance expressed as a ratio in the Lewis relation; therefore, if $im = 0$ (−), then the clothing is completely water vapour impermeable, while if the index $im = 1$ (−), then the clothing is completely water vapour permeable. The water vapour permeability index, *im*, ranges from 0.15 to 0.55 (−) for general clothing, with an average value of 0.38 (−).

An estimation equation for the total water vapour permeability resistance of clothing using the Lewis relation has been proposed by ISO 9920 [31] as shown in the following Eq. 2.25.

$$R_{e_t} = \frac{I_t}{im \cdot L} = \frac{0.06}{im}\left(\frac{I_a}{f_{cl}} + I_{cl}\right) \tag{2.25}$$

where

L: Lewis coefficient (16.65 K/kPa = 2.22 K/mmHg)

If the average water vapour permeability for general clothing $im = 0.38$ (−) is used, the equation can also be expressed as follows

$$R_{e_t} = 0.16 \cdot I_t = 0.16\left(\frac{I_a}{f_{cl}} + I_{cl}\right) \tag{2.26}$$

Both of the water vapour permeability resistances from R_{e_a} and from $R_{e_{cl}}$ can be estimated from the following equations:

$$R_{e_a} = \frac{0.06}{f_{cl} \cdot h_c} \tag{2.27}$$

and

$$R_{e_{cl}} = 0.06\frac{I_{cl}}{im_{cl}}. \tag{2.28}$$

where

h_c: convective heat transfer coefficient (W/m^{2}K)
im_{cl}: water vapour permeability index of fabric of clothing itself (−)

Since im_{cl} of common fabric for clothing is approximately 0.38 (−), Eq. 2.28 can also be expressed as

$$R_{e_{cl}} = 0.18 \cdot I_{cl}. \tag{2.29}$$

Effective water vapour permeability resistance of clothing, R_D, can be estimated as an equivalent thickness of a standard still air layer (d^*), employing the thickness of clothing and its packing factor (v_f) which indicates the ratio of fibre filling within the clothing volume as shown in Eq. 2.30.

$$d^* = \left(1 + 0.052 v_f\right) \cdot d \tag{2.30}$$

$$R_D = \frac{d^*}{D} \tag{2.31}$$

where

d: thickness of clothing (m)

d^*: equivalent thickness of standard still air expressing water vapour permeability resistance of clothing (m)

D: diffusion coefficient of water vapour in air (2.56×10^{-5} m^2/s at 20 °C)

R_D: effective water vapour permeability resistance of clothing (s/m)

v_f: packing factor of clothing (%)

2.2.2.3 Major Factors Affecting Water Vapour Transfer in Clothing

Insensible evaporation occurs continuously from human body surface regardless of ambient temperature; for example, the amount of insensible perspiration is about 40 g per hour in a young adult male [41, 45]. This amount of insensible evaporation has an effect of suppressing a rise in core temperature of 0.4–0.5 °C for a human who is 1.7 m in height and 60 kg in weight. It is, therefore, important to ensure that water vapour transfer of the insensible evaporation must not be interfered by clothing. Thermogenic sweating, which occurs when body temperature rises above a desirable level due to activity or exposure to a hot environment, is important in thermoregulation. It should, therefore, also be ensured that the evaporation of sweat from the body surface is not prevented by clothing. For example, Fig. 2.26 shows changes in tympanic temperature as core (a) and mean weighted skin surface (b) temperature over time for subjects wearing short pants or overalls [41, 46]. As seen in the figure, it was difficult to obtain a cooling effect on the body surface while wearing the overalls, because evaporation of sweat from the body surface was interrupted by the overalls themselves. Figure 2.26 also shows changes in absolute humidity (c) and humidity sensation on the body surface (d) when

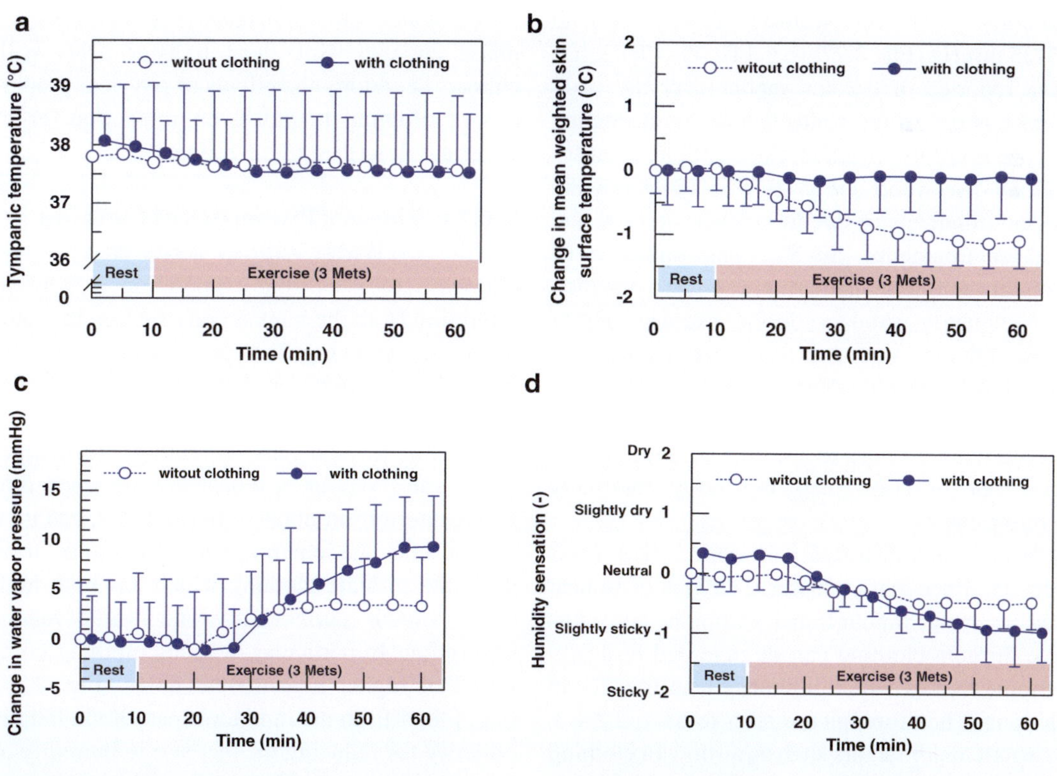

Fig. 2.26 Physiological responses (**a**: core temperature, **b**: change in mean weighted skin surface temperature, **c**: change in water vapour pressure over the body surface) and humidity sensation in the whole body (**d**) in young males with/without clothing exposed to a thermally neutral condition [41, 46]

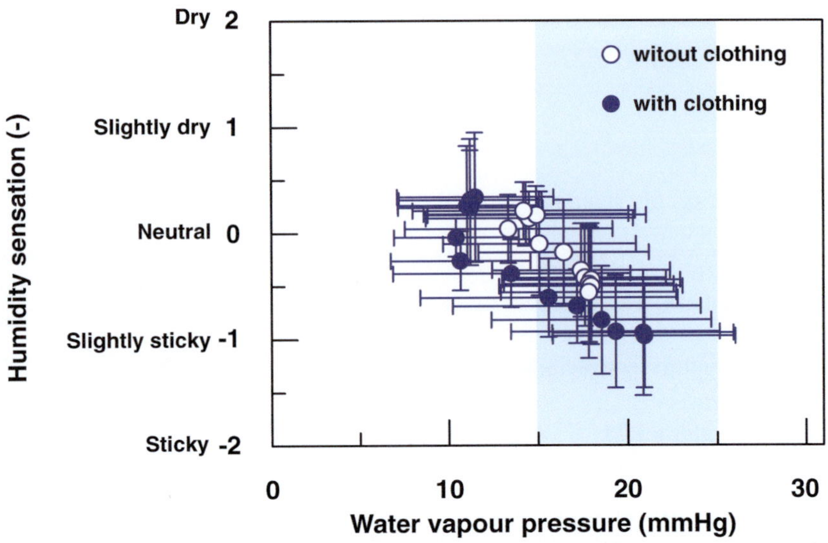

Fig. 2.27 Humidity sensation is plotted against water vapour pressure over the body surface [41, 46]

wearing the short pants or overalls. Humidity sensation increases with an increase in absolute humidity over the body surface. In addition, the body perceives moisture when the absolute humidity over its surface reaches around 15–25 mmHg (Fig. 2.27) (e.g. [41, 47–49]). It has been reported that water vapour over the body surface is one of the major factors for evaluating thermal comfort sensation. For example, it causes thermal discomfort due to stickiness in warm and hot environments or due to coldness in cool and cold environments [50–54]. The influence of humidity sensation on thermal comfort evaluation is particularly stronger during exercise and/or other activities in warm and hot environments. Therefore, smooth water vapour transfer from the body surface through clothing is required not only to maintain core temperature at a favourable level but also to keep the body thermally comfortable.

Water vapour transfer is also called mass transfer. Since mass transfer is analogous to heat transfer, water vapour transfer from the body surface through clothing can be regarded basically as same factors which influence heat transfer in clothing. Therefore, please refer to Sect. 2.2.1.3. in which major factors such as airflow in clothing by air gap size, condition of openings, and body movement on heat transfer in clothing are described. However, there is a significant difference between heat and water vapour transfer. This means water vapour transfer deals with water, which is a substance that undergoes a phase change, while heat transfer deals with energy. Therefore, the effects of physical properties of clothing materials on mass transfer of water are described in the following.

2.2.2.4 Physical Properties of Clothing on Water Vapour Transfer

Pathways of insensible evaporation and sweat migration from the human body surface in clothing are shown in Fig. 2.28. Some water vapour by insensible evaporation and vapourised sweat moves into the environment through openings of clothing or through gaps in clothing. Meanwhile, other water vapour is repeatedly absorbed into and released from fibres of the clothing, and then moves into the environment. Therefore, it is desirable to wear clothing which has both high water vapour permeability and water vapour absorption. In particular, wearing clothing composed of higher hygroscopic fibres (Fig. 2.29) (e.g. [38, 40]) in the innermost part of clothing is effective for reducing the increase in humidity of the microclimate.

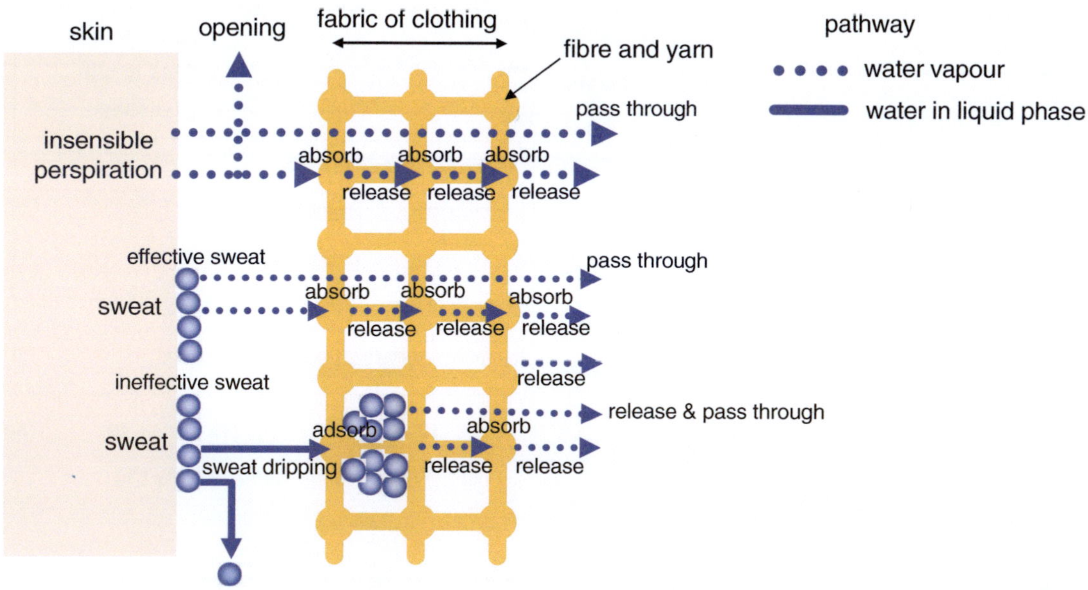

Fig. 2.28 Schematic diagram of pathways of insensible perspiration and sweat from the body surface to the environment through clothing

Water Vapour Absorption

Water vapour absorbability of clothing is determined by the hygroscopicity of fibres that make up the clothing. In general, water vapour absorbability is higher for hydrophilic fibres due to having hydrophilic groups, while it is lower for hydrophobic fibres due to having few hydrophilic groups. Figure 2.29 shows water vapour absorbability of fibres under the standard condition of 20 °C with 65%RH and under a high humidity condition of 20 °C with 95–100%RH. In general, natural fibres (such as cotton, linen, wool, and silk) and recycled fibres (such as rayon and cupra) have such higher hygroscopic capacity that these fibres can easily absorb evaporated water vapour from the body surface. Fibres such as silk, wool, and rayon indicate a higher water vapour absorbability of more than 10% at the standard condition. Their water vapour absorbabilities exceed 25% under the higher humidity condition of nearly 100%RH. Fibres such as cotton and linen also show a high water vapour absorbability of around 8% at the standard condition and show a higher water vapour absorbability of over 25% during exposure to the higher humidity condition of around 100%RH. On the other hand, synthetic fibres such as polyester, acrylic, and nylon show a low water vapour absorbability. Although the water vapour absorbability of nylon fibre is the highest among these synthetic fibres, its ability is only about 1/3 compared to that of wool fibre.

Figure 2.30 shows the results of a series of model experiments in which evaporated water vapour from a simulated skin surface was transferred through a layered clothing system [55]. The vertical axis in the figure shows the difference in absolute humidity between the simulated skin surface and the innermost clothing. If there is a larger difference in absolute humidity between the simulated skin surface and the innermost clothing, it means that an air gap between them was maintained at a lower absolute humidity. The figure shows that the absolute humidity in the gap was kept lower if cotton or wool fibres, which have a high hygroscopic capacity, were employed for the innermost clothing than if polyester or acrylic fibres, which have a non-hygroscopic capacity, were employed. Furthermore, in the case of cotton fibre, storage conditions prior to the experiment affected the absolute humidity difference between the simulated skin surface and the innermost clothing. A

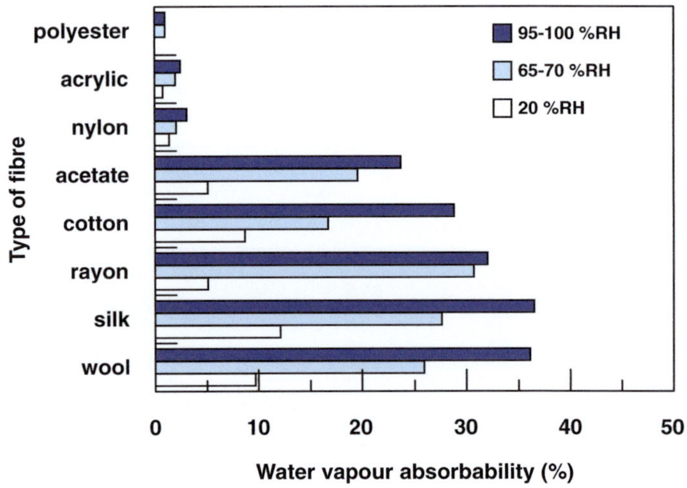

Fig. 2.29 Water vapour absorbability of fibres for clothing during exposure to low to high humidity conditions

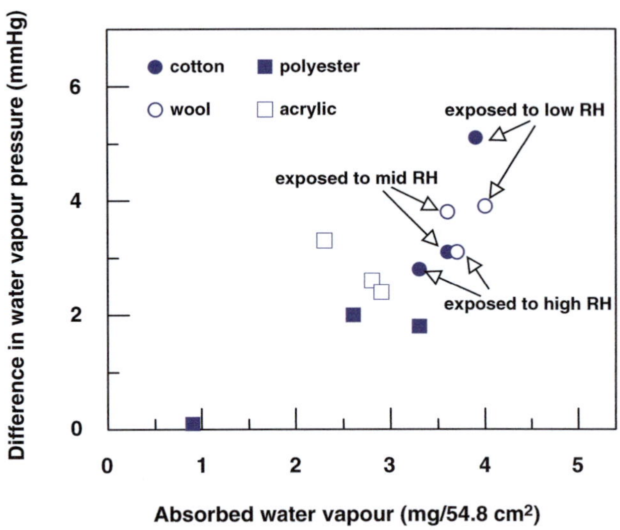

Fig. 2.30 Effects of hygroscopicity and hydrophobicity of fibres upon the absolute humidity of the microclimate within clothing [55]

larger absolute humidity difference was found in the gap between the simulated skin surface and the innermost clothing made of cotton fibre stored in a low humidity condition. On the other hand, a smaller absolute humidity difference was found in the gap between the simulated skin surface and the innermost clothing made of cotton fibre stored in a high humidity condition. This might be because the innermost clothing made of cotton fibre absorbed water vapour during storage depending upon the exposed ambient humidity level. Interestingly, the effect of storage conditions prior to the experiment was not observed for the clothing made of wool fibre. The absolute humidity difference was stably larger in the gap between the simulated skin surface and the innermost clothing made of wool fibre than that of cotton fibre, i.e. the absolute humidity in the gap remained lower for the clothing made of wool fibre than for that of cotton fibre. Because water vapour absorbability for wool fibre is higher than that for cotton fibre. Thus, there is a strong relationship between the water vapour absorbability of innermost clothing and the absolute humidity in the gap between the skin surface and the innermost clothing.

Water Vapour Permeability

Water vapour permeability of clothing refers to the ability of its fabric to allow water in the vapour phase to pass through it. When water vapour permeates through a fabric, it either passes through voids between fibres or yarns that make up the fabric or passes through the fabric with repeated water vapour absorbing into and releasing from the fibres.

For the measurement of water vapour permeability through a fabric of clothing, Japanese Industrial Standards (JIS) provides several methods for evaluating water vapour transfer resistance through a fabric as summarised in Table 2.6; the methods can be without (e.g. JIS L 1099: B method) [56] or with a layer of air on both sides of the fabric (e.g. JIS L 1099: A method) [56]. Intrinsic water vapour permeability resistance of a fabric, as a physical property, can be obtained by means of the former measurement method. On the other hand, water vapour permeability resistance obtained by the latter measurement method is sometimes regarded as relative water vapour permeability resistance. This is because there are air layers on both sides of a specimen in the method in which water vapour permeability resistance of both the air layers is often and largely affected by airflow over the specimen.

Table 2.6 Evaluation methods for water vapour resistance of textile materials for clothing specified in Japanese Industrial Standards L 1099 (2012) [56]

Category		Method
A method	A-1	Calcium chloride method
	A-2	Water method
B method	B-1	Potassium acetate method
	B-2	Another potassium acetate method I
	B-3	Another potassium acetate method II
C method		Sweating hot plate method

Figure 2.31 shows the relationship between the water vapour permeability resistance of a fabric for clothing and its packing factor, which is a ratio of fibres filling in the fabric, as reported by Fourt and Harris [57]. The water vapour permeability resistance increases with an increase in the packing factor as seen in the figure, although the packing factors of fabrics for general clothing are less than 30%. The water vapour permeability resistance of fabrics composed of hydrophilic fibres such as cotton and wool increases gradually with an increase in the packing factor. On the other hand, although the water vapour permeability resistance of fabrics made of hydrophobic fibres such as nylon increases gradually with an increase in the packing factor up to around 30%, the resistance increases exponentially when the packing factor exceeds 30%. It can be noted that the water vapour transfer resistance of fabrics made of hydrophobic fibres is strongly affected by the packing factor.

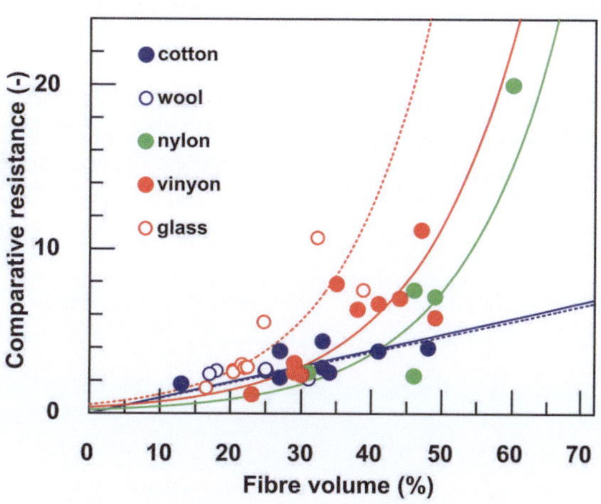

Fig. 2.31 Relation between fibre volume fraction and moisture transfer resistance in fabric for clothing [57]

As shown schematically in Fig. 2.28, if sweat does not evaporate from the skin surface, it remains there as ineffective sweat for thermoregulation. In such a case, the ineffective sweat on the body surface causes an increase in humidity in the microclimate of clothing. It is, therefore, necessary for clothing to absorb water in the liquid phase in order to prevent an increase in humidity in the microclimate due to remaining sweat. Water absorption into clothing is a phenomenon of capillary action caused by surface tension between fibres and yarns. Its water absorbability is highly dependent on the structure of clothing such as woven or knitted fabric, whereas it is not so influenced by the hydrophilicity of clothing's fibres. As well as being highly water absorbent, it is also important and required for clothing to be able to release the absorbed sweat quickly from the clothing to the environment. There are several reasons why absorbed sweat in clothing should be immediately removed; one of the reasons is that the weight of clothing has a significant impact on physical performance. The second reason is if clothing absorbs a large amount of sweat, physical movement is impeded because of an increase in friction between the skin surface and the clothing. The third reason is that it may induce thermal discomfort such as coldness or chilliness when clothing absorbs a lot of sweat at a cold environment because of an increase in sensible heat transfer from the body surface due to the high thermal conductivity of water.

Absorption of Water and Wicking

Water absorption in clothing is a phenomenon in which water is absorbed into microscopic voids between fibres and yarns of clothing fabric. As mentioned above, the phenomenon of water absorption by clothing is a kind of capillary phenomenon, so it is strongly influenced by surface conditions of fibres and yarns and by the woven or knitted fabric's structure of the clothing, rather than the fibres' hydrophilic/hydrophobic property. Figure 2.32 shows water absorbability of several materials for clothing, which expresses how much water in the liquid phase was taken up by the specimen during a certain period of time in terms of height. It can be seen that both woven and knitted fabrics made of cotton exhibit a high water absorbency regardless of their fabrics' structures, because the fibre is entirely hydrophilic. Even though wool fibre is also categorized as one of the hydrophilic fibres because it shows remarkably high water vapour absorbability, it can absorb little water in the liquid phase because its surface is tightly covered with cuticles which are completely hydrophobic regardless of the structure of the woven or knitted fabric. Woven fabric made of polyester fibre shows low water absorbability, because the fibre itself cannot absorb liquid water due to its hydrophobic property. On the other hand, knitted fabric made of polyester fibre shows high water absorbability because of the capillary phenomenon due to its structure. Thus, for hydrophobic fibres, it can be

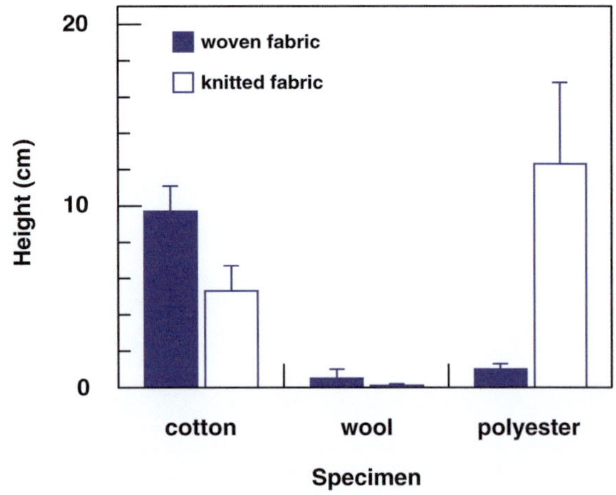

Fig. 2.32 Wicking abilities of cotton, wool, and polyester are measured by Byreck method

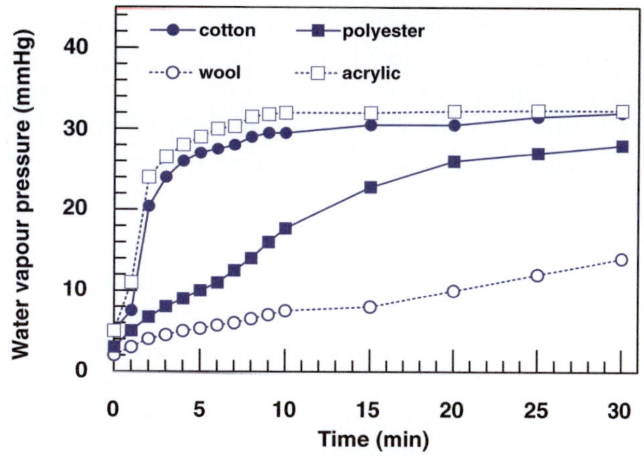

Fig. 2.33 Effects of water vapour absorbability on water vapour pressure of the microclimate in the case where sweat stays on the simulated body surface [58]

noted that water absorption is strongly dependent on the structure of the fabric.

Figure 2.33 shows the water vapour pressure in a simulated microclimate obtained from a series of experiments for a clothing system in which sweat remained on a simulated skin surface [58]. As seen in the figure, the water vapour pressure was kept remarkably higher in the simulated microclimate when cotton fibre was employed for the innermost clothing than when wool fibre was employed. This was because the liquid sweat on the simulated skin surface was absorbed into the innermost clothing made of cotton fibre and then it was quickly and widely spread there. On the other hand, although the sweat in the liquid phase was not absorbed in the innermost clothing made of less water absorbent fibres, some of the sweat evaporated from the simulated skin surface and was absorbed by or passed through the innermost clothing. It can be seen that employing fibre such as wool which has a higher water vapour absorption capacity for the innermost clothing not only prevents to increase in absolute humidity in the microclimate but also prevents the occurrence of wetting the clothing itself.

2.2.3 Adaptable Climates by Clothing

Mecheel and Umbach [59] have derived an estimation equation for the thermally comfortable climate range for a human wearing a clothing ensemble. The estimation equation has been derived from a basis that core temperature is maintained at a favourable level when thermal equilibrium of heat production and dissipation is established for the body. An environmental temperature at which a human can maintain his/her thermal equilibrium while wearing clothing can be estimated using thermal resistance and water vapour transfer resistance of the clothing. The estimation equation is shown as Eq. 2.34:

$$H = M - H_{res} = H_d + H_e \tag{2.32}$$

$$H = \frac{\overline{T}_{sk} - T_a}{R_d} + \frac{w\left(\overline{p}_{sk} - p_a\right)}{R_e} \tag{2.33}$$

$$T_a = \overline{T}_{sk} - R_d \left\{ \frac{w\left(\overline{p}_{sk} - p_a\right)}{R_e} \right\} \tag{2.34}$$

where

H: released heat flux from the body surface (W/m^2)
M: heat production in the body (W/m^2)
H_{res}: released heat flux through respiratory (W/m^2)
H_d: sensible heat flux from the body surface (W/m^2)
H_e:latent heat flux from the body surface (W/m^2)
T_a: ambient temperature (K)
$\overline{T}_{sk}$: mean weighted skin surface temperature (K)
R_d: total thermal resistance in the clothing (m$^2\cdot$K/W)

R_e: total water vapour transfer resistance in the clothing (Pa·m²/W)
w: skin wettedness (−)
$\overline{p}_{sk}$: saturated water vapour pressure at the mean weighted skin surface temperature, $\overline{T}_{sk}$ (Pa)
p_a: environmental water vapour pressure (Pa)

In the equation, an index expressed as w is the skin wettedness (−) proposed by Gagge [60]. It expresses the ratio of the evaporated heat flux from the body by regulatory sweating, H_e (W/m²), to the maximal evaporative heat flux from the body at an actual skin surface temperature of $\overline{T}_{sk}$ for fully wet skin, H_{emax} (W/m²), as shown in Eq. 2.35. So, the skin wettedness should be w = 1.0 (−), if the body surface is evenly and totally wetted with sweat. On the other hand, skin wettedness shows a rate of w = 0.06 (−) during rest under a thermoneutral condition because the insensible water vapour occurs continuously, even if the body surface is not wet with sweat [60].

$$w = \frac{H_e}{H_{emax}} \tag{2.35}$$

where

w: skin wettedness (−)
H_e: evaporated heat flux from the body (W/m²)
H_{emax} : maximal evaporative heat flux from the body with the actual skin surface temperature at $\overline{T}_{sk}$

According to Mecheel and Umbach [59], a lower limit of thermally comfortable environment temperature, $T_{a_{min}}$ (K), for a clothed human can be estimated by Eq. 2.36, in which a temperature of 32 °C has been employed as the mean weighted skin surface at rest in a thermally neutral state. On the other hand, an upper limit of thermally comfortable environmental temperature, $T_{a_{max}}$ (K), can be obtained from Eq. 2.37, in which a higher temperature of 36 °C has been put as the mean weighted skin surface due to an increase in skin blood flow caused by physical activity. They also suggested employing the total thermal resistance, R_d', and the total water vapour transfer resistance, R_e', of clothing during a physical activity to estimate $T_{a_{max}}$.

For estimating $T_{a_{min}}$:

$$T_{a_{min}} = (32 + 273.15) - R_d \left\{ H - \frac{0.06(35.7 \times 133.322 - p_a)}{R_e} \right\} \tag{2.36}$$

For estimating $T_{a_{max}}$:

$$T_{a_{max}} = (36 + 273.15) - R_d' \left\{ H' - \frac{w(46.6 \times 133.322 - p_a)}{R_e'} \right\} \tag{2.37}$$

where

$T_{a_{min}}$: lower limit of thermally comfortable environment temperature (K)
$T_{a_{max}}$: upper limit of thermally comfortable environment temperature (K)
H': released heat flux from the body surface during the work (W/m²)
R_d': total thermal resistance in the clothing during the activity (m²K/W)
R_e': total water vapour transfer resistance in the clothing during the activity (Pa·m²/W)

Figure 2.34 shows the estimated thermally comfortable climate zones while wearing three ethnic clothing: yukata in Japan, kanga in Kenya, and merry-blouse in Papua New Guinea (Fig. 2.35) [10]. In the estimation, the metabolic rates at rest and during physical activity were employed 50 W/m² and 135 W/m², respectively, according to ISO 8996 [61]. In addition, the released heat flux from the body surface was obtained by subtracting the latent heat flux from respiration from the metabolic rate. Rates of skin wettedness, w, of 0.06 (−) and 0.3 (−) were employed for estimating $T_{a_{min}}$ and $T_{a_{max}}$, respectively. These are because the skin wettedness expressed by w becomes 0.06 (−) due to the insensible water vapour evaporation from the body surface during rest under thermoneutral condition as mentioned above and a state of thermal comfort can be kept with a skin wettedness, w, of less than 0.3 (−). Through the estimation method, the thermal comfort climate zones for these three types of clothing are calculated to be 12–23 °C for yukata in Japan, 15–23 °C for kanga in Kenya, and 25–31 °C for merry-blouse in Papua New Guinea.

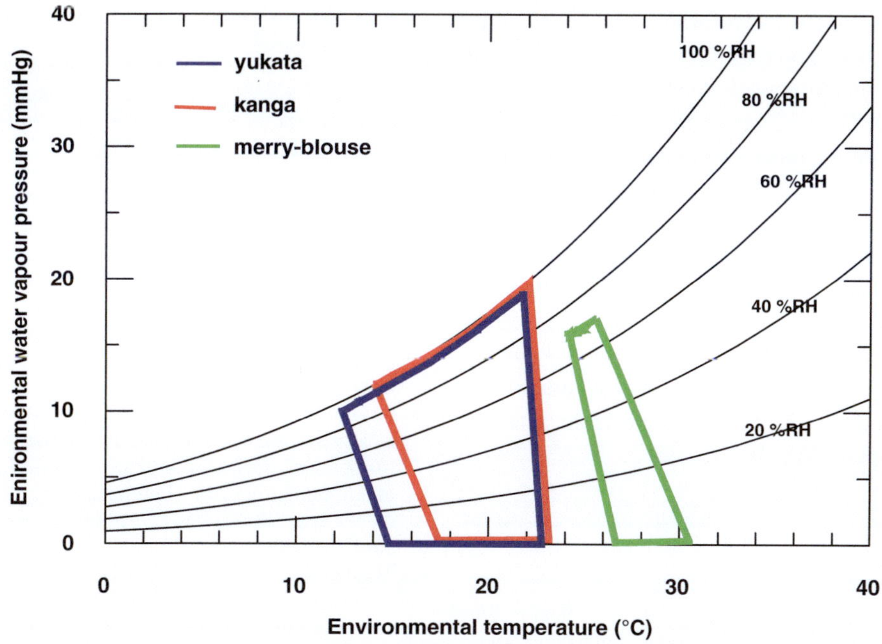

Fig. 2.34 The adaptable range of environments in which thermal comfort can be maintained while wearing Japanese yukata, Kenyan kanga, and Papua New Guinea merry blouse [10]

Fig. 2.35 Pictures of the three traditional ethnic clothing in Japan, Kenya, and Papua New Guinea [10]

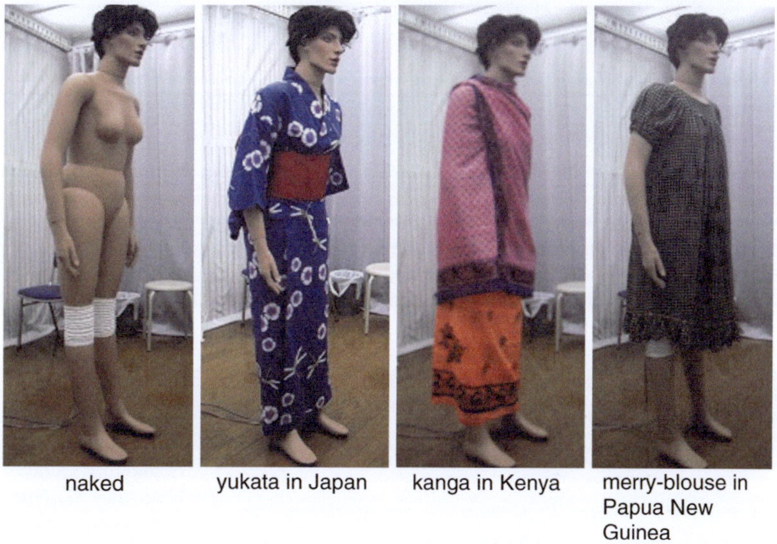

For wearing the merry-blouse, thermally comfortable state can be realised within the climate of 25–31 °C, which is well suited for the climate in Papua New Guinea, where the ambient temperature is slightly higher throughout the year. It is turned out that although wearing the yukata is adapted to the climate from April to June and from September to October in Japan, it is not adapted to the climate from July to August, such as the midsummer season. Although the climate in Kenya varies widely from cool to warm throughout the year, wearing the kanga in Kenya is suitable to keep the body in a thermally comfortable state due to the wide comfort climate zone.

References

1. Fukusako S. Appendix (in Japanese). In: Seki N, editor. Heat transfer. Tokyo: Morikita Publishing; 1996. p. 239–48.
2. Kawabata S. Measurement of anisotropic thermal conductivity of single fiber (in Japanese). J Text Mech Soc Japan. 1986;39:T184–6.
3. Holman JP. Heat transfer. 8th ed. New York: McGraw-Hill; 1997.
4. Sugawara M. Radiation heat transfer (in Japanese). In: Seki N, editor. Heat transfer. Tokyo: Morikita Publishing; 1996. p. 154–81.
5. Fukazawa T, Takao Y, Mitsuno T, Morooka H. Measurement of evaporation and sweat rates in young and elderly females during with/without hoop tension on under bust (in Japanese). Proceedings of the 39th symposium on Human-Environment System 2015; 79–82.
6. Fukazawa T, Takao Y, Mitsuno T. Measurements of insensible perspiration and sweat rates from the body surface during compressing lower chest in Japanese young females (in Japanese). Proceedings of the 41st symposium on Human-Environment System 2017; 185–8.
7. Saito H. Mass transfer (in Japanese). In: Seki N, editor. Heat transfer. Tokyo: Morikita Publishing; 1996. p. 182–203.
8. Tamura T. Climate and clothing. J Hum-Environ Syst. 2016;19:1–11.
9. Kang IH, Tamura T. A study on the developed of an infant-sized movable sweating thermal manikin. J Hum-Environ Syst. 2001;5:49–56.
10. Fukazawa T, Tochihara T. The thermal manikin; a useful and effective device for evaluating human thermal environments. J Hum-Environ Syst. 2015;18:21–8.
11. Kuklane K, Sandsund M, Reiertsen RE, Tochihara T, Fukazawa T, Holmér I. Comparison of thermal manikins of different body shapes and size. Eur J Appl Physiol. 2004;92:683–8.
12. Fukazawa T, Ikeda S, Kim SS, Tochihara Y. Seasonal clothing variation and thermal resistance of clothing ensembles of infants living in Kyushu (in Japanese). J Home Econ Jpn. 2009;60:635–43.
13. Yassin FC, Delanaud S, Szcrupak C, Dubos C, Durand E, Tourneux P. Optimization of the incubator air temperature during LED phototherapy treatment for the preterm infant. Eur J Pediatr. 2021;180:277–81.
14. Delanaud S, Yassin FC, Durand E, Tourneux P, Liber JP. Can mathematical models of body heat exchanges accurately predict thermal stress in premature neonates? Appl Sci. 2019;9:1541. https://doi.org/10.3390/app9081541.
15. Delanaud S, Gossart L, Leclercq M, Libert JP. Use of a novel mathematical model to assess the effectiveness of skin-to-skin care for the prevention of hypothermia in low-birth-weight neonates. Appl Sci. 2023;13:4412. https://doi.org/10.3390/app13074412.
16. Mitsumatsu S, Kondo T, Hori T, Hata M, Sano J. Survey on infants' clothing -a nursery school case study (in Japanese). Jpn J Clo Res. 1999;43:17–23.
17. Araki T, Matsushita K, Tsujita J. Clothing weight of infants in relation to cold-and heat-sensation of their mothers (in Japanese). Jpn J Sch Health. 1981;8:393–9.
18. Fukazawa T, Lee G, Matsuoka T, Kano K, Tochihara Y. Heat and water vapour transfer of protective clothing systems in a cold environment, measured with a newly developed sweating thermal manikin. Eur J Appl Physiol. 2004;92:645–8.
19. Kurazumi Y, Tsuchikawa T, Torii T, Kude T, Yamashita C, Matsubara N, Horikoshi T. Radiative heat transfer areas of the human body in relation to the posture (in Japanese). J Environ Eng AIJ. 2005;598:31–6.
20. Inaba H. Forced convective heat transfer (in Japanese). In: Seki N, editor. Heat transfer. Tokyo: Morikita Publishing; 1996. p. 42–86.
21. Tamura T, Kasahara S. Convective heat transfer coefficients over human body (in Japanese). Proceedings of the 10th symposium on man-thermal environment system -Memorial meeting- 1986; 70–73.
22. Fukazawa T, Kumagai S, Orimoto C, Tochihara Y. A study on mechanical and surface properties and adaptable climate ranges of Kanga clothing systems. Proceedings of the 62nd Japanese Society of Home Economics Conference 2010;46.
23. Fukazawa T, Tochihara Y, Takahara Y. Different impacts of normobaric/hypobaric hypoxia on physiological and subjective responses at a cold environment. J Hum-Environ Syst. 2013;16:11–9.
24. Fukazawa T. Learning from the kanga, Kenya's traditional costume how to maintain thermally comfortable state in hot environments (in Japanese). Jpn J Clo Res. 2011;55:1–4.

25. ISO. ISO 15831 clothing—physiological effects—measurement of thermal insulation by means of a thermal manikin. Geneva: ISO; 2004.
26. Kuklane K, Gao C, Holmér I, Giedraitytė L, Bröde P, Candas V, den Hartog E, Meinander H, Richards M, Havenith G. Calculation of clothing insulation by serial and parallel methods: effects on clothing choice by IREQ and thermal responses in the cold. Int J Occup Saf Ergon. 2007;13:103–16.
27. Fukazawa T, Shibata Y, Ikeda S, Watanabe K, Tochihara Y. Evaluation of thermal resistances of individual clothing and clothing ensembles for infant. Proceedings of the fifth International Conference on Human-Environment System 2016; 20015–1-7.
28. Harada T, Tsuchida K, Maruyama J. Micro climate in clothing and its material (in Japanese). J Text Mech Soc Japan. 1982;35:p350–7.
29. Gagge AP, Burton A, Bazett HC. A practical system of units for the description of the heat exchange of man with his thermal environment. Science. 1941;94:428–30.
30. Ueno S, Sawada S. Measurement of the thermal resistance of firefighters' protective clothing using a movable thermal manikin (in Japanese). J Occup Saf Health. 2008;1:189–96.
31. ISO. ISO 9920 ergonomics of the thermal environment—estimation of thermal insulation and water vapour resistance of a clothing ensemble. Geneva: ISO; 2007.
32. Fujimoto T, Seki N. Effective thermal conductivity and role of radiative transfer in many kinds of clothing materials (in Japanese). J Text Mech Soc Jpn. 1987;40:T13–22.
33. Fujimoto T, Niwa M. Experimental study on effective thermal conductivity of fiber assembly part 1: evaluation of anisotropic effective thermal conductivity and role of radiative heat transfer (in Japanese). J Text Mech Soc Japan. 1989;42:T27–35.
34. Mitsui M. Clothing materials and processing (in Japanese). In: Okada N, editor. Visual textbook of clothing and human life. Tokyo: Kenpakusha; 2010. p. 64–77.
35. Iwasaki F, Tamura T. Studies on thermal resistance of clothing by thermal manikin (part 2) -relationship between clothed area and thermal resistance- (in Japanese). Annual reports of studies, Bunka Women's College. 1985;16:231–9.
36. Fukazawa T, Satsumoto Y, Ishikawa K, Takeuchi M. Effect of air permeability and water vapor permeability upon simultaneous heat and mass transfer in simulated clothing systems (in Japanese). Sen'i Gakkaishi. 1998;54:443–51.
37. Satsumoto Y, Takeuchi M, Hasebe Y, Ishikawa K. The effects of the air space in clothing and the thermal properties of clothing material on heat transfer (in Japanese). Sen'i Gakkaishi. 1990;46:206–15.
38. Fukazawa T. Regulation of microclimate by clothing to be in thermally comfortable state (in Japanese). In: Okada N, editor. Visual textbook of clothing and human life. Tokyo: Kenpakusha; 2010. p. 38–51.
39. Hanada K, Mihira K. Studies on the regional thermal resistance of clothing system Part 6: thermal resistance of clothed manikin accompanied by the water vapor evaporation (in Japanese). Jpn Res Assn Text End-Uses. 1981;22:59–65.
40. Tamura T. Heat and water vapor transfer through clothing (in Japanese). In: Tamura T, editor. Clothing environment. Tokyo: Kenpakusha; 2004. p. 34–50.
41. Fukazawa T. Measurement of human thermal comfort sensation in Japanese young male (in Japanese). Bull Kyoto Univ Edu. 2021;138:105–13.
42. Havenith G, Nilsson HO. Correction of clothing insulation for movement and wind effects, a meta-analysis. Eur J Appl Physiol. 2004;92:636–40.
43. Woodcock AH. Moisture transfer in textile systems, Part I. Textile Res J. 1962;32:628–33.
44. Woodcock AH. Moisture transfer in textile systems, Part II. Textile Res J. 1962;32:719–23.
45. Nakai S. Exercise and hydration (in Japanese). In: Hirata K, Inoue Y, Kondo N, editors. Body temperature: the human thermoregulation system during exercise and its modifying factors. Tokyo: NAP; 2002. p. 129–44.
46. Fukazawa T, Havenith G. Ethnic differences in thermal responses and comfort sensation between Japanese and Caucasian young males under a temperate environment (in Japanese). Proceedings of the 36th symposium on Human-Environment System 2012;165–8.
47. Ushioda H, Nakanishi T, Nakajima T. The relationship between a humid stimulus and a humid sense (in Japanese). Jpn Res Assn Text End-Uses. 2001;42:322–9.
48. Ushioda H, Aoki A, Nakajima T. Factors affecting the evaluation of humidity sensation (in Japanese). Jpn Res Assn Text End-Uses. 1995;36:162–4.
49. Fukazawa T, Havenith G. Differences in comfort perception in relation to local and whole body skin wettedness. Eur J Appl Physiol. 2009;106:15–24.
50. Yamauchi Y, Morooka H, Morooka H. Effect of stimulation strength of clothing humidity on humidity sensation (in Japanese). Sen'i Gakkaishi. 2002;58:61–7.
51. Lee JY, Tochihara Y, Wakabayashi H, Stone EA. Warm or slightly hot? Differences in linguistic dimensions describing perceived thermal sensation. J Physiol Anthropol. 2009;28:37–41.
52. Miura T. Humidity effects on the human health (in Japanese). J Sci Lab. 1971;47:1–11.
53. Gagge AP, Stolwijk JA, Hardy JD. Comfort and thermal sensations and associated physiological responses at various ambient temperatures. Environ Res. 1967;1:1–20.
54. Winslow CEA, Herrington LP, Gagge AP. Physiological reactions and sensations of pleasantness under varying atmospheric conditions. ASHVE Transac. 1939;44:179–94.
55. Fukazawa T. Conditions for health—suitable clothing in cold environments to maintain core temperature at the favorable level (in Japanese). Proc Grant-in-Aid Pub Sci Res Result. 2013;2455009:2–7.
56. JIS. JIS L 1099 Testing methods for water vapour permeability of textiles (in Japanese). JIS. Tokyo; 2012.

57. Fourt L, Harris M. Diffusion of water vapor through textiles. Textile Res J. 1947;17:256–63.
58. Yasuda T, Miyama M, Yasuda H. Dynamic water vapor and heat transport through layered fabrics Part II: effect of the chemical nature of fibers. Textile Res J. 1992;62:227–35.
59. Mecheels J, Umbach KH. Thermophysiologische Eigenschaften von Kleidungssystemen-Thermophysiological properties of clothing systems. Melliand Textilberichte. 1976;57:1029–32.
60. Gagge AP. A new physiological variable associated with sensible and insensible perspiration. Am J Phys. 1937;120:227–87.
61. ISO. ISO 8996 ergonomics of the thermal environment—determination of metabolic rate. Geneva: ISO; 2021.

Thermal Perception and Thermal Sensation

3

Joo-Young Lee

3.1 Thermosensory Systems and TRP Ion Channels

The human body has five primary senses: sight, hearing, taste, smell, and touch. The sense of touch allows us to detect mechanical forces, chemical stimuli, and temperature. Temperature (or heat energy) is perceived at the tips of sensory neuron projections located in the skin [1]. On a molecular level, ion channels in the skin have been identified as gateways that directly respond to different temperature stimuli. The discovery and analysis of temperature-activated transient receptor potential (TRP) ion channels marked the first instance of a sensory ion channel being gated by a physical stimulus [2]. There are four types of heat-activated TRP channels (TRPV1, TRPV2, TRPV3, and TRPV4) and two types of cold-activated TRP channels (TRPM8 and TRPA1). Each of these thermo-TRP channels exhibits unique characteristics, particularly in terms of the distinct temperature thresholds at which they become active. Interestingly, capsaicin activates TRPV1, which is responsive to high temperature (above 42 °C), while menthol activates TRMP8, which is responsive to cold temperatures (below 25 °C). TRPV1 is predominantly expressed in small-to-medium-diameter sensory neurons that are characteristic of nociceptive Aδ- and C-fibers [3, 4]. TRPV1 is also present in the brain, spinal cord, bladder, as well as the skin [2]. In general, innocuous cool temperature is defined as those ranging between 15 and 30 °C, while noxious cold is perceived at temperatures below 15 °C [5, 6]. Both Aδ-fibers and C-fibers are responsive within the range of innocuous cool temperatures, though cold fibers tend to decrease their firing frequency over time or adapt when exposed to a constant temperature [7, 8].

3.2 Warm and Cold Spots in the Skin

A thermoreceptor is a temperature receptor which is a thermosensitive neural element for which its afferent function and response characteristics are electrophysiologically identified [9]. Before the discovery of TRP channels, many researchers studied the distribution of cold and warm spots in the human skin, identifying them as thermoreceptors. It was observed that cold spots are more densely distributed than warm spots [6]. For example, cold spots were found at the following densities: 8.0 spots/cm^2 on the nose, 9–10 spots/cm^2 on the chest, 6.0–7.5 spots/cm^2 on the forearm, 7.4 spots/cm^2 on the back of hand, and 4.5–5.2 spots/cm^2 on the thigh [6]. In contrast, warm spots were found at much lower densities: 1.0 spots/cm^2 on the nose, 0.3 spots/cm^2 on the chest, 0.3–0.4 spots/cm^2 on the forearm, 0.5

J.-Y. Lee (✉)
Department of Fashion and Textiles, College of Human Ecology, Seoul National University, Seoul, South Korea
e-mail: leex3140@snu.ac.kr

Y. Tochihara (ed.), *The Thermal Environment*, https://doi.org/10.1007/978-981-96-3009-7_3

spots/cm^2 on the back of hand, and 0.4 spots/cm^2 on the thigh [6]. No regions of the body exhibited both high warm sensitivity and low cold sensitivity [10]. Due to the technical limitations of measurement instruments at the time, it was challenging to determine whether or not the thermal sensations detected at specific spots on the skin were originated from a single thermoreceptor. It was observed that a single cold or warm spot in the skin does not necessarily correspond to a single thermoreceptor [6]. Thermoelectrical methods have determined that cold receptors are located at an average depth of 0.15–0.17 mm beneath the skin, while warm receptors are found at a greater depth of 0.3–0.6 mm [11, 12]. This suggests that cold receptors, being closer to the surface of the skin, might account for the higher number of cold spots compared to warm spots. Cold receptors are innervated by both thin myelinated fibers (Aδ) and unmyelinated fibers (C-fibers), whereas warm receptors are innervated only by unmyelinated fibers (C-fibers) [6].

3.3 Cutaneous Thermal Threshold and Sensitivity

Human temperature sensations are associated with localized cold or warm spots on the skin. However, these sensations are not purely neurophysiological responses; they also involve psychophysiological factors, a concept known as "phenomenal specificity" [6]. Cutaneous thermal receptors relay information about skin temperature, which not only contribute to autonomic temperature regulation but also influence thermal comfort through the integration of thermal sensations. These sensations depend on both the absolute temperature (T) and the rate of temperature change (dT/dt). Two key terms are used to quantitatively describe cutaneous thermal sensation: thermal threshold and sensitivity. The term "threshold" often refers to the "thermoeffector threshold," which is the level of activity of a potential thermoeffector that must be surpassed for it to become actively involved in temperature regulation. A thermal threshold is defined as the absolute temperature (level) at which a specific thermal sensation is first activated, also referred to as "thermoeffector threshold." For instance, the cutaneous warmth threshold is the skin temperature that initially generates a sensation of warmth.

The term "sensitivity" is commonly used in both statistics and medicine though it has different meanings in each context. For statisticians, sensitivity refers to the probability of a positive test result when the condition is present. For biologists or chemists, sensitivity is the smallest amount of substance in a sample that can be accurately detected. In the context of thermal sensations, thermal sensitivity can be defined as the degree of change in skin temperature—either an increase or decrease—required for a person to first perceive a warm or cool sensation as the skin is warmed or cooled from a thermally neutral state (see "*Method of Limit*").

As previously mentioned, the quantitative measurement of thermal sensory thresholds is a form of psychophysical testing, rather than purely physical or physiological measurements. Several factors can influence the measurement of thermal thresholds, acting as covariates. These factors include the method used (Method of Limit vs. Method of Level), the type of heat applied (contact vs. radiant), the adaptation temperature (cold, neutral, or warm), the rate of temperature change (e.g., 0.1, 0.5, 1, or 2 °C/s), the stimulated surface area, age (young vs. elderly), thermal acclimatization (tropical vs. temperate climates), sex (male vs. female), body mass index (thin vs. obese), exposure to chemicals (alcohol consumption, menthol spray, or capsaicin), skin type (hairy vs. glabrous), initial skin temperature, measurement posture (supine vs. sitting), the force applied during testing (lighter vs. heavier push force), etc. [13]. Due to these influencing factors, it is crucial to clearly identify the exact methodology used in studies before making quantitative comparisons between them. In this chapter, we will discuss the factors that influence cutaneous thermal thresholds based on our published research.

3.3.1 Method of Limit vs. Level

Currently, there are two primary approaches for measuring cutaneous thermal thresholds: the Method of Limit and the Method of Level. We

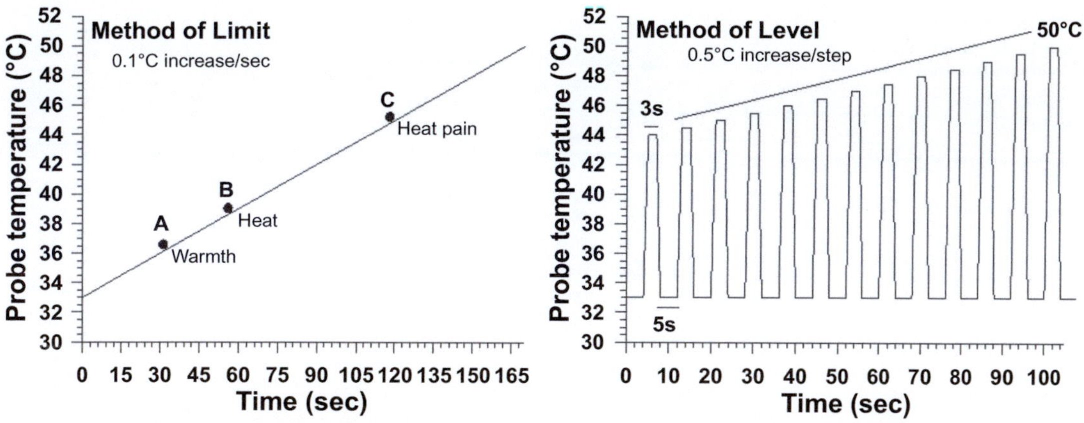

Fig. 3.1 Method of Limit (left) and Method of Level (right) [14]

assessed heat pain sensitivity using both methods [14]. In the left panel of Fig. 3.1, points A, B, and C represent the moments when warmth, heat, and heat pain sensations are perceived over time (Method of Limits). Point C is considered the heat pain threshold in this method. In contrast, the Method of Level involves the application of stimuli with predetermined intensity and duration (Fig. 3.1, right). For instance, this method uses 3-s heat pulses that increase from 44 to 50 °C in 100 s, with 5-s intervals between pulses. Variations in the Method of Levels may include differences in baseline temperature, the starting and final temperatures of the stimuli, the rate of temperature change, interstimulus intervals, stimulus duration, or the gradual increase in stimulus temperature. Reulen and his colleague [15] recommended using a trapezoid-shaped temperature-time graph (where the temperature is maintained for several seconds at the peak) rather than a triangular graph (where the temperature decreases immediately after reaching the peak) for the Method of Level. Our findings indicated that cutaneous heat pain thresholds were, on average, 3.2 ± 2.1 °C higher when measured using the Method of Level compared to the Method of Limit [14].

3.3.2 Type of Heat Applied: Contact vs. Radiant

Since 2010, we have researched cutaneous warmth, hotness, and heat pain thresholds across various body regions using a conductive heat probe (2.5 × 2.5 cm). The knowledge gained from these studies can be applied to optimize temperature settings for heating systems in car seats, active heating wear, and similar applications. Conductive heat is transferred through direct contact, and the rate of transfer depends on three factors: (1) the thermal conductivity of the skin, (2) the temperature difference between the skin and the contact probe, and (3) the thickness of the skin. In contrast, radiant heat is transferred through emitted electromagnetic waves and is influenced by (1) the emissivity of the object and (2) its surface temperature. The use of a 6.25 cm^2 conductive heat exposure area in our previous studies may have led to an underestimation of the true thermal impact experienced from larger radiant heat sources, such as those affecting passengers or drivers of vehicles. We also evaluated body regional differences in cutaneous thermal thresholds in response to radiant heat exposure [16] and compared these findings with results from conductive heat exposure. Our results showed that cutaneous thermal thresholds from radiant heat exposure over a 100 cm^2 area were lower than those from conductive heat exposure of 6.25 cm^2 (Fig. 3.2), likely due to the effect of spatial summation and the different types of heat transfer involved. Further studies on radiant heat with different wavelengths are required, as well as studies on heat with different sizes.

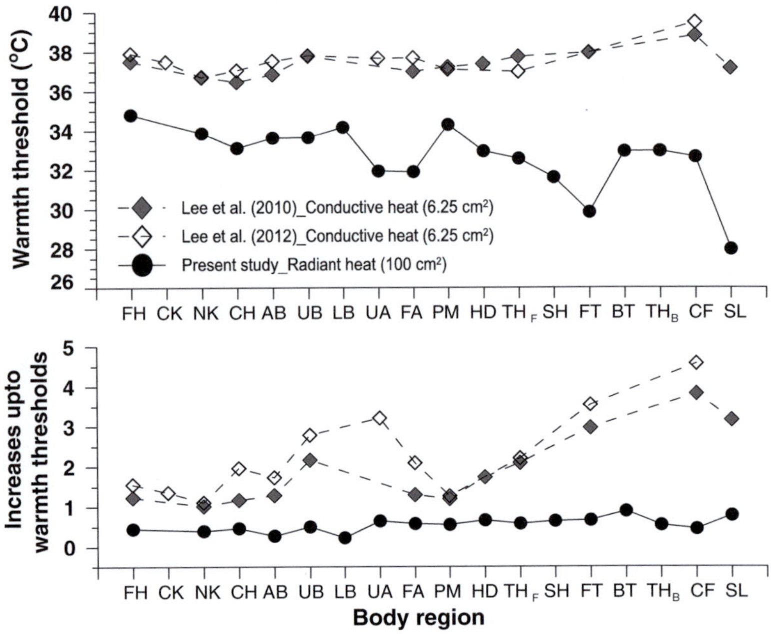

Fig. 3.2 Warmth thresholds and increases in warmth thresholds: Comparison of conductive heat exposure in the previous studies (6.25 cm^2) and radiant heat exposure in the present study (100 cm^2). Abbreviations of *FH*, *CK*, *NK*, *CH*, *UB*, *LB*, *FA*, *PM*, *HD*, TH_F, *SH*, *FT*, *BT*, TH_B, *CF*, and *SL* represent forehead, cheek, neck, chest, upper back, lower back, forearm, palm, hand, thigh (front), shin, foot, buttocks, thigh (back), calf, and sole, respectively [16]

3.3.3 Adaptation Temperature: Cold vs. Neutral

Cutaneous warm sensitivity was blunted in a cool environment compared to a thermoneutral environment [17]. However, this reduction was not significant in young males, while elderly males showed significant reductions in warm sensitivity on the shin and foot [17]. Since neither the head nor the trunk regions exhibited blunted warm sensitivity in both young and elderly males, this suggests that the reduction in a cool environment may have a peripheral origin rather than a central one. Further studies with larger sample sizes are needed to confirm these findings.

3.3.4 Rate of Temperature Change: 0.1, 0.3, or 0.5 °C/s

We investigated the effects of heating rate on cutaneous warm thresholds across 17 body regions in thermoneutral environments. The thermal thresholds were measured at three different heating rates: 0.1, 0.3, and 0.5 °C/s. Significant differences in skin temperature required to detect warmth were observed across 11 body regions, including the sole, depending on the heating rate (Fig. 3.3). However, there was no significant difference in cutaneous warmth thresholds between the heating rates of 0.1 and 0.3 °C/s. In contrast, the difference between heating rates of 0.3 and 0.5 °C/s did affect thermal sensitivity in certain peripheral body regions.

3.3.5 Stimulated Surface Area: Spatial Summation

Early studies using classical methodologies found that human thermal sensitivity varies with the size of the exposed stimulus area: larger surface areas exposed to heat result in lower thermal thresholds [19–22]. The concept of spatial summation was first introduced by Greene and Hardy [23], who reported findings from skin areas ranging from 2.5 to 16 cm^2. Kenshalo and his colleague [24] identified a negative exponential relationship among areas of 1.7, 7.1, and 14.4 cm^2 on the back, forearm, and forehead, respectively. Similarly, Green and Zaharchuk [20] used stimulus areas of 0.64, 1.28, 2.56, and 5.12 cm^2, finding a negative exponential relationship between detection

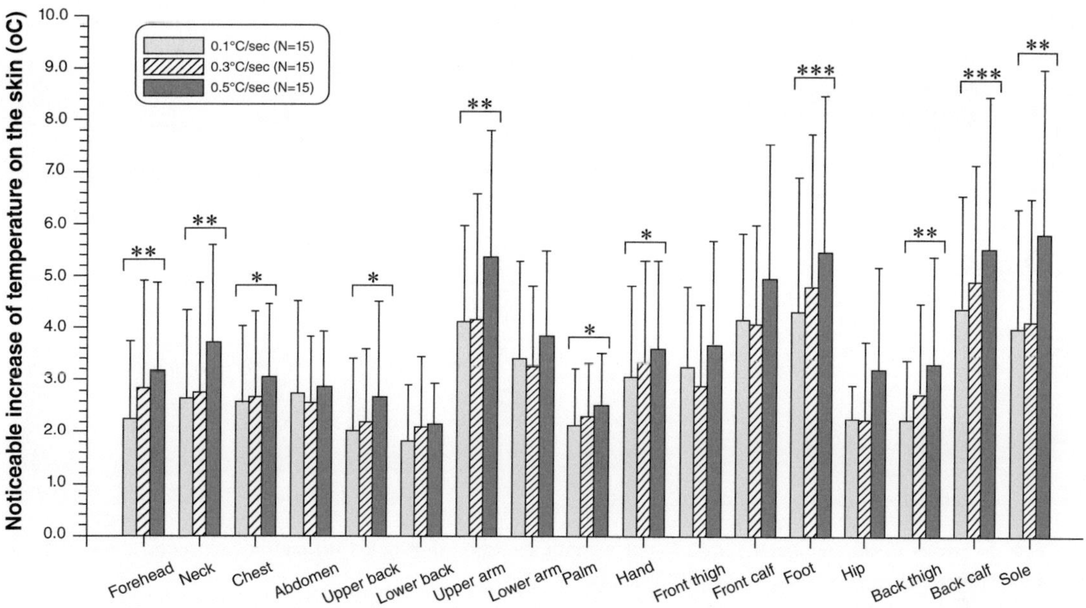

Fig. 3.3 Noticeable increases in skin temperature until the subjects felt their cutaneous warmth on the 17 body regions by the heating speed of 0.1, 0.3, and 0.5 °C/s [18]

thresholds and stimulating areas. Defrin and his colleague [25] observed that the warm-sensation threshold increased as the stimulation area decreased from 9 to 2.25 cm^2 on both the palm and hand. These studies often referenced spatial summation, which refers to the synaptic integration of input signals from multiple, distributed presynaptic neurons [26] and is related to the size of stimulated area or the number of neurons activated. However, most previous studies focused on thermal probes stimulating very small areas (e.g., 1–16 cm^2), covering only a small fraction of the human body surface area (<0.1%BSA). We investigated whether spatial summation of thermal sensitivity occurs with larger areas of radiant heat exposure, ranging from 100 to 400 cm^2 [27]. Our results showed that warmth and heat sensation thresholds were higher for the 100 cm^2 condition compared to the 225 or 400 cm^2 conditions, but no differences were found between the 225 and 400 cm^2 conditions (Fig. 3.4). These findings, along with previous studies, suggest that spatial summation exists but is limited to relatively small stimulus areas, likely smaller than approximately 1% BSA.

3.3.6 Age

From a thermophysiological perspective, aging is associated with diminished autonomic function [28], which may contribute to reduced thermal sensitivity in the elderly. For instance, temperature discrimination on the fingers of young individuals is approximately 0.5 °C for both coolness and warmth, but this increases to 1.0–5.0 °C in older adults [29]. In our study on age-related differences in cutaneous warmth thresholds [17], elderly males showed significantly greater increases in perceived temperature on the hand, shin, and foot compared to young males (Fig. 3.5). Specifically, warmth sensitivity in the young males was 1.3 ± 1.0 °C on the hand, 3.7 ± 2.6 °C on the shin, and 3.4 ± 2.6 °C on the foot, while in the elderly males, it was 3.5 ± 2.6 °C, 7.9 ± 2.7 °C, and 6.8 ± 3.3 °C, respectively. These findings suggest that age-related differences in cutaneous warmth perception are not uniform across the body, particularly in areas such as the hand, shin, and foot. The reduced warmth sensitivity in the extremities of elderly males may indicate a decline in temperature-sensitive affector units with age. This decline in cutaneous warm

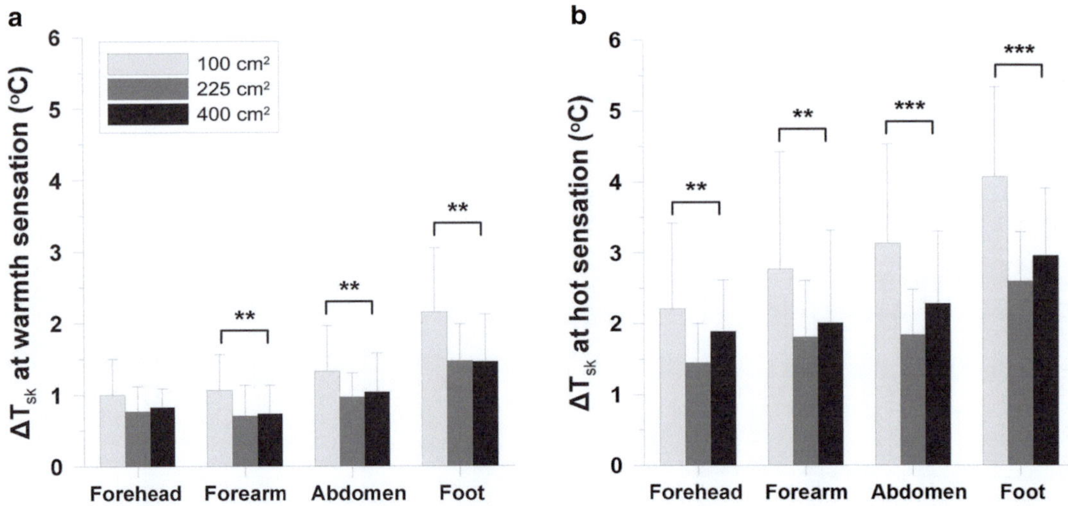

Fig. 3.4 Increases in skin temperatures (ΔT_{sk}) at warmth sensation thresholds (**a**) and hot sensation thresholds (**b**) while being exposed by the three different heating-sized areas [27]

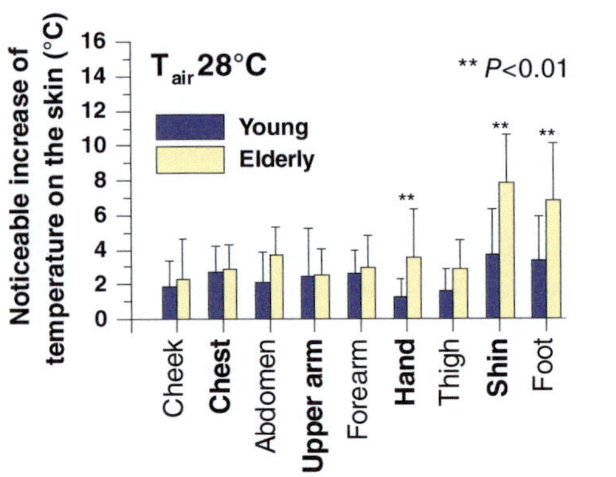

Fig. 3.5 Detectable (noticeable) increment in temperature to perceive warmth on the skin by age [17]

thresholds is likely due to functional deterioration in neuronal pathways, including slower nerve conduction velocities in older individuals compared to young ones [30].

3.3.7 Heat Acclimatization: Tropical vs. Temperate Climates

Adaptation is a common feature of most sensory systems and refers to the decrease in neural responsiveness of primary afferents to a stimulus due to prolonged exposure. We investigated ethnic differences in cutaneous warm and cool sensations between individuals from tropical and temperate regions [13, 31]. Our findings showed that Japanese males were more sensitive to detecting warmth compared to Malaysian males [13]. We concluded that individuals from tropical regions tend to perceive warmth at a higher skin temperature and have a broader inter-threshold sensory zone than those from temperate regions [13].

As the next step, we recruited individuals from tropical regions who were now residing in temperate climates to examine the deacclimatization of cutaneous thermal sensations. The results showed that, on average, these tropical individuals were 3.3, 3.5, 4.2, and 7.3 °C less sensitive to warm, hot, cool, and cold sensations, respectively, compared to temperate natives (Fig. 3.6). This indicates that the tropical individuals' acclimatization to heat, as reflected in their cutaneous thermal thresholds, was maintained despite living in a temperate climate for several years. An interesting finding was that tropical individuals did not show significant regional differences in their cutaneous cool or cold thresholds, whereas clear regional differences were observed in their warm and hot thresholds. Additionally, tropical individuals were less sensitive overall to cutaneous cool, cold, warm, and hot sensations compared to temperate natives.

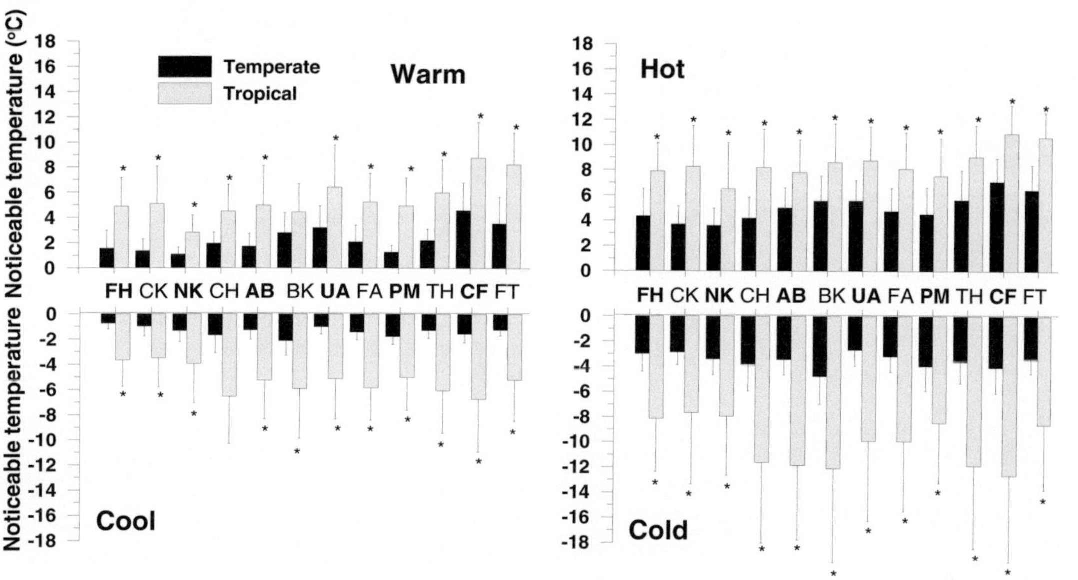

Fig. 3.6 Temperatures at which cutaneous warm, hot, cool, and cold sensations were noticeable in the tropical and temperate males (*FH* forehead, *CK* cheek, *NK* neck, *CH* chest, *AB* abdomen, *BK* upper back, *UA* upper arm, *FA* forearm, *PM* palm, *TH* thigh, *CF* calf, *FT* foot) (*$P < 0.05$) [31]

3.3.8 Body Mass Index

The obese group exhibited reduced thermal sensitivity compared to the nonobese group, particularly in the abdomen where excess subcutaneous fat was present [32]. However, obese individuals might experience increased sensitivity to pain, as pro-inflammatory cytokines produced by adipose tissue can stimulate nociceptors [33]. Anthropometric factors such as total body fat (%BF), subcutaneous fat, and body mass index (BMI) may influence cutaneous thermal sensitivity. We investigated relationships between body fat, BMI, and cutaneous thermal thresholds [34]. We found that the overall hot and heat pain thresholds were higher in the overweight group (42.9 ± 2.3 °C and 46.4 ± 1.7 °C) compared to the lean group (40.2 ± 0.7 °C and 42.6 ± 0.8 °C). The overweight group was approximately 3–4 °C less sensitive to detecting overall hotness and heat pain than the lean group. Body mass index was positively correlated with overall hotness thresholds ($r = 0.592$, $P < 0.05$, Fig. 3.7a) or overall heat pain thresholds ($r = 0.771$, $P < 0.05$, Fig. 3.7b). Total body fat was positively correlated with overall pain threshold (Fig. 3.7c). In addition, we observed that warmth and hotness thresholds on the abdomen had significant relationships with body mass index, indicating that the overweight individuals are less sensitive to detecting radiant heat on the abdomen [16].

Tashani and his colleague [35] compared body parts with different subcutaneous fat distribution and found that the thenar eminence, an area with less fat content, had lower warm thresholds compared to the suprailiac area, which has more fat. Both current and previous studies suggest that subcutaneous fat is one of the key factors determining cutaneous thermal thresholds. Price and his colleague [32] proposed that excess subcutaneous fat might stretch the skin, thereby reducing local sensitivity by decreasing the density of thermal receptors. In obese rats, a significant reduction in intraepidermal nerve fiber density was observed, along with slower sensory nerve conduction and thermal hypoalgesia [36].

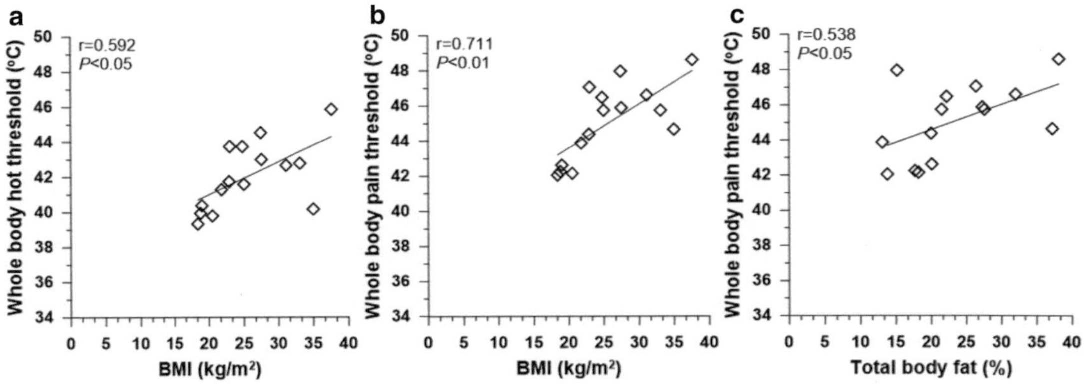

Fig. 3.7 Relationships between overall body thermal thresholds and body mass index (BMI) or total body fat (%) [34]

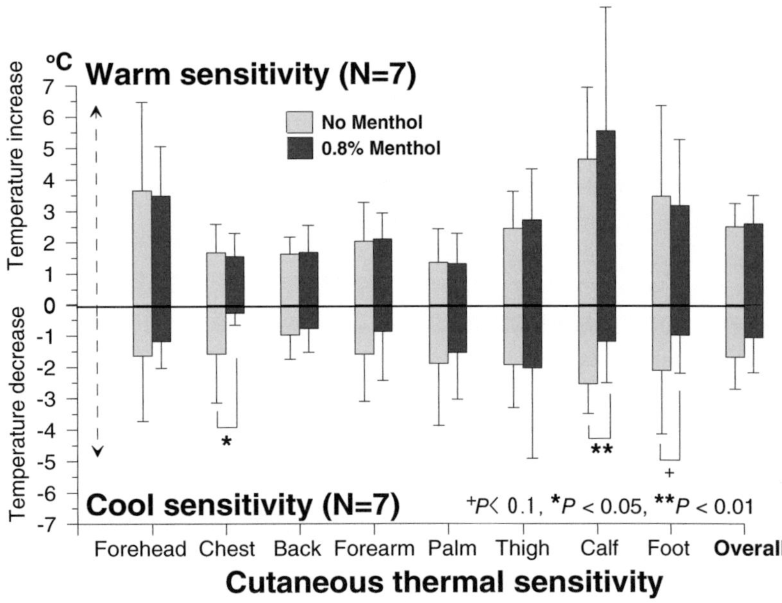

Fig. 3.8 Thermal sensitivity with/without 0.8% menthol application in a thermoneutral environment [37]

3.3.9 Influences of Chemical: Menthol Application

Menthol is a well-known substance with specific effects on thermoreceptors. Goldscheider (in 1898) demonstrated that the cold sensation caused by menthol is not due to evaporative cooling but rather to a chemical action [6]. Neurophysiologists have since identified that TRPM8 is activated by both menthol and cold stimuli. In a practical study on cooling strategies, we examined the effect of 0.8% menthol on cutaneous thermal sensitivity in a thermoneutral environment, as well as on body temperature and thermal sensation under heat stress [37]. We found that menthol significantly affected cutaneous cold sensitivity in a thermoneutral environment but did not alter warm sensitivity (Fig. 3.8). Notably, the application of 0.8% menthol impaired thermoregulatory functions during exercise under heat stress. Specifically, menthol caused an earlier and greater increase in rectal temperature, higher skin temperature, and a reduced total sweat rate. These effects are likely due to peripheral vasoconstriction and delayed sweating. Despite subjects feeling cooler with menthol application compared to the control, it may be risky for individuals undergoing prolonged heat stress.

3.4 Thermal Sensation and Thermal Comfort

Thermal sensation is one of the fundamental variables used to determine thermal comfort, which is a major factor driving behavioral thermoregulation in daily life. Thermal sensation is a perception experienced rationally, while thermal comfort is a form of cognition experienced affectively. Thermal sensation is based on terms like "warm" and "cold," whereas thermal comfort connotes satisfaction, enjoyment, or pleasure. We discussed several issues related to thermal sensation and thermal comfort from both conceptual and methodological perspectives.

3.4.1 Thermal Sensation and Language Issues

Human thermal sensation is one of the ways we directly perceive our thermal environments. Thermal sensation involves two modalities: warmth and cold, which are closely linked to thermoreceptors such as warm and cold spots on the skin and within the core body. The American Society of Heating, Refrigerating and Air-Conditioning Engineers (ASHRAE) defines "thermal sensation" as a conscious feeling, commonly categorized as "cold," "cool," "slightly cool," "neutral," "slightly warm," "warm," and "hot" [38]. Thermal sensation is a universal human experience, but the way it is expressed through language varies across cultures [39]. Typically, scales used to measure thermal sensation are formatted as categorical scales [40, 41]. The wording of these scales depends on the vocabulary choices of each language, which can introduce bias [41]. We have published studies on scale construction and language [39, 42–44].

The first concern relates to the term of "warm" in Korean and Japanese languages [39]. When constructing thermal sensation scales, some Korean and Japanese studies have placed the descriptor "slightly hot" between "warm" and "hot" (Type B in Table 3.1). This arrangement differs from the scales used by ASHRAE or ISO 10551 [41], which do not include "slightly hot"

Table 3.1 Categorical thermal sensation scales

Type A	Type B	Type C	Type D	
			Very, very hot	10
				9
			Very hot	8
				7
			Hot	6
(Extremely hot)	(Extremely hot)			5
Very hot	Very hot	Very, very warm	Uncomfortably warm	4
Hot	Hot	Very warm		3
Warm	Slightly hot	Warm	Comfortably warm	2
Slightly warm	Warm	Slightly warm		1
Neutral	Neutral	Neutral	Neutral	0
Slightly cool	Cool	Slightly cold		−1
Cool	Slightly cool	Cold	Comfortably cool	−2
Cold	Cold	Very cold		−3
Very cold	Very cold	Very, very cold	Uncomfortably cool	−4
(Extremely cold)	(Extremely hot)			−5
			Cold	−6
				−7
			Very cold	−8
				−9
			Very, very cold	−10

Type A—ISO 10551 [41]; Type B—modified ISO scale [45]; Type C—modified scale from ISO 10551. This scale shows the variation of two opposite words, cold and warm [46]; Type D [47, 48]

(Type A in Table 3.1). We observed notable cultural differences in how thermal sensations are linguistically expressed, particularly concerning thermal descriptor "warm" in Korean. In Korean, the term "ttatthada" (warm) often conveys a sense of thermal comfort or thermal satisfaction. As a result, phrases like "comfortably warm" (Type D in Table 3.1) may seem redundant expression for Koreans, while "uncomfortably warm" can appear contradictory (an oxymoron) to Korean speakers. Consequently, Koreans rarely use the term "warm" during hot summers. Instead, "slightly hot" is more naturally used to describe thermal sensation between "neutral" and "hot" in hot environments. A comprehensive survey supported these findings, revealing that 80% of 988 Korean respondents associated "warm" with thermal comfort, whereas 54% associated "slightly hot" with thermal discomfort [43]. This indicates that "warm" and "slightly hot" in Korean are not sequentially aligned on a thermal sensation scale during hot summers [43]. To address these linguistic differences, the use of visual analog scales or pictorial scales has been suggested as a potential solution. However, there remain additional methodological and conceptual issues beyond language translation, which are discussed in Sect. 3.4.3.

We further highlighted that linguistic differences present a significant challenge in the application of the PMV-PPD model [39]. The predicted mean vote (PMV), a general thermal sensation index, is based on the ISO/ASHRAE 7-point thermal sensation scale. The predicted percentage of dissatisfied (PPD) index is derived from the PMV [49–50]. For instance, when the PMV predicts a score of 2 (warm), the PPD is expected to be 75% of the population exposed to the environment. However, when applying this prediction to Koreans, the PPD would be significantly lower than 75%. Similar issues have been observed in Japanese contexts [51]. Different words have different linguistic dimensions, and it should not be assumed that intervals between adjacent descriptors are uniform. These differences may introduce bias when interpreting the relationship between thermal sensation and environmental temperatures. Therefore, it is necessary to adjust the current fixed descriptors to reflect the typical linguistic usage of each language. We recommend presenting both the original English descriptors alongside modified descriptors that correspond to the typical terms used in each language on a scale.

Building on the studies mentioned above, we proposed a new concept of thermal acclimatization in language, which may reflect "linguistic thermal acclimatization" among tropical populations [44]. To test this hypothesis, we investigated whether there are semantic differences in how thermal sensation is expressed between tropical populations (Indonesians) and temperate populations (Japanese). The results indicated that for Indonesians, the term most closely associated with thermal comfort was "cool" (75%), followed by "slightly cool" (7%), "slightly cold" (5%), and "cold" (5%) (Fig. 3.9). This evidence suggests that linguistic heat acclimatization among Indonesians is reflected on the cooler end of thermal descriptors, rather than the warmer end, indicating a preference for coolness in their consistently hot climate. This linguistic heat acclimatization can be understood through the concept of intersubjectivity. We concluded that for Indonesians, the term "cool" connotes a thermally comfortable feeling, and linguistic heat acclimatization occurs at a cognitive level. Further research is needed to explore linguistic cold acclimatization.

3.4.2 Thermal Comfort

A significant number of studies on thermal comfort have been conducted by environmental engineers, primarily focused on indoor climates such as heating and air-conditioning. From this perspective, Fanger [49] defined thermal comfort as the condition of mind in which a person expresses satisfaction with the thermal environment, aligning with ASHRAE Standard 55. In contrast, thermal physiologists define thermal comfort as a state of subjective indifference to the thermal environment [9]. In this section, we examine the differences between these two definitions, focus-

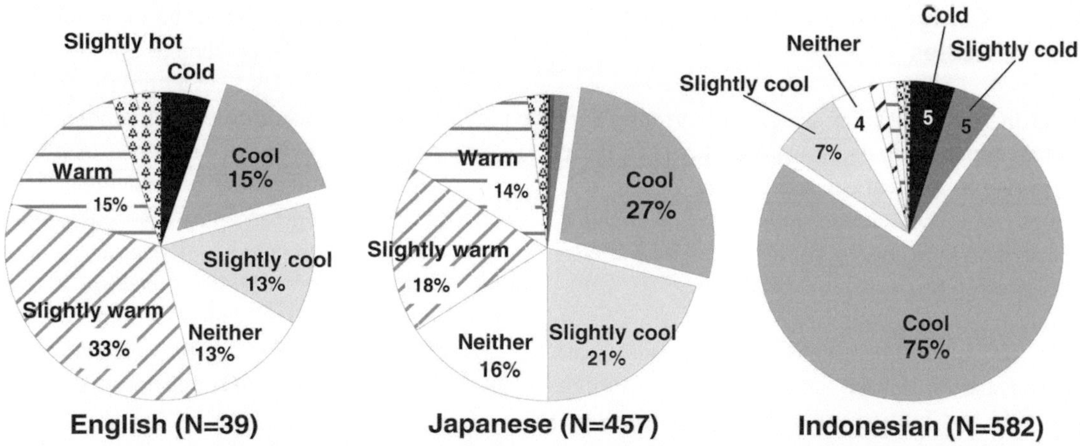

Fig. 3.9 Percentages of descriptors on the closest description of a feeling of thermal comfort [44]

ing on the concept of "thermal satisfaction" and "thermal indifference."

Environmental engineers have sought to create optimal thermal comfort for groups of indoor occupants. However, it is impossible to satisfy everyone simultaneously at the same room temperature due to significant physiological, psychological, and social variations among individuals. These variations include factors such as age, sex, body shape, ethnic differences, thermal acclimatization, diet, clothing, activities, circadian rhythm, menstrual cycle, thermal transients, draught, asymmetric thermal stress, expectations, and duration of exposure [49]. The concept of an optimal thermal environment refers to a thermal condition where the highest possible percentage of the group is in thermal comfort. Based on this idea, Fanger [49] introduced the concept of "predicted percent dissatisfied (PPD)." ISO 10551 [41] provides a thermal comfort scale with five categories: comfortable (score 0), slightly uncomfortable (score 1), uncomfortable (score 2), very uncomfortable (score 3), and extremely uncomfortable (score 4). Strictly speaking, this is not a thermal comfort scale but a thermal discomfort scale. The ISO/ASHRAE scale does not differentiate between levels of thermal satisfaction, while thermal dissatisfaction is categorized into four levels. Both the PPD and the ISO/ASHRAE scale are useful for evaluating thermal dissatisfaction in indoor environments.

Subjective indifference is more closely aligned with thermal neutrality than thermal comfort. Thermal neutrality refers to a condition where an individual has no preference for either warmer or cooler surroundings. A bipolar scale incorporating this concept can be constructed as follows: very uncomfortable (score −3), uncomfortable (score −2), slightly uncomfortable (score −1), indifference (neither; score 0), slightly comfortable (score 1), comfortable (score 2), and very comfortable (score 3). This bipolar scale can be particularly useful for evaluating thermal (dis) comfort in individuals exposed to extreme or dynamic thermal environments.

The difference between the ISO scale and the bipolar scale, with indifference in the middle, may be rooted in philosophical perspectives. For example, darkness is not an entity in itself but rather the absence of light, while light is an entity. Similarly, in philosophy, evil is not an existence but the absence of good. A lack of something can be graded, whereas something that exists is complete in itself. If we consider discomfort as the absence of comfort, the graded levels of discomfort (with no graded level of comfort) on the ISO/ASHRAE scale appear reasonable. However, the choice between a unipolar or bipolar scale should depend on the research objective. We recommend using the bipolar scale to evaluate thermal comfort when subjects are exposed to extreme thermal environments or engaged in physical activity.

3.4.3 Categorical and Visual Analog Scales

Most thermal sensation scales are currently formatted as categorical descriptors. However, there is growing interest in combining categorical scales with visual analog scale (VAS) to enhance sensitivity. This approach introduces both methodological and conceptual challenges. We compared a 9-point categorical scale (9-pts CS), a VAS, and a combined scale that merges the categorical scale with VAS (combined scale) through a survey and human trials during exercise [42] (Fig. 3.10).

We found that the combined scale was more closely related to indoor air temperature compared to the VAS or 9-pts CS. In particular, subjects' responses on the VAS clustered significantly around the thermal neutral zone. We explored various methodological and conceptual issues related to using the combined scale, including term definitions, the use of descriptors, the number of categories, scoring length, unipolar vs. bipolar construction, language translation, central and anchor terms, orientation, and color [42].

ISO 10551 (2019) discussed similar issues on thermal sensation scale construction. (1) A 7-degree scale should be applied in the case of environments judged to be close to neutrality, whereas a 9-degree scale should be applied in the case of environments judged to be more intense. (2) Full scales should be presented, even in cases of surroundings located only in a limited range of conditions. (3) The wording of all the degrees of a scale provides a frame of reference useful to those asked to verbalize their instantaneous experience. (4) For a unipolar scale, a 4-degree one-pole scale can be extended to 5 degrees, with a point of origin indicating the absence of the effect. (5) For bipolar scales, a symmetrical 7-degree two-pole scale can be extended to 9 degrees, comprising a central indifference point.

3.4.4 Air Temperatures Judged as Hot and Cold Weather in the Mind

Many studies by environmental engineers have focused on achieving optimal thermal comfort in indoor environments, including heating, cooling, and air-conditioning. According to IUPS [9], there are three key air temperature zones used to determine optimal indoor conditions: (1) Zone of thermal indifference: this is the range of ambient temperatures, along with specified water vapor pressure, air velocity, and radiant exchange, where 80% of active individuals do not complain about the thermal environment. (2) Zone of thermal comfort: this range includes ambient tem-

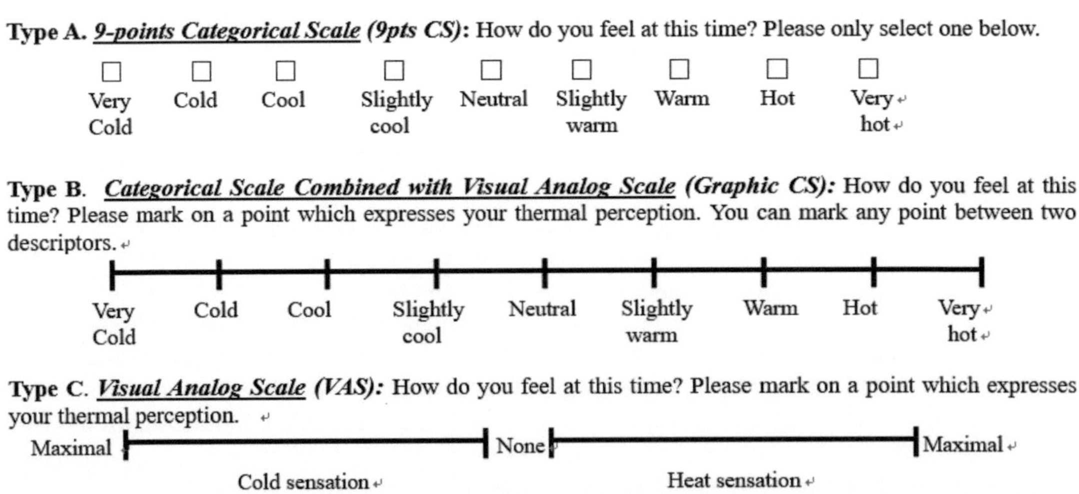

Fig. 3.10 Three types of thermal sensation scales which were utilized for the questionnaire survey (Type A, B, and C) [42]

peratures, mean radiant temperature, humidity, and air movement, where individuals in specific clothing feel indifferent to the thermal environment for an extended period. (3) Thermoneutral zone (TNZ): this is the range of ambient temperatures at which temperature regulation is maintained solely through sensible heat loss, without requiring changes in metabolic heat production or evaporative heat loss. The common element among these zones is that they all refer to ambient temperature ranges.

The thermoneutral zone, or thermal comfort zone, is influenced by factors such as sex, season, clothing insulation, metabolic rate, and ethnicity. Fanger [49] defined the thermoneutral zone as 24 °C with 0.85 clo, and 20 °C with 1.4 clo. Preferred indoor air temperatures vary by region: 24 °C in the USA, 18 °C in the UK, 12 °C in USSR [52], and 32 °C in Iraq [53]. Tanabe [54] found that females preferred an indoor air temperature of 26.3 °C (0.6 clo) in summer and 25.3 °C (0.6 clo) in winter. In contrast, young Korean males preferred an indoor air temperature of 23 °C year-round [55]. Ellis [56] reported that the optimal air temperature for residents in Singapore, wearing typical tropical clothing (0.4 clo), was 27 °C. Ballantyne and his colleagues [57] identified an optimal air temperature of 25.8 °C for individuals resting in still air with 80% relative humidity and typical clothing.

We investigated the range of air temperatures that people perceived as hot and cold weather in their minds. The difference between these two temperatures can be considered as an acceptable range of air temperature in mind (Fig. 3.11). For Korean young males in Seoul, air temperatures above 29 °C were cognized as hot weather, and temperatures below 7 °C were considered as cold weather [55]. This indicates that the acceptable range of air temperatures in their perception is between 7 and 29 °C. In Busan, Korean young males identified temperatures above 30 °C as hot in summer and below 1 °C as cold in winter [58]. We hypothesized that the acceptable temperature range would shift upward for people in tropical climates and downward for those in colder climates. This hypothesis was supported by studies conducted by Tochihara and his colleagues [44] and Kim and his colleagues [59]. For Indonesians, hot weather was most commonly defined as temperatures above 32 °C and cold weather as below 20 °C, making their acceptable range between 20 and 32 °C. In contrast, Japanese individuals identified hot weather as above 29 °C and cold weather as below 15 °C, with their acceptable range between 15 and 29 °C [44]. Elderly Chinese-Koreans living in northern China considered temperatures above 24 °C as hot weather and below −16 °C as cold weather, reflecting an even broader range [59]. These acceptable

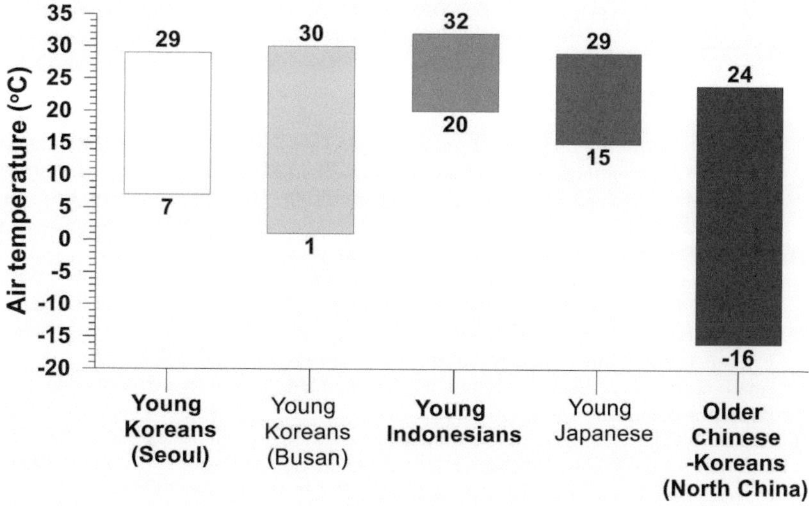

Fig. 3.11 The inter-threshold zone of air temperatures judged as hot and cold weather in the mind

temperature ranges reflect geographical characteristics, with broader ranges potentially offering better adaptation to extreme thermal environments.

References

1. Smith CUM. Biology of sensory systems. West Sussex: Wiley; 2000.
2. Dhaka A, Viswanath V, Patapoutian A. TRP ion channels and temperature sensation. Annu Rev Neurosci. 2006;29:135–61.
3. Caterina MJ, Schumacher MA, Tominaga M, Rosen TA, Levine JD, Julius D. The capsaicin receptor: a heat-activated ion channel in the pain pathway. Nature. 1997;389(6653):816–24.
4. Tominaga M, Caterina MJ, Malmberg AB, Rosen TA, Gilbert H, Skinner K, et al. The cloned capsaicin receptor integrates multiple pain-producing stimuli. Neuron. 1998;21(3):531–43.
5. Davis KD, Pope GE. Noxious cold evokes multiple sensations with distinct time courses. Pain. 2002;98(1–2):179–85.
6. Hensel H. Thermoreception and temperature regulation. London: Academic Press; 1981.
7. Campero M, Serra J, Bostock H, Ochoa JL. Slowly conducting afferents activated by innocuous low temperature in human skin. J Physiol. 2001;535(3):855–65.
8. Darian-Smith I, Johnson KO, Dykes R. "Cold" fiber population innervating palmar and digital skin of the monkey: responses to cooling pulses. J Neurophysiol. 1973;36(2):325–46.
9. IUPS Thermal Commission. Glossary of terms for thermal physiology. Jpn J Physiol. 2001;51(2):245–80.
10. Hensel H. Thermal sensations and thermal receptors in man. Springfield: Thomas; 1982.
11. Bazett HC, McGlone B, Brocklehurst RJ. The temperature in the tissues which accompany temperature sensations. J Physiol (London). 1930;69:88–112.
12. Bazett HC, McGlone B. Experiments on the mechanism of stimulation of end-organs for cold. Am J Phys. 1930;93:632.
13. Lee JY, Saat M, Chou C, Hashiguchi N, Wijayanto T, Wakabayashi H, Tochihara Y. Cutaneous warm and cool sensation thresholds and the inter-threshold zone in Malaysian and Japanese males. J Therm Biol. 2010;35(2):70–6.
14. Park S, Roh SH, Lee JY. Body regional heat pain thresholds in young males using the method of limit and level. Eur J Appl Physiol. 2019;119:771–80.
15. Reulen JPH, Lansbergen MDI, Verstraete E, Spaans F. Comparison of thermal thresholds tests to assess small nerve fiber function: limits vs. levels. Clin Neurophysiol. 2003;114(3):556–63.
16. Kwon JY, Cho K, Jung D, Lee JY. Cutaneous warmth and hotness thresholds to radiation heat exposure at a distance of 10 cm from 17 body regions. J Therm Biol. 2023;115:103611.
17. Tochihara Y, Kumamoto T, Lee JY, Hashiguchi N. Age-related differences in cutaneous warm sensation thresholds of human males in thermoneutral and cool environments. J Therm Biol. 2011;36:105–11.
18. Jung DH, Cho GY, Gorski A, Lee JY. Influences of the heating rate on cutaneous warm thresholds on 17 body regions. J KSLES. 2023;30(1):110–20.
19. Defrin R, Givon R, Raz N, Urca G. Spatial summation and spatial discrimination of pain sensation. Pain. 2006;126(1–3, 123):–31.
20. Green BG, Zaharchuk R. Spatial variation in sensitivity as a factor in measurements of spatial summation of warmth and cold. Somatosens Mot Res. 2001;18:181–90.
21. Pertovaara A, Kauppila T, Hämäläinen MM. Influence of skin temperature on heat pain threshold in humans. Exp Brain Res. 1996;107(3):497–503.
22. Price DD, McHaffie JG, Larson MA. Spatial summation of heat-induced pain: influence of stimulus area and spatial separation of stimuli on perceived pain sensation intensity and unpleasantness. J Neurophysiol. 1989;62:1270–9.
23. Greene LC, Hardy JD. Spatial summation of pain. J Appl Physiol. 1958;13:457–64.
24. Kenshalo DR, Decker T, Hamilton A. Spatial summation on the forehead, forearm, and back produced by radiant and conducted heat. J Comp Physiol Psychol. 1967;63:510–5.
25. Defrin R, Petrini L, Arendt-Nielsen L. Spatial summation of thermal sensations depends on skin type and skin sensitivity. Exp Brain Res. 2009;198:29–36.
26. Etherington SJ, Atkinson SE, Stuart GJ, Williams SR. Synaptic integration. eLS. 2010.
27. Roh SH, Moon JH, Lee JY. Spatial summation of thermal sensitivity is limited to small areas: comparisons of the forehead, forearm, abdomen, and foot. J Therm Biol. 2023;115:103627.
28. Collins KJ, Dore C, Exton-Smith AN, Fox RH, MacDonald IC, Woodward PM. Accidental hypothermia and impaired temperature homoeostasis in the elderly. Br Med J. 1977;1(6057):353–6.
29. Cowburn EJ, Foxm RH. A technique for studying thermal perception. J Physiol. 1974;239:63–87.
30. Verdú E, Ceballos D, Vilches JJ, Navarro X. Influence of aging on peripheral nerve function and regeneration. J Peripher Nerv Syst. 2000;5(4):191–208.
31. Lee JY, Bakri I, Toramoto S, Tochihara Y. Cutaneous thermal thresholds of tropical indigenes residing in Japan. J Therm Biol. 2011;36:461–8.
32. Price RC, Asenjo JF, Christou NV, Backman SB, Schweinhardt P. The role of excess subcutaneous fat in pain and sensory sensitivity in obesity. Eur J Pain. 2013;17(9):1316–26.
33. McVinnie DS. Obesity and pain. Br J Pain. 2013;7:163–70.
34. Roh SH, Moon JH, Kang CH, Lee JY. Relationships between cutaneous thermal thresholds and individual body characteristics: body mass index and regional subcutaneous fat. Manuscript submitted for publication 2024.

35. Tashani O, Astita R, Sharp D, Johnson M. Body mass index and distribution of body fat can influence sensory detection and pain sensitivity. Eur J Pain. 2017;21(7):1186–96.
36. Davidson EP, Coppey LJ, Calcutt NA, Oltman CL, Yorek MA. Diet-induced obesity in Sprague-Dawley rats causes microvascular and neural dysfunction. Diabetes Metab Res Rev. 2010;26(4):306–18.
37. Lee JY, Nakao K, Bakri I, Tochihara Y. Body regional influences of L-menthol application on the alleviation of heat strain while wearing firefighter's' protective clothing. Eur J Appl Physiol. 2012;122(6):2171–83.
38. ASHRAE. ASHRAE Standard, ANSI/ASHRAE Standard 55-2004: thermal environmental conditions for human occupancy. Atlanta: American Society of Heating, Refrigerating, and Air-Conditioning Engineers; 2004.
39. Lee JY, Tochihara Y, Wakabayashi H, Stone EA. Warm or slightly hot? Differences in linguistic dimensions describing perceived thermal sensation. J Physiol Anthropol. 2009;28(1):37–41.
40. Bedford T. Environmental warmth and human comfort. Br J Appl Phys. 1950;1(2):33–8.
41. ISO 10551. Ergonomics of the physical environment—subjective judgement scales for assessing physical environments. Geneva: International Standards Organization; 2019.
42. Lee JY, Stone EA, Wakabayashi H, Tochihara Y. Issues in combining Likert and visual analog scales for the assessment of perceived thermal sensation: methodological and conceptual considerations. Appl Ergon. 2010;41(2):282–90.
43. Lee JY, Tochihara Y. Linguistic dimensions in descriptors expressing thermal sensation in Korean: warm projects thermal comfort. Int J Biometeorol. 2010;54(4):357–64.
44. Tochihara Y, Lee JY, Wakabayashi H, Wijayanto T, Bakri I, Parsons K. The use of language to express thermal sensation suggests heat acclimatization by Indonesian people. Int J Biometeorol. 2012;56:1055–64.
45. Maeda T, Kobayashi T, Tanaka K, Sato A, Kaneko SY, Tanaka M. Seasonal differences in physiological and psychological responses to hot and cold environments in the elderly and young males. In: Tochihara Y, Ohnaka T, editors. Environmental ergonomics, Elsevier ergonomic book series, vol. 3. Amsterdam: Elsevier; 2005. p. 35–41.
46. Holmér I. Cold but comfortable? Application of comfort criteria to cold environments. Indoor Air. 2004;14(Suppl 7):27–31.
47. Havenith G, Heus R. A test battery related to ergonomics of protective clothing. Appl Ergon. 2004;35:3–20.
48. Heus R, Daanen HAM, Havenith G. Physiological criteria for functioning of hands in the cold. Appl Ergon. 1995;26:5–13.
49. Fanger PO. Thermal comfort: analysis and applications in environmental engineering. New York: McGraw-Hill Book Company; 1970.
50. ISO 7730. Ergonomics of the thermal environment—analytical determination and interpretation of thermal comfort using calculation of the PMV and PPD indices and local thermal comfort criteria. Geneva: International Standards Organization; 2005.
51. Kaneko T, Horikoshi T. A study on evaluation of Japanese psychological responses to thermal environment by word-choice method with a unipolar scale. J Archit Plann Environ Eng (Transactions of AIJ). 2001;543:93–9.
52. Burton AC, Edholm OG. Man in cold environment. London: Arnold; 1955.
53. Humphreys MA. Field studies of thermal comfort compared and applied. Building Research Establishment. 1975;29.
54. Tanabe S. Thermal comfort requirements of sedentary college-age subjects during the winter season in Japan. J Home Econ. 1990;41(5):437–45.
55. Hyun CS, Roh SH, Kim DH, Son SY, Baek YJ, Kim KR, et al. Comparison of the perception of summer heat wave and thermoregulatory behavior between adult males who live in Seoul and those in Daegu. Korean J Community Living Sci. 2018;29(1):17–32.
56. Ellis F. Tolerable and desirable levels of warmth in warm climates, with special reference to the requirements of men in the royal navy. Ann R Coll Surg Engl. 1953;13(6):369–91.
57. Ballantyne ER, Barned JR, Spencer JW. Environment assessment of acclimatized caucasian subjects at Port Moresby, Papua New Guinea. In: Proceedings 3rd Aust. Bldg. Research Congress, Div. of Bldg. Research. CSIRO, Melbourne, Reprint No.400. 1967.
58. Ju HY, Kim KR, Lee JY. Heat stress and thermoregulatory behavior of Seoul and Busan residents: comparisons of 2021 and 2022. J KSLES. 2023;30(3):1–14.
59. Kim D, Jin CZ, Min J, Oh JW, Lee JY. Self-identified thermal perception and wearing habits of elderly Korean Chinese in Yanji city, Jilin, China. J KSLES. 2022;29(6):613–24.

Thermoregulation

4

Su-Young Son

4.1 Mechanism of Thermoregulation

4.1.1 Body Heat Balance

Humans are endothermic and homeothermic. They can generate heat and maintain their body temperature within a narrow range, even under fluctuating environmental conditions, to regulate the chemical reactions in the body. When the body temperature is outside the normal range, the nervous system becomes functionally damaged, and protein denaturation occurs. Therefore, a control mechanism for maintaining body temperature is required.

Body temperature is controlled by the balance between the body's heat gain and loss. Heat gain is a combination of the heat produced in the body and acquired from the environment, whereas heat loss means that heat is lost from the body to the environment. If the amounts of heat gain and heat loss are the same, the body temperature does not change. When heat gain is greater than heat loss, the body temperature increases more than usual, but when it is smaller, the body temperature decreases. Humans produce heat or send heat out of the body in response to the external environment to maintain a constant body temperature.

S.-Y. Son (✉)
Department of Home Economics Education, Teachers College, Kyungpook National University, Daegu, South Korea
e-mail: sonsy@knu.ac.kr

Figure 4.1 shows the relationship between environmental temperature and autonomic thermoregulation. Significantly high or low temperatures cause hyperthermia and hypothermia, respectively, which lead to death caused by heat stroke or cold. Survivable regions are classified into physical and chemical regulatory regions [1–3].

A thermoneutral zone (TNZ) is a range of environmental temperatures within which a person can maintain their core body temperature with the least metabolic effort. When a person is naked, the thermoneutral zone is thought to be around 25–30 °C [4]. In this temperature range, the body temperature can be controlled by adjusting heat loss through changes in the skin's blood flow. Below this temperature, however, even maximal vasoconstriction cannot prevent heat loss, and the body must increase heat production to maintain its temperature. At higher temperatures, even when blood vessels are maximally dilated, the rate of heat production cannot keep pace with the speed of heat loss; therefore, another heat transfer mechanism is required. When the outside temperature is higher than the body temperature, heat enters the body through radiation and conduction. Evaporation through sweating is necessary to release this heat. In the TNZ, the temperature range of the environment where the metabolic rate at rest can be maintained at the lowest level, the lower and upper limits of the temperature are called the lower critical temperature (LCT) and upper critical temperature (UCT), respectively (Fig. 4.1). TNZ

Y. Tochihara (ed.), *The Thermal Environment*, https://doi.org/10.1007/978-981-96-3009-7_4

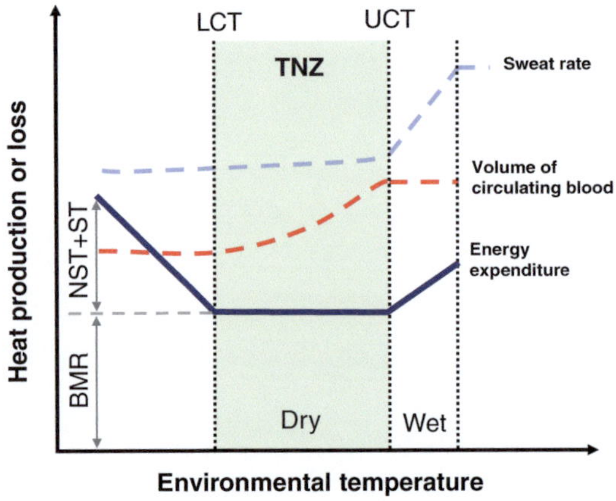

Fig. 4.1 Relationship between environmental temperature and autonomic thermoregulation. (Partially modified from Gordon [1], Henderson and Halsey [2], and Pallubinsky [3]). *TNZ* thermoneutral zone, *LCT* lower critical temperature, *UCT* upper critical temperature, *BMR* basal metabolic rate, *NST* nonshivering thermogenesis, *ST* shivering thermogenesis

differs according to insulation, posture, or basal metabolic rate (BMR) [5].

LCT was defined as "the ambient temperature below which the rate of metabolic heat production of a resting thermoregulating tachymetabolic animal must be increased by shivering and/or nonshivering thermogenesis in order to maintain thermal balance" [5]. In cold environments, vasoconstriction and decreased blood flow to the skin cause a decrease in skin temperature. As a result, the temperature difference between the environment and the skin decreases, and the human body attempts to control the heat loss. When the environmental temperature decreases and falls below the LCT, heat production increases, causing goosebumps and shivering. Goosebumps are the movements of the piloerector muscle; that is, the hair on the skin surface creates an air layer between them to maintain body temperature. Muscle contractions cause shivering, increase heat production, and maintain body temperature. UCT is defined as "the ambient temperature above which the rate of evaporative heat loss of a resting thermoregulating animal must be increased (e.g., by thermal tachypnea or by thermal sweating) in order to maintain thermal balance" [5]. When the temperature of the external environment increases, vasodilation of the skin surface increases blood flow and skin temperature, and the amount of heat loss increases. When the ambient temperature exceeds UCT, that is, when the temperature increases further, it causes sweating. Sweating decreases the skin temperature by evaporating perspiration from the skin surface once it reaches the heat of vaporization.

The metabolic rate (M) produced by burning proteins, fats, and carbohydrates in the body is transferred via convection (C) and conduction (K) to the body surface from the core of the body and is dissipated via convection (C), conduction (K), radiation (R), and evaporation (E). When heat is dissipated from the body surface to the environment, the amount of thermal radiation from the body surface is received by each thermoreceptor and transmitted to the brain, which is the center of thermoregulation. Based on this, the brain attempts to maintain a constant body temperature by controlling the energy that moves from the core to the surface of the body. The following equation represents heat transport in and out of the human body:

$$M-\left(C+K+R+E+W\right)=S \tag{4.1}$$

where,

M: metabolic rate; *C*: amount of heat loss due to conduction; *K*: amount of heat loss due to convection; *R*: amount of heat loss due to radiation; *E*: amount of heat loss due to evaporation; *W*: workload; *S*: heat storage.

The human core body temperature remains mostly in a constant range, following a circadian rhythm that is lower in the morning and higher in the evening. To maintain homeostasis, it is necessary to balance the amount of heat production with that of heat loss.

$$\text{Heat production} - \text{Heat loss} = S \tag{4.2}$$

$$S=\left(0.83\cdot m/A\right)\left(\Delta Tb/\Delta t\right) \tag{4.3}$$

where,

S: heat storage (W/m^2); *m*: body mass (kg); *A*: body surface area m^2; $\Delta Tb/\Delta t$: mean change of body temperature (°C/h); 0.83 = specific heat of the human body.

Thermal energy is produced during food intake, respiration, and mitochondrial adenosine triphosphate (ATP) production, which regulates body temperature. In the balance of heat production and loss, the metabolic rate refers to the minimum energy required to live as a homeotherm and is the basis for life activities (Fig. 4.2).

Blood flow carries heat and affects vasoconstriction and vasodilation in both cold and hot environments. Thus, the heat flow from the body to the outside is regulated. The brain consumes a lot of energy and corresponds to approximately 20–25% of the basal metabolic rate (or resting metabolic rate), which is far higher than the 8–10% observed in other primates [6, 7]. In cold environments, heat production is promoted through skeletal muscle contraction and shivering in addition to nonshivering thermogenesis, such as that in the brown adipose tissue of newborns.

On the skin's surface, moisture constantly evaporates to help heat loss, and this moisture evaporation activity that humans do not feel is called insensible perspiration. We are estimated to lose approximately 900 mL of water per day, accounting for about 20% of the total heat loss. One-third of this insensible perspiration is related to breathing and two-thirds to the skin.

When the environmental temperature increases, that is, when it becomes hot, sweating begins beyond the stage of insensible perspiration. Sweating is a heat loss effector, and conduction, convection, and radiation contribute to it. Each value increases or decreases depending on the situation of the human body and the surrounding environment; it is an efficient means of heat loss, even in the absence of an air conditioner. When combined with airflow or wind, sweat evaporation is accelerated, further increasing the effect of heat loss. However, in extremely hot environments with high temperatures and humidity, it can backfire and increase the body temperature instead.

The wind is counterproductive to heat loss at 100% relative humidity and temperatures around 36 °C. Humidity affects how well humans tolerate a hot environment. For example, under extremely dry conditions, humans can tolerate temperatures up to 130 °C for more than 20 min, whereas at high humidity, they can only tolerate up to 46 °C for a few minutes [4].

4.1.2 Autonomic Thermoregulation and Behavioral Thermoregulation

Thermoreceptors detect changes in temperature. This information is transmitted to the thermoregulatory center, behavioral or autonomy control is activated, and the thermoreceptors accept these changes. This process is known as negative feedback control.

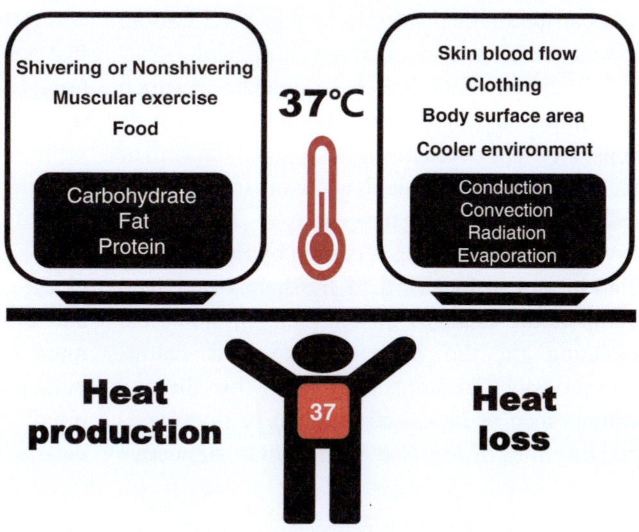

Fig. 4.2 Balance between heat production and heat loss mechanisms

4.1.2.1 Autonomic Thermoregulation

Autonomic thermoregulation refers to involuntary heat production and heat loss reactions (such as vasomotor activity, sweating, nonconductive heat production by brown adipose tissue, metabolism, and piloerection). Autonomic thermoregulation can be classified into the central nervous and endocrine systems. In autonomic thermoregulation, the skin vascular reactions (vasodilation and vasoconstriction) are heat loss reactions. During regulation by the sympathetic nervous system, α1 receptors in vascular smooth muscle cells (spermatocytes) accept adrenaline and cause smooth muscles to contract, resulting in blood vessel contraction. In the endocrine system, adrenaline is released from the adrenal medulla and transported throughout the body via blood, causing vasoconstriction.

In a hot environment, cutaneous blood vessels dilate to increase heat transfer from the core of the body to the surface and promote heat loss from the body to the environment. In contrast, in cold environments, the cutaneous blood vessels are constricted to control heat production. The cutaneous vascular system is divided into capillaries and arteriovenous anastomoses. Sweating is generated by the sympathetic nerves and acetylcholine is released from the sympathetic terminals. In addition, acetylcholine, which is accepted by muscarinic acetylcholine receptors in sweat glands, induces sweating. Moisture evaporation from the skin is called transpiration and can be divided into insensible perspiration, thermogenic sweating, and psychological sweating. Insensible perspiration is the state in which sweating is not perceived. However, it involves constant evaporation of moisture from the skin or the respiratory tract, which accounts for a considerable portion of the body's heat loss. Psychological sweating is caused by mental stimulation and is localized to the hands and feet. Temperature changes also affect psychological sweating on the body surface and palms. Thermogenic sweating is caused by thermal stimuli such as increased ambient or body temperature due to exercise. The human subjective perception of thermal comfort refers to a state in which no sweating activity exists.

In cold environments, a thermogenic response occurs through skeletal muscle contraction. Heat production is accompanied by shivering and involves both the somatic and sympathetic nervous systems. In contrast, nonshivering thermogenesis, which involves the endocrine system, occurs in brown adipose tissue or internal organs. Individual differences in the autonomic thermoregulatory response are affected by sex, age, physique, body composition, circadian rhythm, female menstrual cycle, and individual lifestyle (such as exercise, drinking, or smoking habits). Thermal acclimatization or adaptation has also been observed in autonomic thermoregulatory responses. Adaptation to high or low temperatures for an extended period increases the ability to regulate body temperature via activated sweating. In a study on Korean breath-hold women divers called Haenyeo, the older groups, which had longer adaptation in cold seawater, showed more stable core temperatures, greater activated sweat glands, and better warm-perception thresholds on the extremities compared to non-Haenyeo and younger Haenyeo groups [8].

4.1.2.2 Behavioral Thermoregulation

Behavioral thermoregulation refers to physical activity, biorhythms, seasonal changes, and other voluntary behaviors that maintain a thermally comfortable state. Thermoregulatory behaviors that are complex for all living organisms except humans in modern society include using heating or cooling devices to regulate body temperature, wearing clothes, exercising, ingesting food, and creating ventilation using a fan [9]. In particular, wearing clothes is a behavioral thermoregulatory response that appears only in humans. Wearing clothes creates a microclimate between the skin and clothing, known as the clothing microclimate. This microscopic layer of air surrounds the human body, and the physical properties of this microspace affect comfort (see Chap. 2 for more details).

4.2 Heat Production and Heat Loss

4.2.1 Heat Production

The body contains three major nutrients (protein, fat, and carbohydrates) that are the primary energy sources from food intake. The amount of heat production, that is, the heat generated by food digestion/absorption or oxidation/combustion of oxygen by breathing, is used as fuel for carrying out life activities such as maintaining body temperature and muscle activity. This process is called energy metabolism, and the energy obtained at this time is called the metabolic rate. Approximately 20% of the energy metabolism is converted into other energy sources in the body and used, and the remaining energy is converted into thermal energy and radiated out of the body. Most of the energy metabolism is carried throughout the body by blood flow, which maintains body temperature and is radiated; moreover, it is radiated through the skin or respiratory system, which is in contact with the external environment. Energy metabolism can be measured using a respiratory gas analysis system; furthermore, it can be estimated from the respiratory quotient (RQ), which is defined as the volume of carbon dioxide (V_{CO2}) released over the volume of oxygen consumed (V_{O2}) and absorbed during respiration [10].

Factors affecting energy metabolism include sleep, age, sex, and environmental temperature. In addition, metabolic rates increase with an elevation in any of the following factors: height, weight, body surface area, growth, pregnancy, menstruation, food intake, muscle activity, emotional stress, ambient temperature, and various hormones. The factors related to heat production include basal metabolism, exercise, hormone levels, muscle activity, and shivering.

Basal metabolism refers to the minimum energy consumption required to maintain life in an awake state. BMR is measured in a stable lying posture in a pleasant environment and after meals, in a state where there is no energy consumption for digestion and absorption after a certain period of time has elapsed. Although there are individual differences, the BMR of a 60-kg human is approximately 1300–1600 kcal. Approximately 30% of the BMR is used for the digestive system, such as liver function, while the rest is used for skeletal muscle, life support, and body temperature maintenance. BMR, according to weight, shows high values in infancy and childhood and then decreases. Females have lower BMR values than males do, which is due to the lower metabolic activity and higher fat content in females than in males. This difference in BMR by sex is larger in adulthood and smaller in childhood and senescence. Notably, in the same-sex and same-age groups, the BMR is proportional to the body surface area and decreases during sleep and climate acclimation, resulting in a large value in winter and a smaller one in summer. Exercise increases skeletal muscle contraction, requiring more oxygen intake for stabilization and increasing energy metabolism. The total energy expenditure during exercise reflects its intensity or workload and adds the energy metabolism required for the activity to the BMR.

Relative metabolism rate (RMR) indicates the amount of exercise. RMR is the metabolism during resting conditions, defined by a combination of assumptions of physiological homeostasis. The difference between RMR and BMR is that BMR measurements must meet total physiological equilibrium. In contrast, RMR measurements are more accessible and can represent an estimation of daily energy expenditure [11]. RMR is widely used in industrial labor and sports medicine.

$$\mathrm{RMR} = \left(\frac{\text{Active Metabolism} - \text{Resting}}{\text{Metabolism}} \right) / \,\mathrm{BMR} \tag{4.4}$$

The metabolic equivalent of task (MET), which represents the proportion of energy metabolism during exercise and rest, is used as an exercise indicator, and 1 MET is characterized by a stable energy metabolism. This is an indicator based on the standard male's metabolic rate at rest (1 MET = 50 kcal/(h·m^2) = 58.2 W/m^2 = 3.5 mL/kg/min). The American College of Sports Medicine describes exercise intensity using MET. In a stable state, it becomes 1, while RMR is 0. The formula for calculating METs is as follows:

$$\text{MET} = \text{Active Metabolism} / \text{Resting Metabolism} \quad (4.5)$$

MET may be represented by utilizing the amount of oxygen consumption.

$$\text{MET} = \text{Oxygen consumption during exercise} / \text{Oxygen intake during rest} \quad (4.6)$$

Skeletal muscle metabolism accounts for 75% of the total metabolism, and skeletal muscle contraction produces a large amount of energy. Therefore, metabolism during exercise increases with the intensity of muscle activity. Typical metabolic rates and METs for various activities are listed in Table 4.1.

Energy metabolism may increase owing to food intake, which is called diet-induced thermogenesis. This depends on the amount or ratio of nutrients in the energy sources of carbohydrates, proteins, and fats consumed during food intake. Metabolism may increase owing to the effects of hormones, such as strained sympathetic nerves and adrenaline secretion. Secretion of thyroid hormones, adrenal corticosteroids, adrenaline, and noradrenaline stimulates basal metabolism. In females, changes in body temperature appear under the influence of hormones such as estrogen and progesterone.

Metabolism caused by skeletal muscle accounts for approximately 25% of the BMR; however, muscle movement can change this ratio. In addition, shivering generated due to cold stimulation is a heat production mechanism caused by muscle contraction. It is an efficient method of heat production. Nonshivering thermogenesis is a chemical process that does not involve skeletal muscle contraction. It is caused by an increase in norepinephrine levels in the peripheral sympathetic nerves (see Chap. 5 for more details).

4.2.2 Heat Loss

Heat moves from inside the body to the outside via blood flow and the respiratory system. Additionally, it moves from the body surface to the ambient environment and is dissipated. This is called heat loss. Heat in the body is carried to each region by the blood. The air inhaled through respiration is warmed, humidified, and exhaled, and the heat is transferred to the external body. This heat loss is smaller than the one deriving from the body surface but must be considered in low-pressure environments. Notably, the thickness of the subcutaneous fat layer influences heat loss. Individuals with obesity with more subcutaneous fat have relatively low skin temperatures, whereas lean people have higher skin temperatures. During heat transfer from the body surface to the ambient environment, heat is dissipated externally by conduction, convection, radiation, and evaporation. Environmental conditions, clothing, body surface area, skin blood flow, sweating, and insensible perspiration are factors involved in heat loss.

4.2.2.1 Conduction (K)

Thermal conduction is the transfer of heat to an object in direct contact with the body surface. At this time, the heat moves from a high to a low temperature region. If the temperature of an object is lower than the skin temperature, heat from the skin is transferred to the object. The heat dissipated by thermal conduction is proportional to the temperature gradient, that is, the temperature change per unit length (Fourier's law). Even under the same environmental conditions, the thermal conductivities of the materials are different; therefore, the degree of coolness when touching is different. For example, the body loses more heat in water than in air when both water and air have similar temperatures because water has higher thermal conductivity than air does.

Table 4.1 Metabolic rates and METs for various physical activities

Physical activity	Metabolic rate (W/m^2)	METs
Resting		
Sleeping	40	0.7
Reclining	45	0.8
At rest, sitting	55	1.0
At rest, standing	70	1.2
Walking		
2 km/h	110	1.9
3 km/h	140	2.4
4 km/h	165	2.8
5 km/h	200	3.4
Office		
Reading, seated	55	1.0
Writing	60	1.0
Typing	65	1.1
Filing, seated	70	1.2
Filing, standing	80	1.4
Walking about	100	1.7
Lifting/packing	120	2.1
Driving/flying		
Car	60–115	1.0–2.0
Aircraft, routine	70	1.2
Aircraft, instrument landing	105	1.8
Aircraft, combat	140	2.4
Heavy vehicle	185	3.2
Miscellaneous occupational activities		
Cooking	95–115	1.6–2.0
Housecleaning	115–200	2.0–3.4
Seated, heavy limb movement	130	2.2
Machine work		
Sawing (table saw)	105	1.8
Light (electrical industry)	115–140	2.0–2.4
Heavy	235	4.0
Handling about 50 kg bags	235	4.0
Pick-and-shovel work	235–280	4.0–4.8
Miscellaneous leisure activities		
Dancing, social	140–255	2.4–4.4
Calisthenics/exercise	175–235	3.0–4.0
Tennis, singles	210–270	3.6–4.0
Basketball	290–440	5.0–7.6
Wrestling, competitive	410–505	7.0–8.7

Partially modified from ASHRAE Handbook [12] and ISO 8996 [13]

4.2.2.2 Convection (C)

Convection refers to the movement of heat through air or water. There are two types of convections: natural and forced. Air in contact with the surface of the human body in a stationary state becomes warm and light, generating an updraft. This phenomenon is known as natural convection. When natural convection is ignored by motion or wind, heat dissipation is promoted by forced convection. Convection continuously supplies cold air to facilitate the exchange of conduction heat. For example, cold air can be generated from wind using an electric or folding fan to promote heat loss.

4.2.2.3 Radiation (R)

Radiation refers to a phenomenon in which thermal radiation transfers energy between objects. Thermal radiation is the process of electromagnetic radiation emitted by a heated surface in all directions. Thermal radiation does not require an intervening medium to carry it. The solar radiant heat warms the earth we live in, and this movement of heat rays is called radiation. Moreover, it is possible to warm the human body with radiant heat from heating devices. The rate at which heat is transferred is determined by the temperature of the surface; thus, if the body's surface is warmer than the surrounding environment, the body loses heat. At this time, the rate of heat loss is determined by the difference between the body surface and the ambient temperature.

4.2.2.4 Evaporation (E)

When water evaporates from the skin surface or respiratory tract, the heat of vaporization is removed from the body. This heat transfer is called evaporation. Evaporation is one of the main processes that causes heat loss in the body, and over 580 kcal/L of heat is used for evaporation (heat from skin converts sweat to water vapor). During moisture evaporation, the heat required for evaporation is conducted from the body surface, and the body surface is cooled. Evaporation from the skin surface can be divided into sweating and insensible perspiration.

The heat dissipated via conduction, convection, and radiation on the body surface is called dry-process heat loss or sensible heat transfer, and heat dissipated by evaporation is called wet-process heat loss or insensible heat transfer [14]. Heat loss can be damaging if the ambient temperature is lower than the skin temperature. When the ambient temperature increases, sensible heat transfer decreases, whereas heat transfer by evaporation increases. If the ambient temperature is lower than the average skin temperature, it causes dry-process heat loss and insensible perspiration. However, if the environmental temperature is higher than the average skin temperature, heat from the environment is introduced into the body, and the only effective heat dissipation pathway is the evaporation of sweat.

4.3 Body Temperature

The human body is divided into the core (inside the body), such as the head and organs, which are important from the perspective of body temperature control, and the shell, such as the subcutaneous tissue and skin (Fig. 4.3). Since the metabolism that supports life activities occurs mainly in the core, its temperature remains constant at about 37 °C regardless of changes in the external environment. This is homeostasis and the temperature at which life occurs in the body. Maintaining body temperature homeostasis is important for determining the reaction path and speed of metabolism. In general terms, body temperature means core temperature. The temperature displayed on the shell of the body is referred to as shell temperature. Because the shell temperature is deeply involved in heat loss from the body surface, it fluctuates significantly according to the ambient temperature. When the shell temperature is specified, it is considered the skin temperature.

Compared to the core temperature, which is maintained around the body's organs, the temperature of the skin surface or peripheral parts of the body is affected by the ambient temperature. The human body is composed of a core and a shell. In hot and cold environments, the core temperature is relatively constant; however, the shell temperature fluctuates depending on the external environment. During exposure to a warm environment, the core and shell temperatures become similar, with the two compartments of body temperature regulation maintaining the same temperature. When exposed to a cold environment, the region in the human body that exhibits the same temperature as the core is very small, showing a significant temperature difference between the two compartments. In other words, the outer shell becomes thinner in summer and thicker in winter (Fig. 4.3).

Figure 4.4 is the result of a study on the effect of ambient temperature on the core and skin temperature. The study involved exposing Japanese male college students, wearing only underwear, to various temperature conditions (17–34 °C). The core temperature was not affected by the low

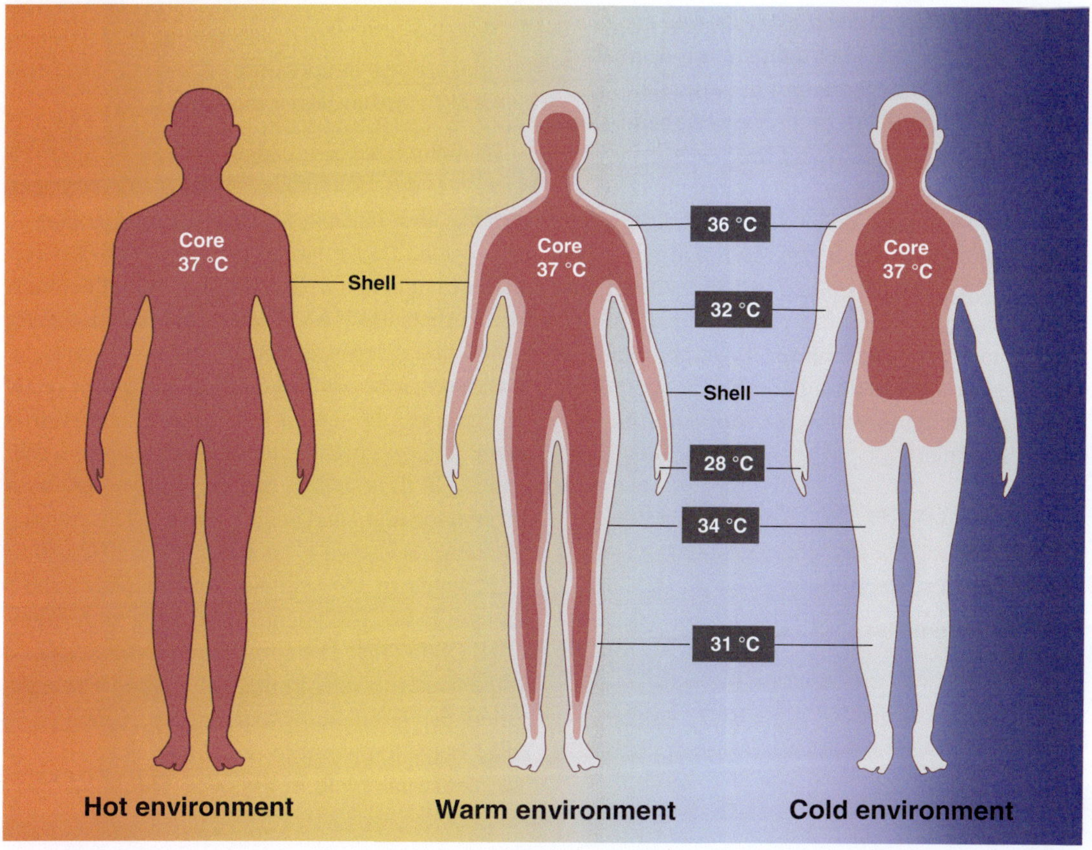

Fig. 4.3 Distribution of body temperature in a hot and cold environment

Fig. 4.4 Temperatures of the rectum and skin on each part of the body in various environmental temperature conditions. (Adapted from Fukazawa [15])

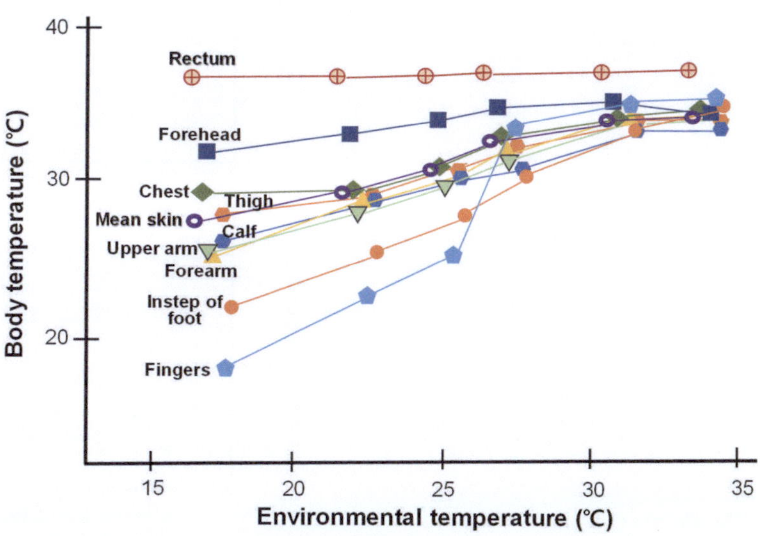

ambient temperature and mostly showed constant values; however, the skin temperature decreased with the ambient temperature. In particular, there was a significant decrease in the peripheral skin temperature in the hands and feet.

Therefore, the average body temperature of an individual should be defined by considering the values obtained through core and skin temperatures, which differ depending on the environmental temperature. When the rectal temperature is used as the core temperature, it can be obtained by changing the weighting coefficient according to the outside temperature, as shown in the following formula:

Cold environment $$T_b = 0.6T_{re} + 0.4T_{sk} \quad (4.7)$$

Neutral environment $$T_b = 0.65T_{re} + 0.35T_{sk} \quad (4.8)$$

Hot environment $$T_b = 0.8T_{re} + 0.2T_{sk} \quad (4.9)$$

where,

T_b: mean body temperature, T_{re}: rectal temperature, T_{sk}: mean skin temperature.

Humans have a circadian rhythm of approximately 1 day. Circadian rhythms are physical, mental, and behavioral changes that follow a 24-h cycle. These natural processes respond primarily to light and dark conditions and affect most living organisms [16]. While a person periodically repeats the sleep-wake cycle, the body temperature also has its own rhythm. In the case of rectal temperature, the lowest level can be observed at dawn and gradually increases thereafter. After peaking in the evening, the temperature decreases through the night and in the morning. This fluctuation range is known to be about 0.7–1.3 °C.

Females in the sexually mature phase show changes in body temperature because of the menstrual cycle [17]. The core temperature measured in the wake-up state immediately after waking is called the basal body temperature. Changes in basal body temperature are shown according to the menstrual cycle in Fig. 4.5. The basal body

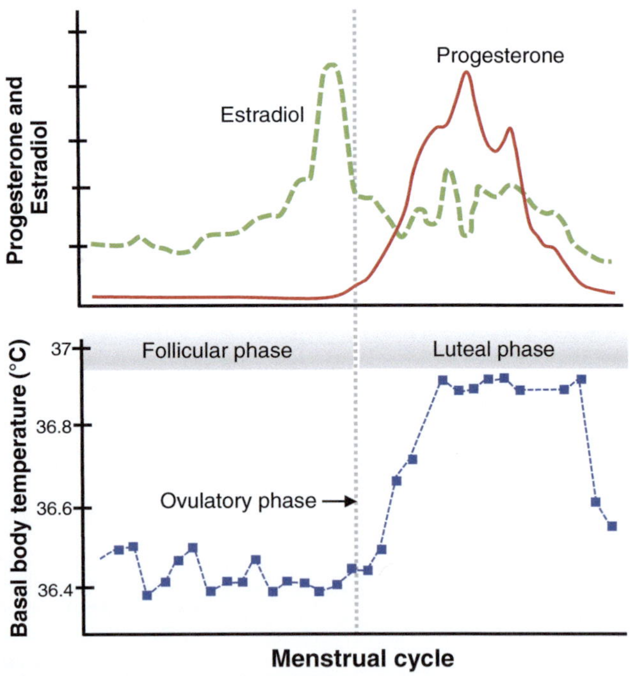

Fig. 4.5 Changes in basal body temperature according to the menstrual cycle regulated by hormones. (The figure is adapted from Baker et al. [18])

temperature is divided into low- and high-temperature parts. From the menstrual to the ovulatory phase, the basal body temperature is lower, and at the ovulatory phase, an additional decrease in temperature occurs. After the ovulatory phase, it rises by about 0.5 °C and remains at a higher temperature for about 14 days. The low-temperature state is repeated after the menstrual phase. Estradiol, a female hormone, lowers core temperatures by controlling heat production and promoting heat loss. However, conflicting results have reported that progesterone promotes heat production and prevents heat loss. According to the menstrual cycle, this change in hormone levels affects basal body temperature changes.

Body temperature changes with age. The axillary temperatures at each age are listed in Table 4.2. Axillary temperature is >37 °C during the neonatal period and gradually decreases, indicating a stable value until about 10 years of age [19]. In the case of older adults aged >65 years, the axillary temperature decreased by 0.23 °C from the average value of the temperatures at 10–50 years of age [20].

4.3.1 Core Temperature

The core temperatures include rectal, oral (sublingual), axillary, esophageal, and tympanic. However, these temperatures are not equal. They are divided and used according to the measurement purpose. Rectal temperature was the standard for calculating the body heat capacity, and axillary, oral, and rectal temperatures were used as fever indicators. On average, the axillary temperature is slightly lower, while the rectal temperature, which reaches 37 °C, is slightly higher than the axillary temperature. In addition, the tympanic temperature, which reflects the brain temperature, is an indicator of the thermoregulation mechanism. The core temperature remains relatively constant; however, hyperthermia and hypothermia can occur in extremely hot and cold environments. The symptoms according to changes in the core temperature are shown in Fig. 4.6. The core temperature is affected by the following factors, which need to be carefully considered: circadian rhythm, female hormone

Table 4.2 Axillary temperature by age

Age group	Axillary temperature
0–1 year	37.09 °C
2–3 years	37.08 °C
4–5 years	37.12 °C
6–9 years	37.06 °C
>10 years	37.01 °C
10–50 years	36.89 ± 0.34 °C
>65 years	36.66 ± 0.42 °C

Adapted from Yamanaka [19] and Iriki et al. [20]

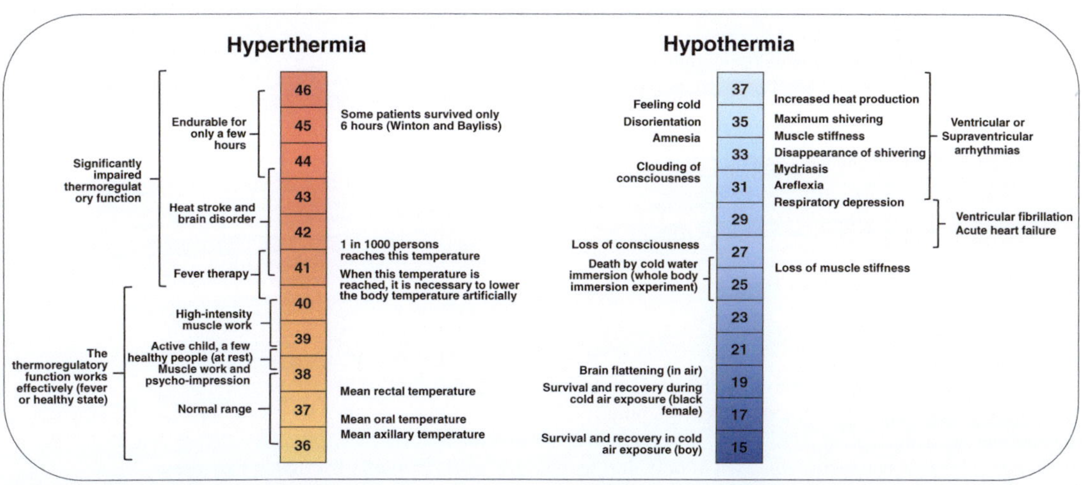

Fig. 4.6 Symptoms because of changes in core temperature. (Adapted from Tochihara [21])

cycle or pregnancy status, age, exercise, bathing, eating, mental stress, differences in the surrounding environment such as temperature or humidity, and measurement device status.

4.3.1.1 Core Temperature Monitoring Methods

Core temperature is an important index for evaluating thermoregulatory functions. Although this measurement is difficult, it is helpful because the appropriate measurement is experimentally verified. There are five representative measurement sites: the axilla, oral cavity, rectum, esophagus, and tympanum. Axillary, oral, and tympanic temperatures are easily affected by the external environment. Therefore, when measuring these core temperatures, it is important to consider insulation (blocking the outside air). The characteristics of each measurement are listed in Table 4.3.

Axillary Temperature

The thermometer is held at a 45° angle and inserted tightly under the armpit. After maintaining this state for 10 min, a stable value is obtained. On measuring several Japanese individuals, an average value of 36.63 ± 0.40 °C was determined as the axillary temperature [23]. This is the simplest measurement method, and subjects experience fewer psychological and physical burdens.

Oral Temperature

The thermometer is placed under the tongue, and the mouth is closed lightly. The displayed value is noted after approximately 5 min. Differences in the measurement spot under the tongue are slight, and individuals may experience minor discomfort. The thermometer must be disinfected before measurement, and there is a risk of damage to the sensor during the measurement. The average oral temperature for Japanese individuals is 36.77 ± 0.34 °C [23].

Rectal Temperature

A thin long sensor is inserted through the anus and is primarily used in experimental research owing to its reliability and stability. Individuals may experience discomfort. Again, the thermometer must be disinfected before measurement, and there is a risk of sensor damage during the measurement. Generally, the average temperature is 36.6–38 °C, 0.4–0.7 °C higher than the axillary temperature, and 0.2–0.4 °C higher than the oral temperature [21]. The measured values may vary depending on the sensor insertion depth. According to a comparative study of the measured values based on insertion length, there was no significant difference in the measurement results when the rectal temperature sensor was inserted to a depth of >10 cm. However, a sensor at a depth of 16 cm was recommended for the highest and most stable temperature measure-

Table 4.3 Characteristics of each core temperature measurement method

Temperature	Difficulty of measurement[a]	Continuous measurement[b]	Obstruction of work[c]	Discomfort[d]	Riskiness[e]
Axillary	0	D	1	0	0
Oral	0	D	1	1	1
Rectal	1	C	0	2	1
Esophageal	2	C	2	3	2
Tympanic (contact type)	2	C	2	3	2
Tympanic (infrared type)	0	D	1	0	0

Adapted and partially modified from Tochihara [21] and ISO 9886 [22]

[a] 0: simple, 1: skill required, 2: medical skills required

[b] C: continuous measurement possible, D: noncontinuous measurement

[c] 0: no limit, 1: work obstruction may occur only during measurements, 2: work obstruction may occur

[d] 0: neutral, 1: slightly uncomfortable, 2: psychologically and probably physically uncomfortable

[e] 0: none, 1: potentially dangerous, 2: may be dangerous if unskilled

ment [24]. Furthermore, a wireless method for measuring rectal temperature using an ingestible telemetric temperature pill allows for remote monitoring of core temperature. Subjects should ingest an activated temperature pill at least 6 h before use. These pills are effective for continuous core temperature measurements and can detect minor changes [25]. Being noninvasive, they place a minimal burden on the subject. The core temperature measured using this method typically differs by no more than 0.3 °C from measurements taken using direct probe insertion [26].

Esophageal Temperature

The esophageal temperature reflects the temperature of the aortic blood flow in the heart. A dedicated thermometer is inserted into the esophagus from the nostril. The sensor should be inserted at a distance of approximately 40 cm. The measurement requires medical skills and monitoring, imposing a significant psychological and physical burden on the individual. As usual, the thermometer must be disinfected before measurement, and there is a risk of damage to the sensor during the procedure.

Tympanic Temperature

The tympanic temperature reflects the temperature of the internal carotid artery blood flow through the hypothalamus, which is the thermoregulatory center. Direct measurement of the tympanic temperature is logically and technically complicated. There is no discomfort in the infrared tympanic thermometers, but they are easily affected by the external environmental temperature. Recently, infrared tympanic temperature sensors have been developed and used for the real-time remote monitoring of workers wearing protective clothing in nuclear facilities [27] (Fig. 4.7). As the tympanic temperature shows differences from the rectal temperature, it is impossible to be sure of the absolute core temperature measurement. However, measuring tympanic temperature during sleep is more practical and convenient than measuring rectal temperature. The tympanic temperature can be used to monitor real-time changes in body temperature during sleep or monthly changes in body temperature to predict a woman's ovulation phase [28].

4.3.2 Skin Temperature

The relationship between skin temperature and physiological and sensory conditions is shown in Fig. 4.8. In addition, the degree of variability in the skin temperature varies depending on the body part. Specifically, when exposed to a cold environment, skin temperature decreases in the head and trunk, and a significant decrease in temperature is observed in the limbs or peripheral parts located far from the trunk. The mean skin

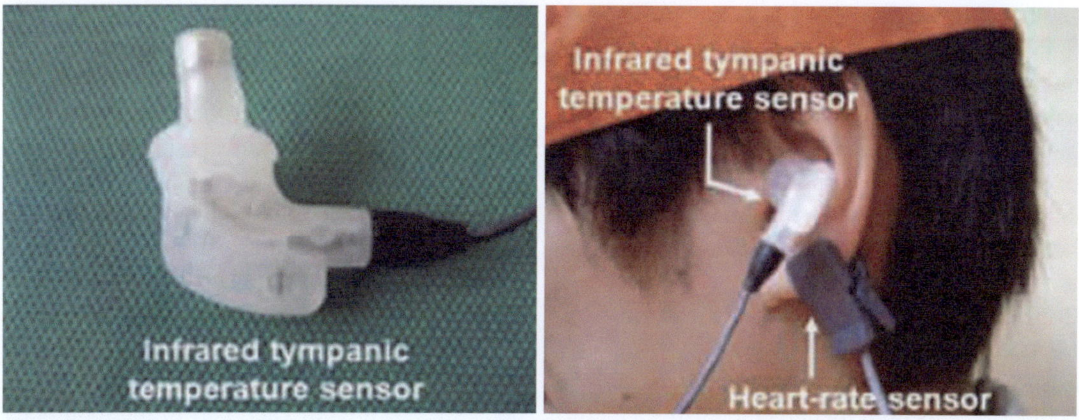

Fig. 4.7 Ear-plug type infrared tympanic temperature sensors and wearing condition. (The figure is adapted from Takahashi et al. [27])

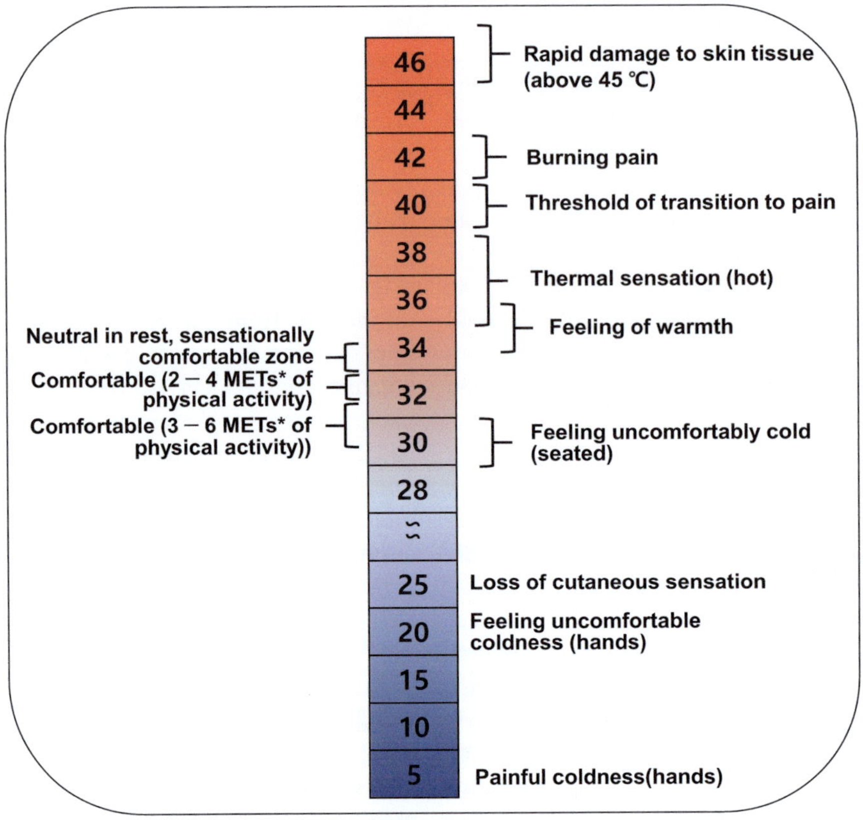

Fig. 4.8 Skin temperature and its physiological condition and sensation. (Adapted from Tamura and Hirata [29]). *MET* metabolic equivalent of task

temperature is obtained from the skin temperature of each body part and the area ratio of the corresponding body part.

4.3.2.1 Skin Temperature Monitoring Methods

Skin temperature is the boundary between the human body and the environment. Moreover, it is a factor that regulates heat exchange between the human body and the environment and is an index of thermoregulatory responses. In addition, it is an indispensable indicator that influences thermal sensation in thermal physiology research.

Contact and Noncontact-Type Measurement

For skin temperature measurement, there is a contact-type measurement method using a thermistor sensor, a copper-constantan thermometer, and a noncontact-type measurement method such as thermography. A regional body part suitable for measurement purposes is selected for contact-type measurements, and a sensor is attached to the skin. The choice of the measurement point is important for determining the mean skin temperature.

Mean Skin Temperature Measurement

The mean skin temperature is the average skin temperature of countless points distributed throughout the body. However, because it is impossible to measure all of these, the body is divided into several parts showing similar temperatures, which are measured and expressed as regional skin temperatures. In addition, to calculate the mean skin temperature, the ratio of the body surface area of each part to the total body surface area is multiplied by the temperature, and all results are summed. Several previous studies have published formulae for calculating mean skin temperature (Table 4.4). The mean skin temperature can be considered closer to the accurate average skin temperature as the number of measurement points increases. However, there are disadvantages in that the measurement requires time and effort and is challenging to measure outdoors. On the other hand, if there are few measurement points, it is possible to measure them easily, but there is a high possibility of error.

Infrared Radiation Measurement

An infrared radiation thermometer measures the skin temperature using a noncontact measurement method. On the surface of any object whose surface temperature is above absolute zero (−273 °C), infrared radiant energy (R) proportional to the fourth power of the absolute temperature of the surface (T_s [K]) is emitted.

$$R = s \cdot \sigma T_s^4 \ (\text{Stefan - Boltzmann Law}) \quad (4.10)$$

where,

R: infrared radiant energy; s: emissivity of an object ($0 \leqq s \leqq 1$); σ: $5.67 * 10^{-8}$ [W/m^2 · K^4]

Human skin is presented as almost black when the emissivity (s) is 0.99–0.98. Depending on the scan rate, it is possible to obtain temperature distribution figures in black-and-white or color. This system is called thermography, and the figures obtained from this system are called thermograms. Thermograms of differences in skin temperature under a hot (34 °C) and cold (15 °C) environment are shown in Fig. 4.9.

Table 4.4 Number of measuring points and the coefficient of mean skin temperature

	Hardy and Dubois	Hardy and Dubois	ISO 9886	Ramanathan	Burton
Number of measuring points	12	7	8	4	
Forehead	0.07	0.07	0.07		
Cheek					
Scapula			0.175		
Chest	0.0875		0.175	0.3	0.5
Upper arm			0.07	0.3	
Abdomen	0.0875	0.35			
Lower arm	0.14	0.14	0.07		0.14
Back of hand	0.05	0.05	0.05		
Thigh	0.095	0.19	0.19	0.2	0.36
Calf	0.065	0.13	0.2	0.2	
Instep of foot	0.07	0.07			
Back	0.0875				
Waist	0.0875				
Back of thigh	0.095				
Back of calf	0.065				

Adapted and partially modified from Hardy and Dubois [30], ISO 9886 [22], Ramanathan [31], and Burton [32]

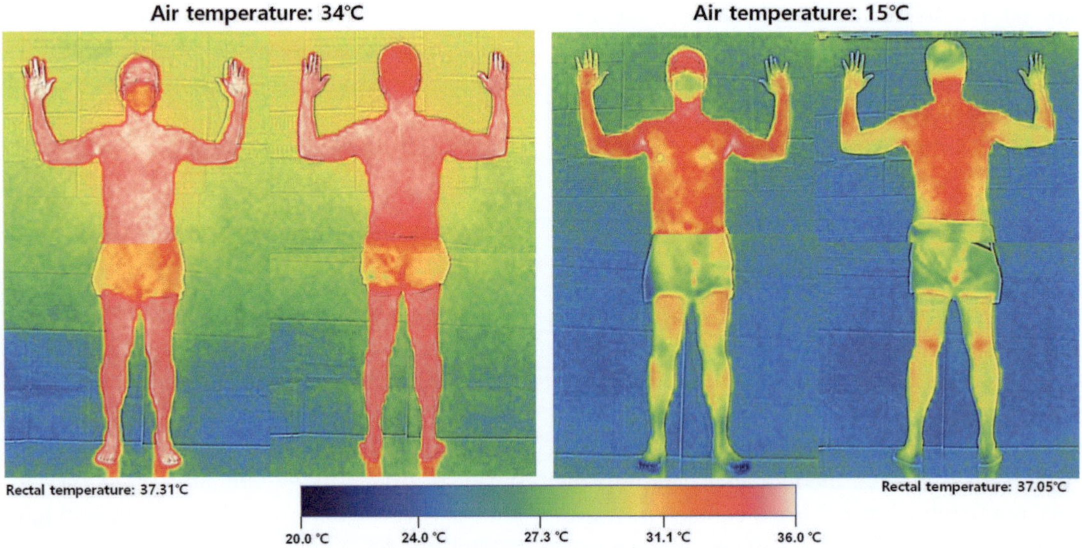

Fig. 4.9 Thermograms of differences in skin temperature under a hot (34 °C) and cold (15 °C) environment. The subject is a 38-year-old male

References

1. Gordon CJ. Thermal physiology of laboratory mice: defining thermoneutrality. J Therm Biol. 2012;37(8):654–85.
2. Henderson MET, Halsey LG. The metabolic upper critical temperature of the human thermoneutral zone. J Therm Biol. 2022;103380
3. Pallubinsky H, Schellen L, van Marken Lichtenbelt WD. Exploring the human thermoneutral zone–a dynamic approach. J Therm Biol. 2019;79:199–208.
4. Widmaier EP, Raff H, Strang KT. Vander's human physiology: the mechanisms of body function (hardcover, 14, revised). Mcgraw-Hill College; 2017.
5. Edition T. Glossary of terms for thermal physiology. J Therm Biol. 2003;28:75–106.
6. Leonard WR, Robertson ML, Snodgrass JJ, Kuzawa CW. Metabolic correlates of hominid brain evolution. Comp Biochem Phys A. 2003;136:5–15.
7. Leonard WR, Robertson ML. Evolutionary perspectives on human nutrition: the influence of brain and body size on diet and metabolism. Am J Hum Biol. 1994;6:77–88.
8. Lee JY, Park J, Kim S. Cold adaptation, aging, and Korean women divers Haenyeo. J Physiol Anthropol. 2017;36(1):1–13.
9. Mota-Rojas D, Titto CG, Orihuela A, Martínez-Burnes J, Gómez-Prado J, Torres-Bernal F, et al. Physiological and behavioral mechanisms of thermoregulation in mammals. Animals. 2021;11(6):1733.
10. Patel H, Kerndt CC, Bhardwaj A. Physiology, respiratory quotient. In: StatPearls. Treasure Island; 2023. https://www.ncbi.nlm.nih.gov/books/NBK531494/. Accessed 13 Feb 2023.
11. Ravussin E, Burnand B, Schutz Y, Jéquier E. Twenty-four-hour energy expenditure and resting metabolic rate in obese, moderately obese, and control subjects. AJCN. 1984;35(3):566–73.
12. ASHRAE. ASHRAE handbook—fundamentals. Norcross: American Society of Heating, Refrigerating and Air-Conditioning Engineers; 2021.
13. ISO. ISO 8996 Ergonomics of the thermal environment—determination of metabolic rate. Geneva: International Organization for Standardization; 2021.
14. Romanovsky AA. The thermoregulation system and how it works. Handb Clin Neurol. 2018;156:3–43.
15. Fukazawa T. Cold, hot, and health (in Japanese). In: Sugai K, Morooka H, Mitsuno T, Hiyabayashi Y, editors. Apparel and health: from the basics to evolving clothing. Tokyo: Inoueshoin; 2012. p. 20.
16. National institute of general medical sciences. Circadian rhythms. In: Fact sheets about science and health. National Institute of General Medical Sciences. 2020. https://nigms.nih.gov/education/fact-sheets/Documents/fact-sheet-circadian-rhythms.pdf. Accessed Oct 2020.
17. Cuevas IF, López NE. Effects of menstrual cycle on skin temperature. In: Thermohuman. 2021. https://thermohuman.com/2021/04/24/effects-of-menstrual-cycle-on-skin-temperature/. Accessed 24 Apr 2021.
18. Baker FC, Siboza F, Fuller A. Temperature regulation in women: effects of the menstrual cycle. Temperature. 2020;7(3):226–62.
19. Yamanaka T. Approaches for 0-year-old children in primary care (in Japanese). Jpn J Pediatr Med. 1991;32(12):1341–5.

20. Iriki M, Kosaka M, Murakami N, Murada S. Studies on the axillary temperature of the Japanese aged (in Japanese). Nippon Ronen Igakkai Zasshi. Jpn J Geriatr. 1975;12(3):172–7.
21. Tochihara Y. Thermal environment (in Japanese). In: Tochihara Y, editor. Artificial environment design handbook. Tokyo: Maruzen; 2007. p. 33–5.
22. ISO. ISO 9886. Ergonomics—evaluation of thermal strain by physiological measurements. 2004 International Organization for Standardization, Geneva; 2004.
23. Iriki M, Tsuchiya K, Kinno I, Uchino K, Kawashima Y. The oral temperature of healthy Japanese (in Japanese). Jpn J Biometeorl. 1988;25(3):163–71.
24. Lee JY, Wakabayashi H, Wijayanto T, Tochihara Y. Differences in rectal temperatures measured at depths of 4–19 cm from the anal sphincter during exercise and rest. Eur J Appl Physiol. 2010;109:73–80.
25. Bongers CC, Hopman MT, Eijsvogels TM. Using an ingestible telemetric temperature pill to assess gastrointestinal temperature during exercise. J Vis Exp. 2015;(104):53258.
26. Gosselin J, Béliveau J, Hamel M, Casa D, Hosokawa Y, Morais JA, Goulet ED. Wireless measurement of rectal temperature during exercise: comparing an ingestible thermometric telemetric pill used as a suppository against a conventional rectal probe. J Therm Biol. 2019;83:112–8.
27. Takahashi N, Lee JY, Wakabayashi H, Tochihara Y. Development and operational results of a real-time remote biological information monitoring device for the workers wearing protective clothes at a nuclear facility. Hoken Butsuri. 2011;46(2):140–7.
28. Ko Y, Jung JY, Kim HT, Lee JY. Auditory canal temperature measurement using a wearable device during sleep: comparisons with rectal temperatures at 6, 10, and 14 cm depths. J Therm Biol. 2019;85:102410.
29. Tamura T, Hirata K. Thermal comfort (in Japanese). In: Tamura T, Kobayashi T, Hirata K, editors. Science of clothing environment. Tokyo: Kenpakusha; 2004. p. 14.
30. Hardy JD, Du Bois EF, Soderstrom GF. The technic of measuring radiation and convection: one figure. J Nutr. 1938;15(5):461–75.
31. Ramanathan NL. A new weighting system for mean surface temperature of the human body. Appl Physiol. 1964;19(3):531–3.
32. Burton AC. Human calorimetry: II. The average temperature of the tissues of the body: three figures. J Nutr. 1935;9(3):261–80.

Physiological Responses and Performance in Cold Environments

5

Hitoshi Wakabayashi

5.1 Physiological Responses to Cold Environments

To maintain core body temperature (T_{co}) within a certain range in response to changes in external environmental temperature, the human body produces autonomic thermoregulatory responses such as vasoconstriction/vasodilation, metabolic heat production, and sweating. The point at which the environmental temperature decreases and cold-induced thermogenesis (CIT) begins is called the thermogenic threshold (lower critical temperature), and the point at which the temperature rises and thermoregulatory sweat response begins is the sweating threshold (upper critical temperature). The temperature range located between the thresholds of CIT and the sweating, where heat dissipation is mainly regulated by the skin vasoconstriction/vasodilation response (cutaneous vasomotor regulation zone), is called the inter-thresholds zone or the thermoneutral zone. The center regulating a balance between heat dissipation and heat production through these thermoregulatory responses exists in the hypothalamus of the diencephalon. It controls cutaneous blood vessels, brown adipose tissue (BAT), white adipose tissue, liver, skeletal muscle (SM), sweat glands, and other effector organs in a complex manner, via the sympathetic nervous system, somatic nervous system, and endocrine system.

In this chapter, we will first introduce CIT and cold-induced vasoconstriction, which are autonomous thermoregulatory responses to cold environment below the thermogenic threshold, that is, in the metabolic regulation zone (chemical control zone). In addition to the general thermoregulatory responses for maintaining T_{co}, this chapter includes literature review on the cold-induced vasodilation (CIVD), which is observed when the hands and fingertips are immersed in cold water. Moreover, we introduce the cold shock responses, which are the ventilatory and cardiovascular initial responses when the whole body is immersed in cold water.

5.1.1 Thermoregulatory Response in Cold

5.1.1.1 Shivering and Nonshivering Thermogenesis

In cold environment, for maintaining T_{co}, human enhances CIT, which is divided into nonshivering thermogenesis (NST) and shivering thermogenesis (ST). A schematic diagram shows the relationship between thermal environment and components of metabolic heat production with major thermogenetic tissues for each component (Fig. 5.1).

H. Wakabayashi (✉)
Division of Environmental Engineering, Faculty of Engineering, Hokkaido University, Sapporo, Japan
e-mail: wakabayashi@eng.hokudai.ac.jp

Y. Tochihara (ed.), *The Thermal Environment*, https://doi.org/10.1007/978-981-96-3009-7_5

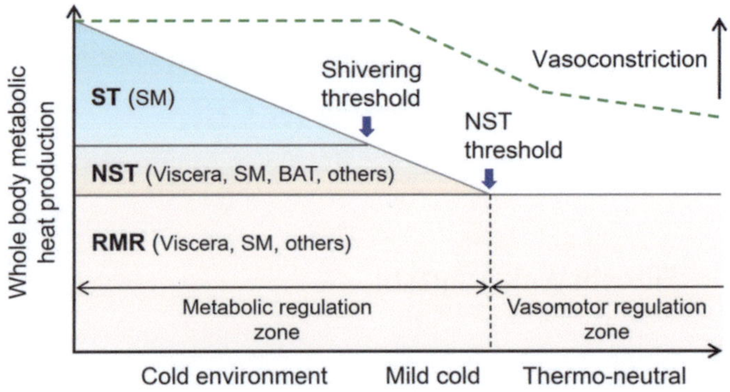

Fig. 5.1 Scheme of autonomic thermoregulatory response in cold environment—nonshivering and shivering components of cold-induced thermogenesis with major thermoeffector organs. *ST* shivering thermogenesis, *NST* nonshivering thermogenesis, *RMR* resting metabolic rate, *SM* skeletal muscle, *BAT* brown adipose tissue

In the thermoneutral environment, basal metabolism and resting metabolism occur in visceral tissues such as the liver and SM. Basal metabolic rate (BMR) and resting metabolic rate (RMR) are classified as obligatory NST and are sometimes distinguished from adaptive thermogenesis such as cold-induced NST. In mild cold environments below thermoneutral zone, cold-induced NST occurs in tissues such as BAT, SM, and internal organs. In this scheme of CIT (Fig. 5.1), the thermogenic threshold at which cold-induced NST begins to occur is specifically defined as the NST threshold. Then, in a cold environment, ST is generated in the SM, and the threshold at which shivering begins is defined as the shivering threshold. Previously, the thermogenic threshold of the human thermoregulatory response had been often used synonymously with the shivering threshold [1, 2]. But in the past decade, since the rediscovery of human BAT activity in 2009 [3–5], researchers made some revisions on the model distinguishing NST and ST components separately with each threshold [6].

There are large individual variations in NST enhancement during mild cold exposure, and it was reported to range from a few percent to 30% of RMR in young and middle-aged people [6]. Our research group also evaluated NST in a climatic chamber at 18.6 °C in young Japanese male participants and reported individual variations in NST; the higher the BAT activity, the greater the increase in NST, ranging from almost no change to around 30% of RMR (13.5% on average) [7]. In a study with larger number of participants, it was reported that in a mild cold environment of 19 °C, NST increased by approximately 5% of RMR in individuals with low BAT activity and approximately 18% in individuals with high BAT activity [8].

ST during cold exposure refers to heat production due to involuntary muscle tremor for thermoregulation induced by a decrease in skin and core body temperature. The maximum ST reaches approximately five times the RMR and 40–50% of the maximal oxygen uptake [9]. The amount of heat generated by ST is considerably larger than that by NST, since the SM is widely distributed throughout the body. Electromyography (EMG) is often used to objectively evaluate the onset of shivering and shivering intensity. At the early phase in cold exposure, muscle tonus and continuous mild tremors occur, and recruitment of low-threshold motor units (slow-twitch muscle fibers) is observed in the EMG waveform [10]. Afterward, burst-like high-intensity shivering caused by higher-threshold motor units (fast-twitch muscle fibers) occurs [9, 10], which is observed mixed with continuous low-intensity shivering. As the intensity of shivering increases, the frequency of burst-like shivering increases, and carbohydrate metabolism is enhanced [9]. There are individual differences in the location of muscle tremor, but shivering intensity evaluated by EMG is higher in the trunk and lower in the extremities [9, 10].

Although the schematic diagram divides CIT into components of NST and ST, demonstrating the major thermoeffectors for them, the contribution ratio of BAT and SM for NST is not specified (Fig. 5.1). Regarding the contribution of BAT to the NST, some studies have reported that the contribution of BAT is negligible [11, 12], while others have reported a higher contribution of BAT [13, 14]. In addition, because BAT activity exhibits inter- and intra-individual variations, such as group differences, individual differences, and seasonal and diurnal variations [15], it is not possible to uniformly determine the average contribution of each tissue to NST. In the next paragraph, we will introduce previous reports that examined the contribution of each tissue to NST according to individual differences in BAT activity and its relationship with other thermoregulatory responses.

5.1.1.2 Inter- and Intra-Individual Variation in the BAT Activity

Before explaining group and individual differences in BAT activity, the standard BAT evaluation methods (FDG-PET/CT) used in various countries and research groups are briefly outlined here. Currently, FDG-PET/CT that combines positron emission tomography (PET) and computed tomography (CT) to image the uptake of ^{18}F-fluorodeoxyglucose (^{18}F-FDG) into BAT during mild cold exposure is most commonly used [16]. The mean standard uptake value (SUV_{mean}) and maximum standard uptake value (SUV_{max}) in the region of interest, where BAT is present, are analyzed and used as an index of BAT activity and as a threshold for determining BAT positive or negative individuals. Based on the BAT activity during mild cold exposure conducted in several countries, it has been reported that the detection rate of BAT-positive individuals differs depending on the populations [17]. Compared to the high detection rate of 95% in the Netherlands (Maastricht) and 70% in the Finnish (Turku) population, it was relatively low as 41% in the Japanese (Sapporo) [18] and 36% in the American (Detroit) populations. Although these population differences include the influence of the environments where participants live and the different protocols, in detail, used in each research group, there is a precious research comparing South Asian and European populations living in the same region, using the same evaluation technique, conducted by a research group [19]. In this study, they evaluated BAT activity in South Asian (both parents are from South Asia) and Dutch Caucasian populations born and grown up in the Netherlands. Although no group differences were found in SUV_{mean} or SUV_{max}, BAT volume was smaller in South Asian than in Caucasian group [19]. Another study identified haplotypes of the uncoupling protein 1 (UCP1) gene in a group of participants who showed significant NST enhancement during mild cold exposure in a climate chamber [20]. They then analyzed the frequency of the haplotype based on the 1000 Genomes Project database and found that the haplotype appears more frequently in populations from higher latitudes (areas with lower average temperatures) [20]. In other words, the frequency of NST-enhanced UCP1 haplotypes is highest in European populations, and the frequency decreases in the order of Asian and African populations, suggesting that BAT activity and NST enhancement may show a similar trend.

As for individual differences in BAT activity within a certain population, early studies have reported a relationship between obesity and BAT activity [3–5], and it is well known that obese people have low BAT activity. It was pointed out that there was a relationship between type 2 diabetes and BAT activity [21], as well as a link between the high prevalence of type 2 diabetes and low BAT activity, especially in South Asian populations [19]. BAT activity was clearly lower in type 2 diabetic patients when compared with young healthy individuals [22, 23], whereas there was no difference when compared with healthy individuals of the same generation [22]. In addition, it is well known that BAT activity decreases with growth and aging [3, 18], and there are also reports of differences depending on outdoor temperature and gender differences [3, 21].

Variations in BAT activity are observed even within the same individual, such as seasonal acclimatization and circadian rhythm [15]. Seasonal acclimatization in BAT activity is well known to be higher in winter, when the outdoor temperature is low, compared to that in summer [4, 24]. In addition to the seasonal acclimatization, it has been reported that experimental cold acclimation through the repeated short-term cold exposures enhanced BAT activity in healthy young participants [8, 14, 25, 26], type 2 diabetic patients, and obese individuals as well [23, 27]. Furthermore, in recent study, we found diurnal variation in BAT activity during mild cold exposure [28]. The NST assessed by the expired gas and supraclavicular skin temperature as a surrogate measure for the BAT thermogenesis tended to be higher in the morning and lower at night in those with high BAT activity, whereas no diurnal differences were observed in those with low BAT activity [28]. This suggests a diurnal variation in BAT activity [28].

As described above, since there are inter-individual and intra-individual variations in BAT activity [15], the contribution of BAT to NST would also vary and fluctuate. According to the variation in BAT activity, the SM and other organs would have diversity in the degree of contribution as thermogenic effectors. In the following paragraphs, we will introduce research on the metabolic contributions of multiple organs to the CIT including NST and ST.

5.1.1.3 BAT and Skeletal Muscle Multiorgan Contributions to CIT

BAT activity and detection rate are particularly low in Asian compared to European populations [17, 19]. Thus, there would be a greater variation in the BAT activity and multiorgan metabolic contribution to CIT in Asian populations. Therefore, we investigated the relationship between CIT during mild to moderate cold exposure and amount of BAT and SM in young Japanese male [7]. Starting from thermoneutral baseline measurement of RMR at 28 °C in an artificial climate chamber, room temperature was controlled at mild cold environment of 18.6 °C for 90 min to induce NST, and then temperature was further lowered to 11.6 °C to initiate shivering. During the protocol, expired gas was measured for evaluating metabolic thermogenesis, the shivering intensity of the pectoralis major muscle was evaluated using EMG technique, and T_{co} and circulatory responses, such as skin temperature (T_{sk}) and skin blood flow, were continuously measured. The main results showed a positive correlation between NST and detectable volume of bioactive BAT, which was assessed using FDG-PET/CT technique, but no relationship with SM mass (Fig. 5.2). The contribution of BAT to NST is consistent with the findings of previous BAT studies [5, 8, 11, 12]. On the other hand, although there was a strong positive correlation between SM mass and RMR, no contribution of SM to NST was observed (Fig. 5.2). This is consistent with the finding that lean body mass was positively correlated with RMR but not NST in supplement data from a previous study [8]. These results suggest that the contribution of SM to NST may be smaller than that of BAT; however, there is a limitation in the evaluation of SM contribution to NST only based on the estimated tissue weight without any measurement of bioactive muscle metabolism.

Several research groups have argued for the high contribution of SM to thermogenesis during mild cold exposure, based on the assessment of tissue metabolism using FDG-PET and [^{15}O]O_2-PET [12, 29]. A study evaluating FDG uptake in multiple organs during mild cold exposure (including minimal shivering) found that although glucose uptake per unit volume was higher in BAT than in SM, total glucose uptake considering tissue volume was much greater in the SM [29]. However, the total glucose uptake shown in this study included the RMR component, and the contribution of both tissues to the NST component was not evaluated. In our study, the total metabolic rate combined with RMR and NST during mild cold exposure was positively correlated with SM mass ($r = 0.683$, $P < 0.01$), but this was mostly due to the component of RMR which was strongly correlated with SM mass. In a study using [^{15}O]O_2-PET to evaluate oxygen uptake in each tissue during mild cold stimulation, tissue-specific NST (difference in

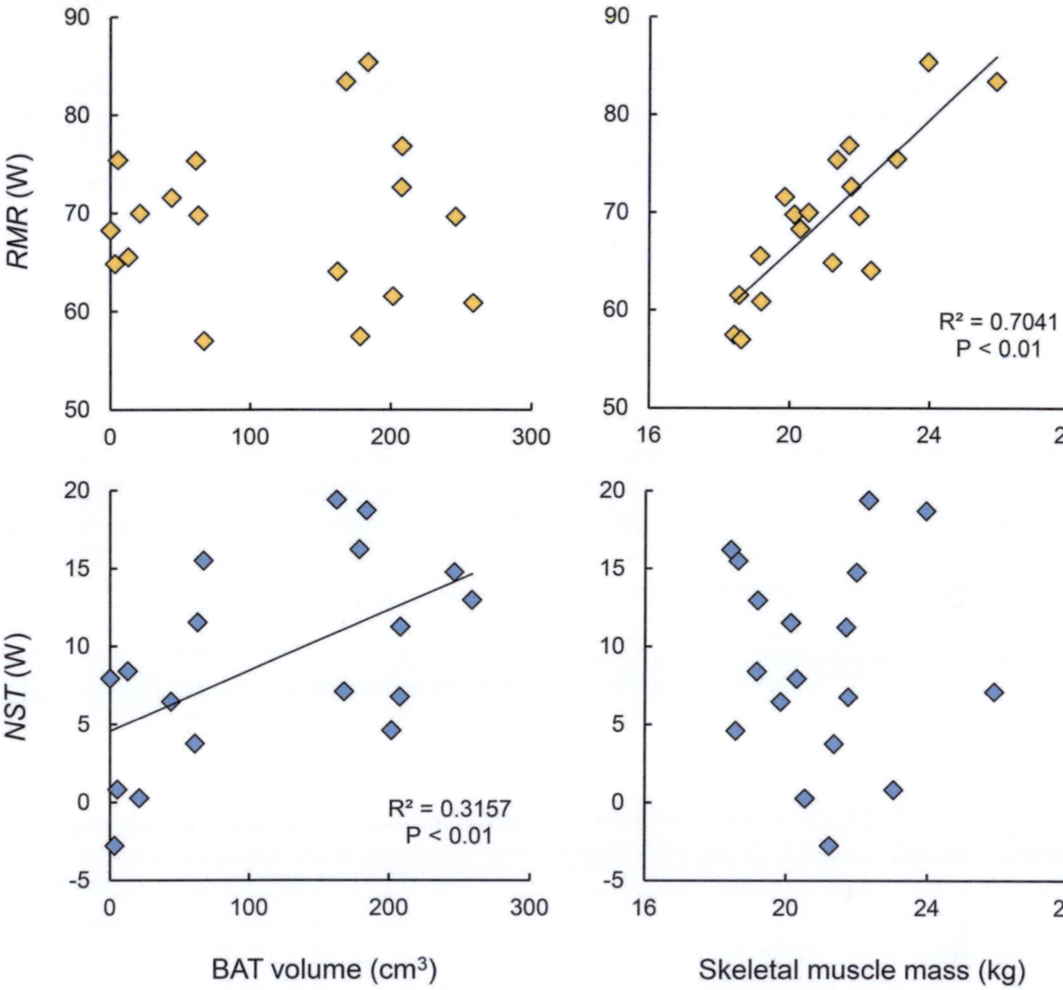

Fig. 5.2 Metabolic heat production during cold exposure as a function of BAT volume and skeletal muscle mass. *RMR* resting metabolic rate, *NST* nonshivering thermogenesis, *BAT* brown adipose tissue, *SM* skeletal muscle. (Modified from Wakabayashi et al. [7])

the oxygen uptake during cold stimulation and at room temperature of 22 °C) in BAT around the cervical and upper thoracic region was only about 1% of total NST in the whole body and does not correlate with total NST. However, when tissue-specific NST from SMs around the cervical and upper thoracic region was combined, there was a positive correlation with whole body NST [12]. These results suggested that the contribution of BAT to total NST was small, and that SM was the main thermogenic tissue for NST. Similarly, some studies have reported that energy expenditure in BAT is approximately 5% of total whole-body NST based on evaluation of tissue oxygen uptake using [^{15}O]O_2-PET technique [11]. Although the evaluation of the contribution of each tissue metabolism to CIT using [^{15}O]O_2-PET is useful and important, it is pointed out that this technique might lead to underestimate BAT-specific metabolism due to the short half-life of [^{15}O] of about 2 min and limited area of view [13]. Indeed, according to the study using FDG-PET/CT [14], it was suggested that more contribution of BAT to the NST, with large individual variation from almost 0 to 16% of RMR, accounting for 90% of NST [13].

Regarding the contribution of SM to NST according to BAT activity, specificity can be seen in the adaptive phenomena of each tissue's metabolism caused by short-term repeated cold exposure. In studies of young healthy participants with high BAT activity, repeated cold exposure showed an increase in NST and enhancement of BAT activity (increased FDG uptake), whereas no adaptive change was observed in FDG uptake into the SM [25] and parameters of SM respiration such as mitochondrial density and respiration [14]. On the other hand, studies targeting type 2 diabetic patients and obese individuals with low BAT activity have shown that repeated cold exposure not only increases BAT activity but also increases glucose uptake into the SM [23, 27]. Considering these findings, in populations with low BAT activity, since the contribution of SM to NST is relatively high, tissue-specific metabolic adaptation due to repeated cold exposure might be enhanced in the SM. Further research is needed to clarify the degree of contribution of each organ to NST, as well as its diversity and fluctuations.

In our study, in addition to examining the BAT and SM contribution to NST during 90-min mild cold exposure (18.6 °C), we also examined a relationship between NST and shivering induced by further reduction of the ambient temperature to 11.6 °C [7]. When participants were divided into high and low BAT activity groups, the high BAT group showed higher metabolic rate during the mild cold exposure phase as described above (Fig. 5.3a). During the subsequent phase of shivering onset, the shivering intensity (%MVC) [30] of the pectoralis major muscle assessed by EMG rose earlier in the low BAT activity group compared to the high BAT group (Fig. 5.3b). After 90 min of mild cold exposure, when the environmental temperature was lowered to 11.6 °C, although no relationship was observed between the shivering onset time and the amount of BAT (Fig. 5.4a), a negative correlation was shown between the rectal temperature at 90 min and the time of shivering onset (Fig. 5.4b). This could be because the larger the amount of BAT, the higher the NST, the higher the T_{co} was maintained (Fig. 5.4c), and as a result, the onset of shivering was delayed. Although a direct relationship between BAT activity and ST by SM was not clearly observed, there might be an indirect correlation between BAT-mediated NST and SM-mediated ST for maintaining T_{co}.

In the study comparing CIT between South Asian and Caucasian populations, an individualized cooling protocol was used to detect the shivering threshold and then adjusted the cold stimulus (temperature of water perfusion suit) for assessing BAT activity at maximum NST [19]. As mentioned above, the South Asian group showed lower BAT content and less NST; in addition, it was confirmed that the shivering threshold temperature was about 2 °C higher on average than the Caucasian population [19]. In

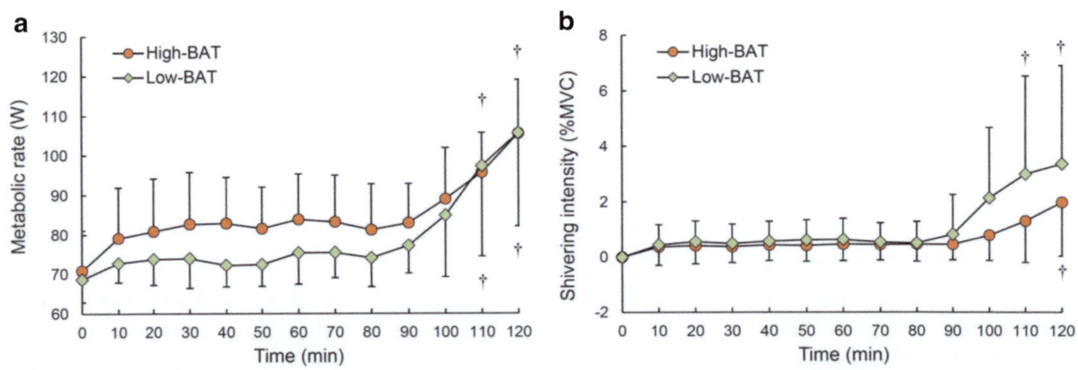

Fig. 5.3 Time course of metabolic rate (**a**) and shivering intensity (**b**) during gradual cold exposure in the high and low BAT activity groups. The values are mean ± SD. † Significant difference compared with values at the end of mild cold exposure (90 min) in each group. (Modified from Wakabayashi et al. [7])

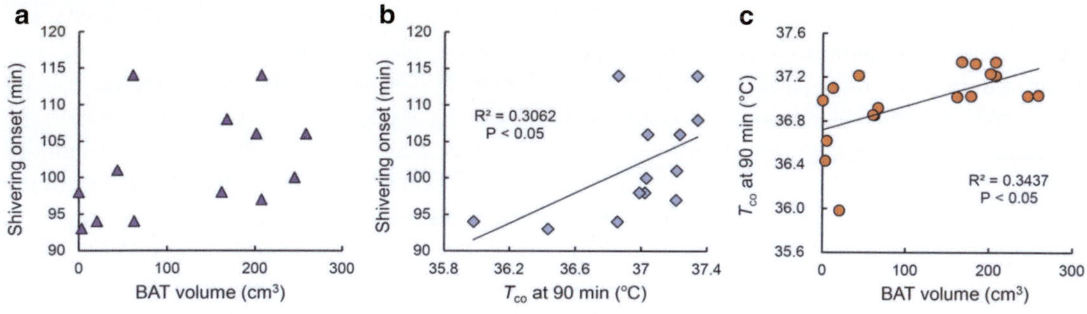

Fig. 5.4 Relationship between BAT volume, core body temperature, and shivering onset time. (Modified from Wakabayashi et al. [7])

other words, in South Asian populations with low BAT activity, relatively mild cold stimulation causes shivering onset for compensating less NST. Similarly, in type 2 diabetic patients with low BAT activity, in addition to higher glucose uptake in SM compared to young healthy participants, they tended to have higher shivering intensity (%MVC) during mild cold exposure [22]. Moreover, there was a tendency for a negative correlation between BAT volume and shivering intensity in all participants, and those with low BAT activity might induce somewhat stronger shivering in SM [22].

These previous studies focused mainly on NST and only the minimal shivering was observed during mild cold exposure [19, 22]. In our study, the shivering phase was only 30 min at 11.6 °C, which was insufficient duration and cold stimulus for initiating strong shivering or reaching steady state [7]. It is necessary to further investigate the multiorgan compensation in metabolic thermogenesis between SM-induced ST and NST according to the variation in the BAT activity.

5.1.1.4 Cold-Induced Vasoconstriction

In order to maintain T_{co} in cold environment, in addition to the CIT for increasing heat production as explained above, cutaneous vasoconstriction is also induced for heat conservation. A number of studies have been conducted on vasomotor thermoregulation including vasodilation and vasoconstriction, which have been summarized in several review articles [31–34]. Focusing on the response to cold environment, a review article briefly overviews the mechanism of the cold-induced vasoconstriction [34]. In the following paragraph, we just provide an overview on a few limited topics in the vasomotor response to cold. It is highly recommended to read the review articles introduced above [31–34] and find the details in the original articles.

The vasomotor regulation is initiated even within the thermoneutral zone where the cold stimulus is milder than the NST threshold (Fig. 5.1). The cold-induced cutaneous vasoconstriction decreases the T_{sk} and reduces the temperature gradient between skin and surrounding environment, which results in suppressing the dry heat loss from the skin surface. When cutaneous vasoconstriction occurs, venous return from the extremities to the heart shifts from superficial to deeper veins [35]. This enables the human body to reduce the heat loss since deeper venous veins are well insulated compared to the superficial veins. In addition, countercurrent heat exchange occurs between the deep veins (accompanying veins) and arteries which are running parallel and intermingled with opposite blood flow direction [36], and the temperature of arterial blood gradually decreases, resulting in lower T_{sk} at the distal extremities. Thus, ununiform T_{sk} distribution, with lower T_{sk} in distal body regions, is often observed in cold environments compared to the uniform T_{sk} in hot environments. From the morphological aspects, since the surface area per volume is far larger in the distal extremities, particularly in fingers and toes, than in the trunk body part, lowering the T_{sk} at the distal extremities is a rational way to lessen dry heat loss.

From the vascular anatomical aspects, the arterio-venous anastomoses (AVAs) are located in the skin of distal body parts, especially in the finger, toe, glabrous (acral) side of hands and feet, and the ears and nose, but not in the trunk body parts [37]. AVAs have a structure of direct connection between small arteries and veins, without a capillary segment for exchanging substances with tissues. The length of AVA is typically from 200 to 500 μm, the inner diameter from 10 to 150 μm, and the thickness of the wall from 40 to 60 μm, with an inner layer of smooth muscle cells, which enables them to close the vessels completely [37]. Since the amount of blood flow through the vessels is proportional to the 4th power of its inner diameter according to the Poiseuille's law, as compared to the ordinal capillaries with 10 μm inner diameter, 256 times greater blood flow should pass though the AVA with 40 μm inner diameter [38], and over 1000 times in AVA with 60 μm diameter [39]. In the case of AVAs with 100 μm diameter, the amount of blood flow could reach 10,000 times greater than that in the capillaries. Those anatomical properties enable AVAs to have very high blood flow and a bunch of heat transfer from body core to skin surface. It has been suggested that the AVAs were regulated by the adrenergic sympathetic nerve activity [37, 39]. Indeed, very heavy innervation of noradrenergic fibers were observed around AVAs [40]. It was supported by the physiological evidence reporting that the fluctuation of the blood velocity in the dorsalis pedis artery was disappeared after epidural block [41]. In addition, the characteristics of rapid vasoconstriction and slower vasodilation process indicate that the vasoconstriction is induced by burst impulses in the adrenergic axons innervating the AVAs, and vasodilation is passively induced by release of the AVA smooth muscle contractions [37].

The AVAs located in acral skin parts play a major role in heat dissipation from the skin within the thermoneutral zone. When the ambient temperature is below the thermogenic threshold (lower edge of the thermoneutral zone), all AVAs remain closed [37]. The effect of moderate local cooling on blood velocity in finger artery was examined with hand immersion in water from 35 to 19 °C [42], and sustained closure of AVAs was observed when the water temperature (T_w) was below 21.5 °C (range 20–23 °C). However, with severe local cooling of finger in 3 °C water, the same research group observed cold-induced vasodilation (CIVD) in the blood velocity in the finger arteries [43]. Based on the time course data, they suggested that dilation of AVAs, which could be due to relaxation of low temperature smooth muscle cells, would be a mechanism eliciting CIVD [43].

Physiological mechanism of cold-induced vasoconstriction in nonglabrous (non-acral) skin parts is briefly summarized as follows [33, 34]. The whole-body cooling induces cutaneous vasoconstriction, even if the site of the skin blood flow measurement is not locally cooled, and this reflex cutaneous vasoconstriction is controlled by the sympathetic nervous system [33]. Noradrenaline (NA) released from the adrenergic nerve endings is a primal neurotransmitter binding with adrenergic receptors (or adrenoceptor) on the vascular smooth muscle to induce vasoconstriction [33, 34]. In addition, neuropeptide Y (NPY) is known as a major cotransmitter for the reflex vasoconstriction when whole-body T_{sk} gradually decreased to 31.7 °C [44]. Another research reported a contribution of adenosine triphosphate (ATP) as a cotransmitter for the reflex cutaneous vasoconstriction observed when whole-body T_{sk} gradually decreased to 30.5 °C [45]. It is generally suggested that NA is responsible for 60% of the reflex cutaneous vasoconstriction [33, 45], and remaining 40% is explained by the cotransmitters such as NPY and ATP, while further studies are required to clarify the contribution of them.

Mechanism of the vasoconstrictor response to local skin cooling varies depending on the duration of the cooling [32]. During the early phase of local cooling (within 10 min), vasoconstriction is mostly regulated by adrenergic neural mechanisms, primarily depending on alpha-2 (α_2) adrenergic receptors [46]. Local cooling increases the reactive oxygen species (ROS) generated from mitochondria in vascular smooth muscle, which enhances Rho-kinase activity [47]. The mitochondrial ROS enhances Rho-kinase activity to elicit translocation of α_{2c} adrenergic receptors to the vascular smooth muscle cell membrane [48]. The increased sensitivity of α_{2c} adrenergic receptors, through the above process, enhances the vasoconstriction, even though NA secretion from the sympathetic nerve is reduced by local cooling [32].

During the late phase of local cooling (more than 10 min), vasoconstriction in nonglabrous (non-acral) skin is regulated by combination of continued activation of adrenergic function and inhibition of nitric oxide (NO) vasodilator pathway [32, 49]. NO is a vasodilator substance synthesized by nitric oxide synthase (NOS) enzyme to relax vascular smooth muscle. This occurs through the process of activating guanylate cyclase, which increases cyclic guanosine monophosphate (cGMP) and activates protein kinase G (PKG), ultimately leading to smooth muscle relaxation. The prolonged local skin cooling inhibits the activity of endothelial NOS enzyme and downstream from the NOS enzymes, which suppresses NO-mediated vasodilation, resulting in vasoconstriction [49]. The inhibition of NOS activity is partly induced by the elevation of ROS from the vascular smooth muscle mitochondria, which enhances Rho-kinase activity [32, 47, 50]. Since the Rho-kinase pathway is a relatively slow event as shown in time course of Rho activity to cooling in human cultured dermal arteriolar vascular smooth muscle [48], the Rho-kinase plays an important role for vasoconstriction during prolonged cooling, both via inhibition of NO pathway and increased sensitivity of α_{2c} adrenergic receptors.

5.1.2 Cold-Induced Vasodilation

5.1.2.1 Outline of Cold-Induced Vasodilation

When fingers or hands are exposed to cold water, cold-induced vasoconstriction reduces finger skin temperature (T_{fing}) and remains low during the early phase of the immersion; after a while, finger blood vessels vasodilate to increase T_{fing}. This response is known as cold-induced vasodilation (CIVD), which was originally found by Lewis in 1930 [51]. The CIVD and vasoconstriction are repeated during the cold exposure to show the hunting reaction in the time course of T_{fing}. The paradoxical vasodilation response in cold fingers has been studied for a century, and several review articles have summarized the findings with perspectives for future research [52–54]. A typical example of the hunting reaction is shown in Fig. 5.5. Since the upregulation of T_{fing} is reasonable for preventing frostbite and nonfreezing cold injury (NFCI), the CIVD response is used as an index of local cold tolerance.

There are several parameters of CIVD response as shown in Fig. 5.5 [52, 53, 55]. The onset time is the time duration from finger immersion to the minimum temperature (T_{min}), which is induced by the initiation of the first CIVD response. The peak time is the time interval from the T_{min} to the first maximal temperature (T_{max}), and the amplitude is the temperature difference between T_{min} and T_{max}. In some cases, the following CIVD waves show a lower nadir or higher peak temperature than T_{min} or T_{max} observed in the first wave. The author believes it is general and reasonable to choose the first T_{min} and T_{max} which are corresponding to the onset time and peak

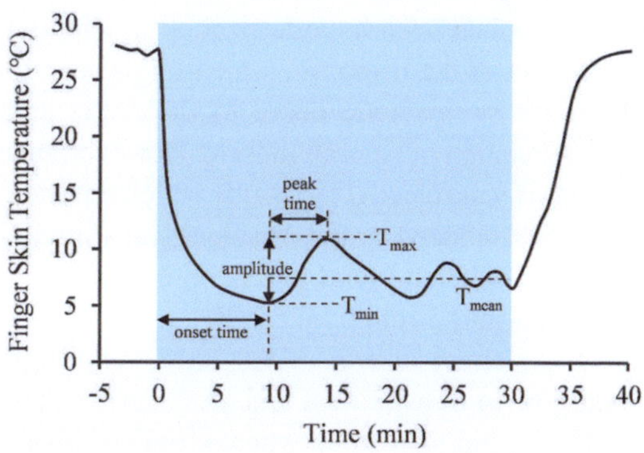

Fig. 5.5 Parameters of CIVD response in time course of finger skin temperature during finger cold immersion. The onset time is the time duration from finger immersion to the minimum temperature (T_{min}). The peak time is the time interval from the T_{min} to the maximal temperature (T_{max}). The amplitude is the temperature difference between T_{min} and T_{max}. The mean finger temperature (T_{mean}) is measured from the CIVD onset to the end of the cold immersion. (Modified from Daanen [52])

time, respectively, while it depends on the studies and CIVD waves. The mean finger temperature (T_{mean}) from the CIVD onset (in practical use, 5 min after immersion) to the end of the cold immersion can be used as a representative parameter for the magnitude of the series of CIVD responses. In addition, area under the curve from the onset of CIVD to the end of the immersion (CIVD index) can reflect both rapidity and magnitude of CIVD response [56]. The numbers of hunting reactions are also used as a parameter of CIVD response. There are participants who do not show any CIVD response.

It should be noted that the changes in the finger skin blood flow measured by laser-Doppler flowmetry precede the changes in T_{fing} during the CIVD response [43, 57, 58]. The time lag between the actual vasomotor response and following T_{fing}, which is generally used as CIVD parameters, should be concerned especially when focusing on the timing of the CIVD response and synchronization with the other factors.

5.1.2.2 Mechanisms of CIVD

In this paragraph, potential physiological mechanism of CIVD response is summarized based on several review articles [52, 54] with some recent updates.

Since the original research on CIVD [51], axon reflex had been hypothesized as a potential mechanism of the response, which might lead to inhibit the sympathetic nerve to the AVAs and cause relaxation of smooth muscle. However, this hypothesis was rejected by a study which reported axon reflex induced by painful electrical stimulation increased skin blood perfusion in warm finger but not in the cold finger [59].

As explained above, AVAs have very high blood flow to transfer heat and are located at the distal and acral skin parts, where CIVD response is mainly observed [37]. In addition, based on the blood velocity measurements in the finger arteries with severe local cooling, it was suggested that the CIVD response reflects relaxation of the AVAs smooth muscle due to the low temperature of the smooth muscle cells (local phenomenon), since sustained vasoconstriction was observed in the noncooled control fingers indicating no change in the sympathetic impulses [43]. Nervous blockade of the neuromuscular junction from adrenergic nerve to the smooth muscle could be induced by low tissue temperature [52, 53]. It was suggested that cessation of transmitter release from adrenergic nerve endings would regulate the CIVD, since no CIVD response was observed in isolated rat tail artery when NA was continuously perfused [60]. Cold-induced vasoconstriction limits blood flow to the finger vessels, resulting in critical low tissue temperature for cessation of NA release from the nerve endings. After a while, smooth muscle relaxes and CIVD occurs to supply blood and heat to the vessels, which warm up the tissue and resume the sympathetic transmitter activity for inducing next vasoconstriction. The repetition of the series of vasomotor responses forms the hunting reaction in the finger blood flow and T_{fing}. In addition to the local phenomenon, it was suggested that sympathetic activity (central phenomenon) could be a major factor for controlling initiation and magnitude of CIVD response, based on the findings of suppressed CIVD response with lower T_{co} and T_{sk} [61, 62], which are associated with high sympathetic activity.

Contribution of some vasodilating substances, such as NO, has been hypothesized as a potential mechanism of CIVD response [52–54]. As far as the author knows, no clear evidence of NO contribution in human CIVD response has been reported, while a study found that cooling potentiated higher NO production to cholinergic stimulation in cutaneous artery dissected from rabbit ear [63]. As explained above, prolonged local skin cooling inhibits the activity of endothelial NOS and downstream of NOS enzymes, which suppresses NO-mediated vasodilation, resulting in vasoconstriction in nonglabrous skin [49]. Recently, several updates have been conducted on the mechanism of the CIVD based on a wavelet analysis on finger skin blood flow measured using laser-Doppler flowmetry [57, 64, 65]. The frequency bands and associated vasomotor function are 0.005–0.01 Hz (endothelial NO-independent activity), 0.01–0.02 Hz (endothelial NO-dependent activity), 0.02–0.06 Hz (neurogenic activity), and 0.06–0.15 Hz (myogenic activity), respectively, with slight variation among studies. During hand immersion in 8 °C water, strong synchronization was observed between time course in the neurogenic activity and laser-Doppler flow, while both endothelial NO-independent and NO-dependent activities were not synchronized with skin blood flow [57]. During finger immersion in 5 °C water, suppression of neurogenic and myogenic activities and increase of endothelial NO-independent activity were observed just prior to CIVD, whereas no change in the endothelial NO-dependent activity [64]. Although there are some discrepancies in the NO-independent endothelial activity between the studies, the minor contribution of NO to the CIVD response and the major role of the neurogenic activity (sympathetic withdrawal) were confirmed consistently [57, 64]. There are studies supporting no contribution of NO for regulating CIVD, based on the effect of acute supplementation of beetroot (red beet) juice with high dose of nitrate, supplying exogenous NO through the sequential reduction process [58, 66]. No effect of beetroot intake was observed in the CIVD parameters in T_{fing} and skin blood flow during hand cold immersion [58, 66], although our study found the vasodilator effect of beetroot enhanced the recovery of T_{fing} and skin blood flow after removing hand from cold water [58].

5.1.2.3 Variations in CIVD

There are population differences in the CIVD response, with the African group having a smaller CIVD response and slower rewarming after hand immersion than those observed in the Caucasian group, while the Asian group shows intermediate response [67]. This is thought to be one of the reasons for the high incidence of frostbite and NFCI in African populations [68].

There have been studies reporting greater CIVD response in populations living in cold regions and working in cold environment such as fishermen, divers, and cold storage workers. Norwegian Lapps and north Norwegian fishermen showed a significantly rapid CIVD onset compared to a control group, while, no difference was observed in T_{min} and T_{max} [69]. Gaspe fishermen showed higher T_{fing}, less pain in finger, and lower blood pressure during hand cold immersion [70]. Ice-chamber workers and swimmers showed higher T_{mean} and T_{min} with short onset time compared to the control [71]. A series of studies were conducted for evaluating CIVD response of several populations living in Hokkaido, the northern island of Japan [72]. In general, Hokkaido natives, who born and live in Hokkaido, showed greater CIVD response with higher T_{mean}, T_{min}, and shorter onset time, compared to the populations moved from the main island of Japan [72]. Recently it was reported that older women wetsuit divers (*Haenyeo* in Korean, *Ama* in Japanese) living in Jeju island in Korea showed higher T_{min} compared to older non-divers [73], whereas, previous study reported that T_{fing} and blood flow was lower in cotton suit divers than in non-divers [74]. In addition, there are studies reporting less CIVD response in the tropical indigenes compared to the temperate indigenes [75, 76]. Based on those findings in the population studies, it had been generally understood (or hypothesized) that the greater CIVD response in populations living or working in cold environment was due to cold acclimatization.

In laboratory-controlled studies, findings on the effect of repeated cold exposure on the CIVD response are inconsistent and differ from the expectation of the population studies. Some of the studies reported enhanced CIVD response after short-term repeated cold exposure [77, 78]. On the other hand, many studies reported no change [79] or even blunted CIVD after short-term cold acclimation [80–85]. Considering those observations in the minor or negative effect of short-term cold acclimation on the CIVD response, currently some review articles concluded that repeated cold exposure does not change the CIVD response [53, 54, 86]. Since repeated cold exposure could induce habituation of pain in the extremities and improve thermal comfort, even with no change (or decrease) in skin temperature of finger or toe [79, 84], it was suggested that the discrepancy between the subjective and physiological response may increase the risk of cold injuries [79]. It is recommended that, for preventing cold injuries, people should not rely on the habituation through the repeated cold exposure [53, 86].

5.1.3 Cold Shock Response

5.1.3.1 Outline of Cold Shock Response

The cold shock response (CSR) evoked during the initial few minutes of cold water immersion is the life-threatening reflex including cardiovascular (tachycardia, arrhythmia, peripheral vasoconstriction, and hypertension) and respiratory (gasping and hyperventilation) components [87–91]. It was reported that the CSR begin in T_w around 25 °C, while it can be controlled consciously at that temperature [92], and respiratory frequency peaks in T_w between 10 and 15 °C as vigorous as that in 5 °C water [92, 93].

A tachycardia within a few seconds of exposure to cold water is one of the most frequently reported cardiac CSR [89]. Exposure to cold water (ice-cold shower) increased cardiac output and peripheral vasoconstriction, resulting in increase of arterial pressures within a few seconds and reach maximal levels within 30–60 s [88]. Several studies have also reported that cold water immersion produces an inspiratory gasp followed by uncontrollable hyperventilation, indicated by a significant fall in the end-tidal partial pressure of carbon dioxide [87, 94]. The cold shock hyperventilation can increase respiratory rate by more than 110% and ventilation volume by more than 600% compared to the pre-immersion baseline [90, 95, 96], which increases the risk of involuntarily water aspiration and drowning [95, 96]. A reduction of maximal breath hold time caused by the inspiratory gasp response has been reported in several studies [96, 97].

The World Health Organization (WHO) estimates that there were more than 2.5 million drowning deaths between 2008 and 2018 [98]. In the case of accidental cold water immersion, the initial CSR could be one of the major causes of near-drowning incidents and drowning death [99], since the initial inspiratory gasping and uncontrollable hyperventilation induce an inhalation of water into the lungs when the head is submersed below the water surface. In addition, during cold water submersion, CSR and diving reflex are simultaneously induced, and both sympathetically mediated tachycardia and parasympathetically mediated bradycardia could make an "autonomic conflict," resulting in the arrhythmias [100, 101]. A policy of "float first" during the initial minutes of accidental cold water immersion to keep the airway above the water surface is reasonable practice for preventing inhalation of water and improve the survival prospects [102, 103]. In addition, the habituation of the CSR by repeated cold water immersion, which is discussed in the following paragraph, has the potential to reduce the number of drownings.

5.1.3.2 Mechanism of the Cold Shock Response

The immediacy of the CSR suggests that the components are driven by stimulation of cutaneous cold receptors on immersion and are mediated by the sympathetic autonomic nervous system [87, 89, 90, 104]. It was reported that a greater CSR was observed in whole skin exposure conditions than in a partially insulated condition with a wetsuit [96, 105], and a greater CSR was induced during one staged immersion than two staged entry [95]. These findings suggested that the CSR is induced by both spatial and temporal summation of regional thermoafferent inputs. A study reported that peripheral cold stimulus to the trunk body part resulted in a greater influence on the ventilatory CSR than when applied to other parts of the body [106].

It has been hypothesized that thermoafferents from the peripheral cold receptors increase ventilatory response by directly stimulating the respiratory center [87, 107]. A chemoreceptor pathway, which is normally activated by higher partial pressures of carbon dioxide and results in an increase in ventilation, does not affect the ventilatory CSR. This was confirmed in a study that found no difference in ventilatory CSR between conditions with and without hyperventilation prior to cold water immersion [93]. The neural pathways of the ventilatory CSR, largely based on animal studies, are summarized in some review articles [90, 91].

The cardiovascular CSR are thought to be due to the reflex stimulation of the sympathetic nervous system as a direct consequence of rapid skin cooling [88, 108]. Johnson et al. [108] reported that the level of plasma noradrenaline from the sympathetic nervous system had almost doubled after 2-min immersion in 10 °C cold water. In contrast, Keatinge et al. [88] found no evidence of the release of catecholamines during 2-min ice-water showers. Cold stimulus to the skin surface of feet during cold immersion was found to increase beta-adrenergic sympathetic activity to the heart and likely inhibited parasympathetic activity, resulting in a tachycardia [109].

The CSR may be affected by anticipatory anxiety and psychological stress, which would be accompanied by a sympathetic overdrive. Several observations of a tachycardia and hyperventilation provide evidence of this anticipatory anxiety response [110, 111]. It was reported that an increase in breath-holding time on immersion in cold water occurred following a psychological skills intervention, while no change was observed in heart rate [97]. It was suggested that the psychological skills improved the ability of voluntary suppression of the cold-induced drive to breathe.

Find more information on the mechanism of CSR summarized in some review articles [90, 91].

5.1.3.3 Habituation of Cold Shock Response

It has been reported that the CSR is reduced (habituated) following repeated immersions in cold water [87, 112, 113]. Recently, a systematic review and meta-analysis was conducted for previous studies on the habituation of CSR [91]. This review article lists eligible studies on habituation of CSR, summarizing research designs and major findings in the cardiovascular and ventilatory components of CSR. In addition, the mechanism of the CSR and habituation of the responses are discussed in the manuscript with citations of previous research including animal studies [91]. Please read the systematic review for understanding the latest summary on the habituation of CSR and find more detail in each original manuscript. Understanding the habituation process of the CSR has the potential to reduce the number of drownings by clinical application of this process to high-risk groups.

The habituation of the cardiorespiratory CSR can be achieved following as few as several short duration (2.5–3 min) cold water immersions [114–117]. It is strongly suggested that a repeated fall in T_{sk} induces the habituation in the CSR, and that this habituation occurs without a reduction in T_{co} [114, 115, 117, 118].

It was reported that six repeated 3-min head-out immersions in 15 °C water was sufficient to significantly decrease the cardiorespiratory CSR in 10 °C water [115]. Even six repeated cold shower exposures induced the habituation of the respiratory CSR during head-out immersion in 10 °C water [118], although habituation of the cardiac responses was not clearly observed. This may suggest that there are different mechanisms involved in producing habituation of the respiratory and cardiac CSR. Furthermore, repeated immersion on one side of the body could reduce the cardiorespiratory CSR when the other side was immersed [114], indicating that the habituation of CSR would occur centrally rather than peripherally. These observations indicated that repeated cold exposures with relatively small cold stimulus (higher T_w or smaller surface area exposed) can partly promote the habituation of the CSR to a greater intensity cold exposure. Also, it is indicated that, especially in the case of a smaller cold stimulus, such as showering, the magnitude of the habituation is determined by the rate of fall in T_{sk} and the body surface area exposed to the cold [118].

There is a study looking at the permanence of habituation of CSR. It was reported that after six times of repeated short duration cold water immersions, there was a gradual loss of habituation, but that the habituation of hyperventilation and tachycardia was retained, in part, for up to 7 and 14 months, respectively [116]. An interesting study reported that the habituation of ventilatory CSR was inhibited by the repeated anxiety inducing task prior to cold water immersion [119].

As explained above, since repeated immersion of one side of body induced habituation of CSR when other side of the body was immersed in cold water, the habituation of CSR would occur centrally rather than peripherally [114]. In animal study, it was reported that no change in static and dynamic characteristics of cold receptors in the nasal area of cats, after 2 months of cold exposure to 5 °C air [120]. However, after 4.7 years long-term continuous cold exposure to 5 °C air, a change in dynamic activity of cold receptors was observed in cats [121]. This is supported by the evidence of a significantly smaller number of cold spots observed in long-term cold-adapted humans, who have lived in polar regions for at least 5 years and constantly worked outdoors [122]. Compared to these studies, the duration for obtaining the habituation of CSR is far too short to induce adaptive change in the cold receptors. Thus, the habituation of CSR would be primarily mediated by central phenomenon rather than peripheral change.

An animal study on cats has suggested that the reflex respiratory responses to cold are mediated at midbrain, and the cerebrum is not essential for the acute responses [123]. However, the involvement of the frontal area of the cerebral cortex on the habituation process has been reported [124, 125]. It was suggested that the frontal cortex is important in developing the habituation of cardiovascular responses, but not in maintaining the established habituation [124]. A study in leucotomized patients, whose frontal cortex had been removed, indicated the involvement of the frontal cortex in the habituation process of the cardiovascular CSR in humans [125]. More recent study reported that the cerebral cortex plays a critical role in top-down inhibitory control on autonomic centers located in the brainstem during innocuous whole body cooling [126]. The precise location and nature of the central changes responsible to the habituation of CSR is still under discussion. Find more information on the mechanism of habituation of CSR summarized in our recent review article [91].

5.2 Cold Adaptation

There are several terms explaining cold adaptation, such as "acclimatization," "acclimation," and "habituation." These are often mix-used, but in this chapter the terms follow the glossary of terms for thermal physiology [127]. Acclimatization is adaptive changes that occur when individuals are exposed to a natural climate including geographical difference or seasonal change [127]. Cold acclimatization is particularly relevant to indigenous populations living in cold regions for a prolonged period, who generally have better cold tolerance. Acclimation is adaptive changes occurring to experimental change in particular climate factors such as temperature in a controlled environment [127]. Cold acclimation is achieved through repeated cold exposures in the laboratory, which generally enhances human thermoregulation and cold tolerance. Habituation is reduction of physiological response or perception to a repeated stimulus [127].

There are several categorizations of cold adaptation based on the differences in thermoregulatory response observed in the cold acclimatized populations compared to the non-acclimatized control groups [86, 128–133]. It has been suggested that the duration and severity of cold exposure and systemic or local cold exposure during the adaptation period are the factors determining the types of cold adaptation [129, 130]. Here, we discuss several types of cold acclimatization and acclimation, based on physiological responses observed in populations who are regularly exposed to cold and studies testing the effect of repeated cold exposure in the laboratory.

5.2.1 Cold Acclimatization

5.2.1.1 Habituation of Metabolic Response (Shivering Thermogenesis)

There have been several populations living in cold environments, such as indigenous hunter-gatherers in central Australia (Indigenous Australians), the Kalahari Desert (San people), and nomadic indigenous Norwegian (Sami people), who showed less metabolic heat production with habituation of shivering during sleep in the cold environment [128, 134, 135]; the names of populations are replaced from the original manuscripts. In those indigenous populations, a simultaneously greater reduction in their T_{co} while sleeping was observed compared to the control group from the temperate climate [128, 134, 135]. This characteristic is called hypothermic acclimatization, which could relate to the less metabolic heat production due to the habituation of shivering.

Habituation of shivering has also been observed in divers who regularly work in cold water. For assessing the shivering threshold temperature in laboratory experiments, the lowest T_w at which a participant can immerse for 3 h without shivering is defined as critical water temperature (T_{cw}) [136]. Studies have shown that the T_{cw} of Korean cotton-suit divers was lower than those of Korean non-divers [136, 137]. The lower T_{cw} implies a reduction in the T_w threshold to onset shivering. A similar response has been reported in several population groups including Asians and Caucasians engaging in diving work in Hawaii [138], with varying amounts of subcutaneous fat. Interestingly, it was reported that after using wetsuits for several years, enabling divers to maintain their T_{co}, the previously observed shift of the shivering threshold was diminished and became similar to non-divers [139]. Greek divers who engaged in sponge diving work with wetsuits also showed a similar shivering threshold as European non-divers [140]. This lack of adaptation of the shivering threshold observed in wetsuit divers suggested the necessity of repeated reduction in T_{co}, and probably T_{sk}, to obtain the habituation of metabolic responses to cold.

5.2.1.2 Metabolic Acclimatization (Nonshivering Thermogenesis)

In an early study, a greater metabolic rate while sleeping in warm environment was observed in Inuit people living in Arctic region compared to the control group [141]. The elevated BMR in Inuit was considered as a kind of metabolic acclimatization. More recent studies also reported the 3–19% greater BMR in the indigenous circumpolar populations compared to the non-indigenous control group [142]. It was reported that cotton-suit Korean divers showed seasonal variation in their BMR, higher in winter and lower in summer [139, 143], whereas wetsuit divers showed no seasonal variation [139]. The increase of BMR is probably due to the adaptation of divers to the cold water environment in winter.

In experimental animal studies, shivering thermogenesis is gradually replaced by the successive increase of NST during repeated cold exposure, and NA injection after the cold acclimation period increases BAT-derived thermogenesis more than before acclimation [144]. It has been recognized in small mammals that adaptation to cold involves an increase of NST activated by NA [145, 146]. In human population studies, a significant but small increase in thermogenesis to NA injection was observed in Korean cotton-suit divers in winter, but not in non-divers [147]. Based on the small increase in thermogenesis, it was suggested that the development of NST was not the main feature of cold acclimatization in divers. In another population, an increase in NST and plasma free fatty acids to infused NA was observed in indigenous people in Hokkaido (Ainu), but not in the control group living in Hokkaido [148]. They suggested the potential development of BAT in the indigenous population in cold northern region in Japan, although it had not yet been clear whether adult humans possessed active BAT at that time.

Recently, after the rediscovery of human BAT activity using FDG-PET technique [3–5], metabolic acclimatization in human adults has become much clear with evidence of the greater NST with higher BAT activity in winter [4, 24]. One of the clear evidence using the FDG-PET/CT technique for population difference in BAT activity was observed in the greater volume of active BAT in the Caucasian European group than in the South Asian group [19]. Recently, based on the higher supraclavicular skin temperature, a surrogate measurement of BAT activity, it was reported that indigenous Siberians (Yakut or Saha people) living in Yakutia showed greater BAT thermogenesis in mild cold environment than the control population living near Chicago [149]. A study focused on population difference in the UCP1 haplotypes and enhancement of NST, based on the metabolic response in a laboratory study and haplotype frequency in the 1000 Genomes Project database [20]. They reported that the frequency of NST-enhanced UCP1 haplotypes was higher in populations from higher latitudes (with lower average temperatures), i.e., the highest in European, followed by Asian and African populations [20]. These recent findings indicate that the activation of BAT is one of the major factors controlling metabolic cold acclimatization in humans, especially in the enhanced NST during mild cold exposure.

5.2.1.3 Insulative Acclimatization

Early studies on indigenous Australians reported lower T_{sk}, as well as the habituation of shivering, during sleep in cold environment compared to the control group [150]. Since the peripheral cooling enabled them to lessen the heat loss from the skin, the insulative acclimatization is the economical thermoregulation with the habituation of shivering and less heat production [134].

There are studies on the insulative cold acclimatization in Korean cotton-suit divers. The level of vasoconstriction and tissue insulation (I_{tissue}) calculated by heat balance equation is assumed to be maximal at the end of 3-hour immersion in T_{cw} without shivering. Rennie et al. [136] calculated maximum I_{tissue} from temperature gradient between T_{co} and T_{cw} and metabolic heat production in participants at steady state after 3-hour immersion in T_{cw}. They reported a similar maximum I_{tissue} for Korean cotton-suit divers and Korean non-divers measured in summer [136]. However, the maximum I_{tissue} for a given subcutaneous fat thickness measured in winter was found to be greater in divers than in non-divers [151], suggesting an adaptive increase in insulation, due to non-subcutaneous fat layer. Divers and non-divers showed a similar blood flow in their limbs during immersion in 30–33 °C water, which is typical range of T_{cw}, however, the heat flux for a given blood flow tended to be lower in divers, suggesting divers might have an adaptive change in their countercurrent heat exchange in their limbs [152].

5.2.2 Cold Acclimation

5.2.2.1 Habituation of Metabolic Response (Shivering Thermogenesis)

In laboratory experimental studies, it has been reported that metabolic heat production is reduced following repeated cold water immersions [112, 153, 154]. A significant delay of shivering onset time, measured in a protocol which gradually decreases T_w, was observed after repeated head out water immersions in 15 °C water [112]. The effect of the adaptation regime on the habituation of metabolic responses was also examined in their study, comparing repeated rest or exercising immersions in cold water [112]. The reduction in metabolic responses to cold was significantly greater after the repeated resting immersions, which induced greater decreases in their T_{co} during the acclimation process, compared to the repeated exercising immersion group [112]. Tipton et al. [113] confirmed that habituation of metabolic response is specific to the repeated reduction of T_{co} induced by cold water immersions during the habituation process. In their series of experiments, one experimental group repeated five times of 45-min immersion on average, resulting in 1.18 °C T_{co} reductions, and conducted testing immersion pre- and post-acclimation process. A significantly blunted metabolic response was observed during the post-testing immersion until T_{co} decreased by 1.18 °C, however, no habituation was observed when their T_{co} decreased further [113]. Budd et al. [153] reported a significant reduction in metabolic response and a delay of shivering onset time, measured by EMG, during 10 °C cold air exposure, following a 10-day repeated cold water immersions lasting 30–60 min [153]. Similarly, a delay in shivering onset time, accompanied with lower T_{co} thresholds, has been reported following repeated cold water immersions [154, 155], while there were individual variations in thermoregulatory response obtained through the cold acclimation and some showed greater metabolic heat production [155].

As shown in the hypothermic acclimatization observed in the cold acclimatized populations sleeping in cold environment [128, 134, 135], a tendency of greater fall in T_{co} during rest in cold water was observed after repeated cold water immersions, which might be associated with the habituation of metabolic response [112]. Another research group also reported a reduction in T_{co} during cold immersion, after repeated cold immersions, with a significant reduction in metabolic response [154]. In addition, significant reduction in resting T_{co} following adaptation regimen has been reported by several investigators [154, 155]. However, several studies did not show any evidence of hypothermic acclimation despite a recognized habituation of metabolic response [113, 153]. This may be explained by an increase in vasoconstriction, as discussed in the section of insulative adaptation, and/or reduction of shivering-induced heat loss may overcome the negative effect of smaller heat production. Recently, the habituation of shivering response after the repeated cold exposure was observed in shivering intensity assessed by EMG technique [26] and subjective report [14], simultaneously with the enhanced BAT activity which is discussed later.

5.2.2.2 Metabolic Acclimation (Nonshivering Thermogenesis)

To compensate for the reduction of metabolic heat production seen on the habituation of shivering response, an increase of NST (metabolic acclimation) after repeated cold exposure had been reported in animal studies [145, 146]. In a field experimental study in humans who stayed at Norwegian cold mountains for 6 weeks, an increase of metabolic rate (metabolic acclimation) with higher T_{sk} (no insulative acclimation) was observed after the acclimation period [156].

Early laboratory-controlled studies, which examined cold adaptation after repeated cold water immersions, found that no difference in the RMR in thermoneutral condition [153], no change in the metabolic response to injection of NA [157], and no significant adaptive change in plasma NA concentration during cold water immersion [158]. During cold air exposures after repeated cold water immersions, there was a significant increase in plasma NA with lower T_{sk} (insulative acclimation) and lower T_{co} (hypothermic acclimation), but no change in metabolic response [159]. Based on these observations, it had been believed that there was no significant adaptation of NST in humans, probably due to the small amount of BAT in human adults compared to small mammals [160].

Recently, after the rediscovery of human BAT activity using FDG-PET technique [3–5], it has been reported that experimental repeated short-term cold exposures enhanced BAT activity in mild cold environment in healthy young people [8, 14, 25, 26], type 2 diabetic patients, and obese individuals [23, 27]. Some of these studies found the evidence of metabolic acclimation with simultaneously elevated NST and BAT activity after cold acclimation [8, 14], while the other studies did not find clear adaptation in the NST. In addition to the enhanced BAT activity after the repeated cold exposure, the habituation of shivering was simultaneously observed in shivering intensity assessed using EMG technique [26] and in the subjective report [14]. These findings suggest that increase of BAT activity and NST (metabolic acclimation) would in part replace the shivering thermogenesis (habituation of shivering) after the cold acclimation in human, while it is still not so clear as in animal studies [144], probably because the amount of BAT in human adults is not enough to totally replace the shivering thermogenesis, and contribution of other thermoeffector tissues and vasomotor response should be concerned.

The findings of the metabolic acclimation (enhanced NST) with the evidence of BAT activation in human adults provide significant progress in the cold adaptation studies. When you compare the flowcharts of cold adaptation in human drawn in a textbook in 1996 [131] and review article in 2016 [133], you can find one question mark for enhanced NST is finally removed throughout the 20 years of studies.

5.2.2.3 Insulative Acclimation

Several laboratory-based studies reported a significant decrease in T_{sk} during cold exposure after repeated cold water immersions [154, 155, 158, 159], which are the insulative acclimation enabling them to reduce the heat loss from the skin. On the other hand, O'Brien et al. [81] showed no significant adaptation in T_{sk} and I_{tissue}, while there was a significant increase in total peripheral resistance after repeated cold immersions [81]. Our research group provided evidence of insulative acclimation after repeated mild cold immersions, with significant reduction in T_{sk} and skin blood flow assessed using laser-Doppler flowmetry [83]. A relatively small number of studies have reported insulative acclimation after repeated cold immersions.

5.2.3 Factors Affecting Cold Adaptation

The variability of cold adaptation has been categorized in several articles and the possible factors affecting cold adaptation have been discussed [86, 128–133]. Originally, Hammel [128] described metabolic, hypothermic, and insulative-hypothermic acclimatization, based on studies of indigenous populations in different climatic regions.

LeBlanc [129] categorized two different types of cold adaptation in humans produced by continuous moderate or intermittent severe cold exposures, applied from the categorization of animal studies. In animal studies, the continuous moderate cold exposure with shivering developed NST (metabolic adaptation), whereas the intermittent severe cold exposure, which exposed a local skin part to colder temperatures without significant shivering, induced a habituation of cardiovascular responses but no development of NST [129]. However, in humans, a population of divers who regularly engaged to cold water submersion and therefore could be categorized in the continuous moderate cold exposure group demonstrated no clear evidence of a replacement of shivering with NST [147]. Inuit and Gaspe fishermen were categorized into the intermittent severe cold exposure group and were found to shiver as much as [161] or even more than the control group [162].

Bittel [155] categorized metabolic and insulative type of cold acclimation based on individual thermoregulatory responses to cold air exposure, obtained after repeated cold water immersions. He defined metabolic acclimation as a higher shivering response after cold adaptation, and insulative acclimation as a lower T_{sk}. From the observations on individual data, the possibility of effect of morphological characteristics and physical fitness on adaptation type was described [155], although it has not been clearly investigated. The effect of morphological characteristics on adaptation type may be due to an indirect consequence of differences in T_{co} reduction during adaptation process which is associated with physical characteristics.

The hypothermic adaptation is thought to be as a result of a reduction in metabolic response to cold after repeated cold exposures [134, 154], whereas, insulative adaptation could maintain T_{co} even with a reduction in the metabolic heat production. Moreover, it is questionable whether the attenuated metabolic response of cold-adapted individuals persists during hypothermia, or whether an unhabituated response returns when T_{co} falls below the levels experienced during the development of the habituation. Tipton et al. [113] tested this question and observed a significant habituation of metabolic response to prolonged cold water immersion, which was specific to T_{co} experienced repeatedly during the acclimation process, however, it returned to unhabituated response when T_{co} fell below these temperatures.

Here we update the potential factors affecting the cold adaptation type with recent findings in the human BAT activity. Compared to the other thermoeffectors such as the SM or white adipose tissue, human BAT has large individual variations in their activity and contribution to the CIT. Thus, the cold adaptation type could be affected by the pre-adaptation state and adaptation potential in the BAT activity in the individual. In young healthy individuals with high pre-adaptation BAT activity, repeated cold exposure induced metabolic acclimation with enhanced NST and BAT activity, whereas no adaptive change was observed in the metabolic parameters in the SM [14, 25]. On the other hand, type 2 diabetic patients and obese individuals with low pre-adaptation BAT activity have shown that repeated cold exposure not only increases BAT activity but also increases glucose uptake into the SM [23, 27]. Considering these findings, depending on the pre-adaptation state and adaptation potential in the BAT activity, the contribution of the other thermoeffectors to the adaptive change might be different.

5.3 Performance in Cold Environments

5.3.1 Physical Performance

A number of research studies have been conducted on various categories of exercise performance in cold such as maximum muscle strength, maximum exercise intensity, energy efficiency, muscular endurance, and manual dexterity, which have been summarized in several review articles [163–166]. In this paragraph, at first, we briefly reviewed the exercise performance impairment in cold environment and the underlying physiological mechanisms, including thermoregulatory, neuromuscular, cardiovascular, and metabolic systems. The scheme of physical performance impairment in cold and physiological mechanism in multiple systems is summarized in Fig. 5.6, which is modified from our short review [167]. Then, we introduce our recent studies on the greater contribution of anaerobic metabolism during exercise in cold [168, 169] and new training method with hypothermic SM [170].

5.3.1.1 Brief Review on Physical Performance in Cold

Regarding maximum muscle strength, Clarke et al. [171] reported a small or no effect of muscle temperature (T_m) on maximal voluntary contraction (MVC) during isometric handgrip within T_m range of 27–40 °C, but a reduction of 60% when T_m reached around 20 °C. Similarly, several studies have reported a reduction in the isometric MVC, when T_m dropped below 27 °C [171, 172]. Based on these observations, in general, although there are some variations between studies and muscle groups, T_m of around 27 °C would be a critical temperature for inducing significant impairment in isometric maximal voluntary contraction [164–166].

On the other hand, it was suggested that maximal muscle performance in dynamic exercise is generally more impaired with relatively mild

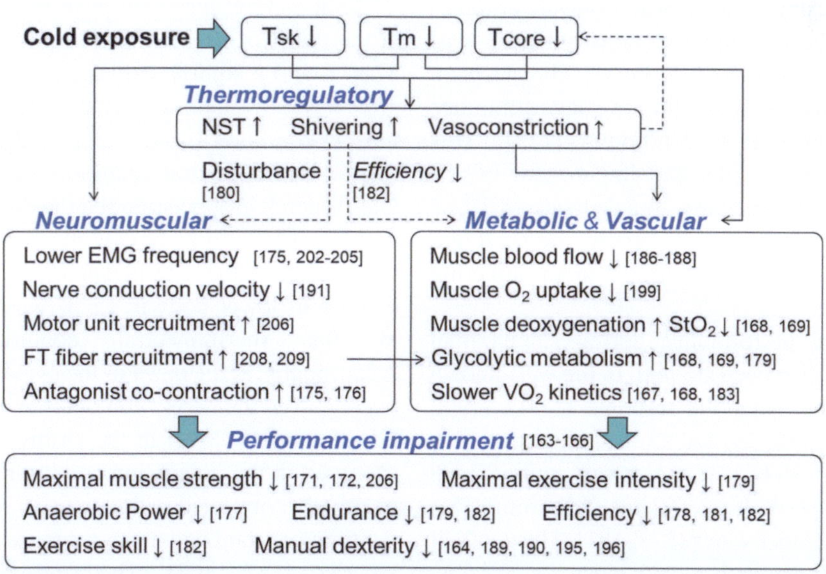

Fig. 5.6 Scheme of physical performance impairment in cold and physiological mechanisms in multiple systems. (Modified from Wakabayashi et al. [167])

cooling compared to the isometric contractions [164, 173, 174]. Bergh and Ekblom [173] reported greater decrease of peak torque per °C reduction of T_m in dynamic knee extension exercise compared to the isometric condition within T_m range of 30–39 °C. Similarly, the flight time and take-off velocity of maximal drop jump was decreased in a dose-dependent manner with degree of cooling [175], which in part might be due to the co-contraction of the agonist and antagonist muscle pair after cooling [175, 176].

In addition, a few studies have focused on anaerobic exercise capacity with hypothermic SM [177]. They conducted Wingate anaerobic test (30-sec maximal effort pedaling) after 30-min lower body immersion in 11.5 °C water. As a result, compared to the normothermic condition, the average power output and peak power during the Wingate test were reduced by around 26% and 30%, respectively, with the lower body precooling [177].

Regarding aerobic exercise capacity, Nadel et al. [178] reported that maximum oxygen uptake (VO_{2max}) during swimming in cold (18 °C) water was lower than in warmer water (26 and 34 °C), especially in the participants with less body fat. More recent study conducted incremental exercise tests at three T_w conditions (18, 26, and 34 °C) using an underwater bicycle ergometer [179]. The maximum exercise intensity was significantly lower in 18 °C water, whereas peak oxygen uptake (VO_{2peak}) did not show difference between temperature conditions [179]. This could be explained by the impairment of the mechanical efficiency due to a decrease in T_m, as well as an increase in extra energy demand and disturbance to motor control due to shivering [180].

Decreased performance during submaximal exercise has also been studied. In the well-known studies on the metabolic response to swimming in varying T_w (18, 26, and 33 °C), greater oxygen consumption (VO_2) for a given submaximal swimming speed (less efficient swimming) was observed in colder water [178, 181]. The detail of submaximal swimming efficiency (swimming distance per unit VO_2) and swimming performance (stroke parameters) was analyzed during self-paced endurance swimming in a swimming flume (circulating water tank) controlled at 10, 18, and 25 °C [182]. The less swimming efficiency with shorter stroke length (distance swum per each stroke) and more frequent stroke rate (number of strokes in a minute) were observed in 10 °C water than those in the warmer water conditions [182]. The impairment of swimming efficiency in cold water observed in these studies [178, 181, 182] would be mainly explained by the extra energy expenditure due to shivering associated with a decrease in T_{co}. In addition, the involuntary muscle contraction during shivering would disturb a coordination of the swimming stroke [180], and co-contraction of agonist and antagonist muscles in cold environment [175, 176] might induce the impairment in the stroke parameters [182]. Thus, the lower swimming energy efficiency would be attributable in part to the impaired swimming performance and lower mechanical efficiency due to impaired neuromuscular function in hypothermic SM. Moreover, swimming failure observed in cold water could partly be attributed to the limitation in the endurance capacity due to greater recruitment of glycolytic metabolism in the hypothermic SM, which is explained later in this chapter with our recent findings [168, 169].

Temperature dependence on the VO_2 kinetics at the onset of exercise has been studied [183]. They found a significantly greater time constant after onset of moderate cycle exercise with cold muscle condition (vastus lateralis T_m = 30.2 °C) than in thermoneutral condition (T_m = 36.8 °C). Our research group supported the evidence of the delay in the VO_2 kinetics, as shown in the longer time constant and mean response time after precooling of lower body in 12 °C water compared to the thermoneutral condition [168]. Additionally, at the onset of the cycle exercise, an overshoot in the deoxy hemoglobin and myoglobin (deoxy[Hb+Mb]) in the vastus lateralis was observed in the precooling condition, which was positively correlated with condition difference in the time constant and mean response time of VO_2 kinetics [167, 168]. The transient peak in deoxy[Hb+Mb] after precooling would indicate greater O_2 extraction fraction [184, 185] and less

O_2 delivery to muscle oxygen uptake due to less blood perfusion in the working muscle in cold [186–188]. Based on these findings, it was suggested that the slower VO_2 kinetics in cold would be caused by the limitation of aerobic energy supply due to less blood perfusion in the working muscle.

Regarding manual performance requiring fine motor control, exposure to cold results in a significant loss of manual dexterity, primarily due to peripheral cooling of the hands. Decrements in manual performance have been firmly established in a wide variety of tasks and are directly related to the degree of cooling of structures and tissues in the hand [189]. It is well known that the lowering of hand and finger skin temperature (T_{fing}) is primarily responsible for the impairment of manual performance in subjects exposed to cold [163, 164, 189]. Indeed, when young male volunteers wearing standard protective clothing were exposed to cold air at −5, −15, and −20 °C for 60 min, the relationship between counting performance and hand skin temperature was highly significant as shown in Fig. 5.7 [190]. Heus et al. [164] concluded that a slight decrease in manual dexterity was found at local hand and finger temperatures of 20 °C, but a significant loss of dexterity occurred at hand skin temperatures of 15 °C and below [164]. Heus et al. [164] summarized in their review article that the following factors are responsible for the reduction of manual performance due to cold: (1) Nerve conduction velocity is significantly reduced when nerve temperature falls below 20 °C, and action potential propagation is suppressed at temperatures below 10 °C [191]. (2) Cutaneous receptors such as contact, pressure, and pain sensations are greatly reduced at skin temperatures of 6–8 °C [192]. (3) When the intra-articular temperature falls below 24 °C, which is equivalent to a skin temperature of 15 °C [193], the range of motion of the joint is reduced [194]. A decrease in the viscosity of the synovial fluid as the temperature drops also contributes to the decrease in manual performance. Since the hands and fingers have a large surface area per volume, the surface tissues are cooled in a short time, which can easily lead to a decrease in dexterity. It is known that different cooling rates at the same skin temperature result in different manual performance. Lockhart et al. [195] compared several manual tasks at different cooling rates, referred to the exposure time 5 min (fast cooling) and 50 min (slow cooling) to reach criterion skin temperatures (23.9, 18.3, 12.7, and 7.2 °C). They found that slow cooling to 7.2 °C T_{fing} resulted in greater decrements of manual performance than fast cooling to the same temperature (Fig. 5.8) [195]. It was suggested that prolonged exposure to moderate cold resulted in a greater reduction in manual performance than brief exposure to severe cold.

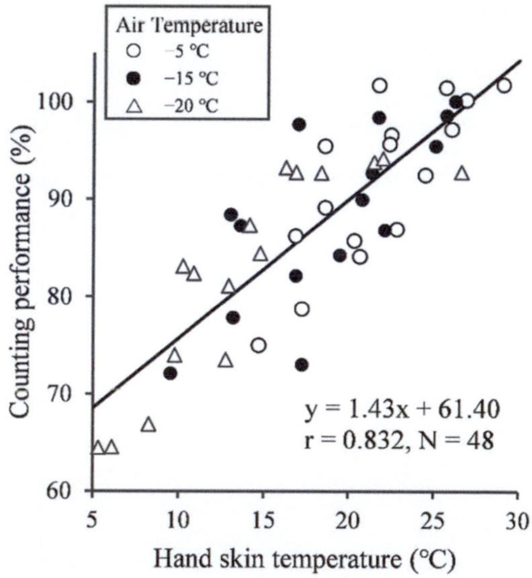

Fig. 5.7 Relationship between counting performance and hand skin temperature during cold exposures at −5, −15, and −20 °C. (Adapted from Tanaka et al. [190])

Although T_{co} and mean T_{sk} appear to play a minor role for dexterity as compared to local skin temperatures of the hand and fingers, the overall thermal condition of the body should also be considered when working in cold. For example, we have reported that the decrease in manual dexterity during cold work was greater at night, even though the T_{fing} is higher than that during daytime [196]. The lower T_{co} at night due to biological rhythms may have contributed to the greater impairment in manual dexterity during night work. Kim et al. [197] investigated the effects of loading work in a cold air at −25 °C on manual performance. Eight male students transported a

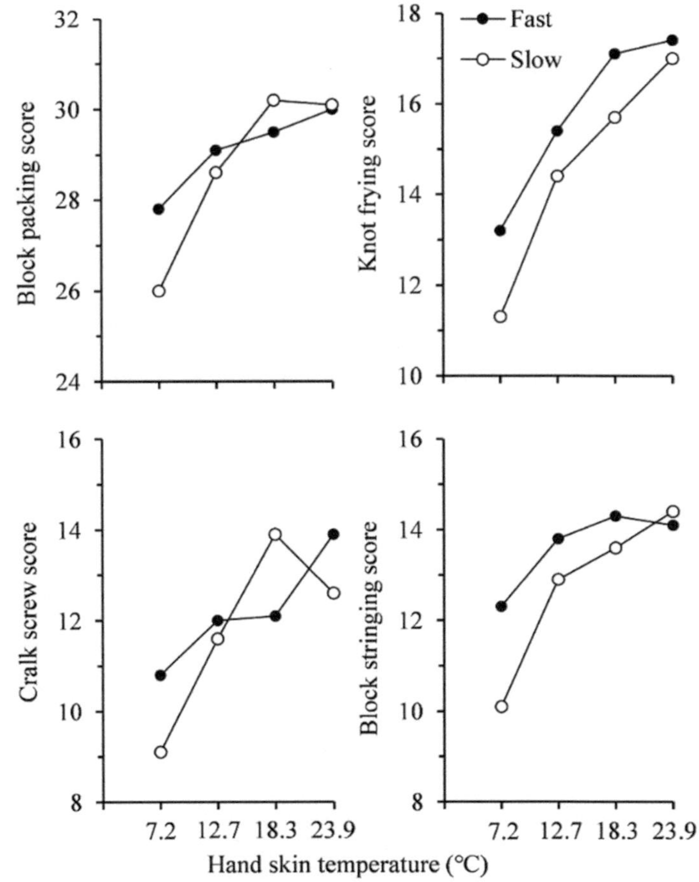

Fig. 5.8 Mean block-packing, block-stringing, Craik screw, and knot-tying scores as a function of hand surface temperature for fast (5 min) and slow (50 min) rates of cooling. (Modified from Lockhart et al. [195])

9 kg or 18 kg object for each 10 min, with resting for 10 min. This protocol was repeated three times, for 90 min of total exposure to extreme cold environment. Although the T_{fing} increased slightly with the loading work, manual performance decreased as the weight of the load increased [197]. Considering those findings, the manual performance in cold environment would have been affected by comprehensive factors other than T_{fing}.

As described in this paragraph, in a cold environment, there are impairments in various exercise performances, from maximal to submaximal exercise, static and dynamic movement, and manual dexterity. A brief summary of the physical performance impairment in cold and potential physiological mechanism in multiple systems are shown in Fig. 5.6 [167].

5.3.1.2 Greater Contribution of Anaerobic Metabolism in Cold

Our research group focuses on muscle deoxygenation and glycolytic metabolism during exercise with hypothermic SM. The blood lactate (La) concentration and muscle oxygenation during submaximal exercise after 30-min lower body cooling in 12 °C water were compared with control condition without precooling [168]. In both cooling and noncooling conditions, the exercise intensity was set to the same absolute workload, which was lactate threshold (LT) level preassessed by an incremental bicycle exercise test on land at thermoneutral environment. The vastus lateralis muscle temperature (T_m), measured using zero heat flux method, was declined to about 23 °C after the lower body immersion, and then T_m was rewarmed to normothermic level in 20 min during the following exercise, due to the heat production from the working muscle. During the first 10 min of the exercise, significantly higher La was observed in cooling condition, but the difference between the two conditions became statistically insignificant after 20 min, when the T_m was rewarmed to normothermic level [168]. This indicated the enhancement of glycolytic metabolism during moderate exercise with hypothermic SM. A greater contribution of anaerobic metabolism was also observed in the increased ventilation equivalent (VE/VO_2) during submaximal incremental exercise in cold water (18 °C) compared to that in warm water [179].

Focusing on the SM metabolism, a greater increase in deoxy hemoglobin and myoglobin (deoxy[Hb+Mb]) in the vastus lateralis was observed during exercise after lower body cooling [168]. The increase of deoxy[Hb+Mb] is considered as an index that reflects the lower O_2 delivery relative to O_2 consumption in the SM [184, 198]. Thus, the greater muscle deoxygenation in the cooling condition would be attributed to the less blood flow in the hypothermic SM [186–188], which would be due to the vasoconstriction as a thermoregulatory response to cold. The less O_2 supply to the working muscle in cold would lead to the enhancement of the anaerobic metabolism. Our research group previously reported that forearm muscle oxygen consumption during handgrip exercise, which was estimated by the oxy hemoglobin and myoglobin (oxy[Hb+Mb]) curve during arterial occlusion, was decreased with the reduction of the muscle temperature [199]. The suppression in the aerobic metabolism in hypothermic SM might compensatorily increase the contribution of the anaerobic metabolism for maintaining the same exercise intensity as in the normothermic condition.

Recently, Sakaue et al. [169] supported the finding of the greater anaerobic contribution in cold by comparing physiological responses during 30-min exercise in cold water (15 °C) and thermoneutral water (33 °C) following 30-min resting immersion in each temperature condition. This protocol could maintain their T_m below the critical temperature inducing impairment of maximal force and muscle endurance [164, 171, 172] during most of the 30-min exercise session. In our previous report using the precooling protocol, a significant increase in the La was only observed in the first 10 min of exercise, since the T_m was rewarmed to normothermic level in 20 min [168]. Whereas the protocol of exercising in cold water using an underwater ergometer enabled them to maintain significantly greater La during the whole 30 min of exercise phase compared to the thermoneutral condition [169]. The overall phase average of La during 30-min exercise was reanalyzed and drawn in Fig. 5.9a. Additionally, a significantly higher respiratory exchange ratio (RER) was observed during the

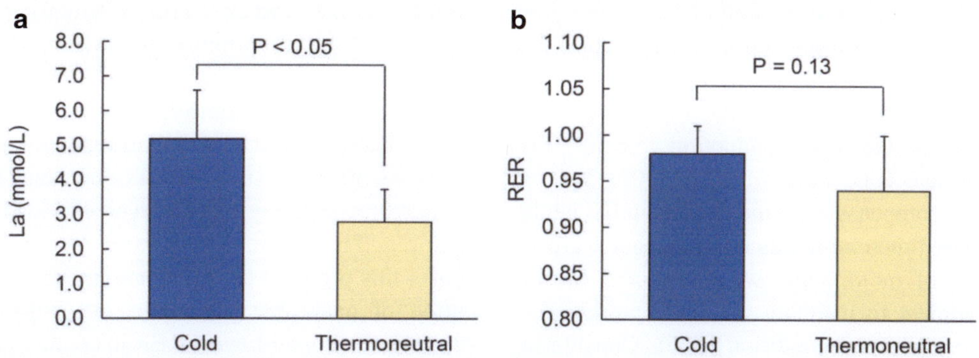

Fig. 5.9 Overall phase average values of La (**a**) and RER (**b**) during 30-min cycling exercise in cold and thermoneutral water conditions. *La* blood lactate concentration, *RER* respiratory exchange ratio. Values are mean + SD. The overall phase average values were reanalyzed from the original time course data in Sakaue et al. [169]

first 5 min of exercise in cold water [169], and this tendency was continued during the 30-min exercise although it was not statistically significant (Fig. 5.9b). These results demonstrated evidence of the greater recruitment of glycolytic metabolism in cold water during exercise even at the LT level intensity predetermined in the thermoneutral water. This means, even at the same mechanical workload in both conditions, physiological strain is greater in the cooling condition due to the limited blood perfusion and impairment in the neuromuscular function. Regarding the SM metabolism, significantly lower oxy[Hb+Mb] and lower tissue O_2 saturation (StO_2) in the vastus lateralis were observed during exercise in cold water [169].

The potential physiological mechanisms of the greater contribution of anaerobic metabolism during exercise in cold are summarized as follows. At first, as explained above, the greater muscle deoxygenation and anaerobic metabolism during exercise in cold would be attributed to the reduction of blood perfusion in the working muscle [186–188], which would be due to the vasoconstriction as a thermoregulatory response to cold. Additionally, a leftward shift of oxyhemoglobin dissociation curve in lower temperature [200, 201], known as the Bohr effect, might reduce O_2 extraction in the active muscle. Due to the suppression of both O_2 delivery and O_2 utilization in the active muscle in cold water, hypoxic-like environment would be constructed at the level of the working SM. Regarding the neuromuscular system, a lower shift of electromyogram frequency with lower muscle temperature has been consistently reported during exercise in cold [175, 202–205], which would be associated with reduction of the nerve conduction velocity [191] and muscle conduction velocity [202, 204]. In order to compensate for the decline in the neuromuscular function, it was reported that a greater number of motor units were recruited in cold environment to maintain the same workload in the thermoneutral condition [206]. Considering the greater cold sensitivity and lower power output of slow-twitch fibers [207], recruitment of fast-twitch fibers would be increased in cold environment even at moderate intensity exercise, as reported in studies in fishes swimming in cold water [208, 209].

5.3.1.3 New Exercise Training with Hypothermic Skeletal Muscle

Based on our findings of greater anaerobic contribution during exercise in cold [168, 169] and potential physiological mechanisms, we proposed a new training method "hypothermic skeletal muscle training," which utilizes the physiological strain in cold as an extra physiological load. This training can provide a hypoxic-like environment at the SM level due to the cold-induced vasoconstriction [186–188] and leftward shift of oxyhemoglobin dissociation curve in lower temperature [200, 201]. In addition to the less O_2 supply to the working muscle in cold, impairment in the neuromuscular function of the hypothermic SM would integratively enhance recruitment of more motor units [206], which could include higher threshold fast-twitch fibers [208, 209], and glycolytic metabolism even at low to moderate mechanical intensity [168, 169]. Based on these physiological mechanisms, we hypothesized that the hypothermic SM training even with low to moderate workload would further improve the anaerobic power.

Our research group conducted a highly original intervention study for assessing the effect of the hypothermic SM training [170]. Participants were divided into cooling and thermoneutral control groups which conducted 4-week exercise training (3 times a week, 12 times in total) in cold water (15 °C) and thermoneutral water (33 °C), respectively. In a training session, each cooling and thermoneutral control group conducted 30-min exercise in cold and thermoneutral water following 30-min resting immersion in each temperature, respectively, at a LT level workload predetermined in thermoneutral water. As explained above, this protocol could induce greater contribution of anaerobic glycolytic metabolism as shown in the higher La and RER in cooling group [169]. Before and after the training period, participants conducted performance tests for assessing LT level workload in thermoneutral water, maximal oxygen uptake (VO_{2max}) using a bicycle ergometer on land, and Wingate test for evaluating anaerobic power.

As a result of the training intervention, no training effect in the aerobic metabolism was observed in the LT level and VO_{2max} in both cooling and thermoneutral control groups. Whereas training in cold water even at the moderate workload significantly improved the anaerobic power [170]. In the Wingate test, greater average power and peak power were observed post-training period compared to pre-intervention only in the cooling group (Fig. 5.10a, b). Additionally, pre-to-post percent change in the time to peak power was significantly shorter in cooling group compared to the thermoneutral control group (−11.5 ± 17.6% and 15.6 ± 28.0% in cooling and control groups, respectively, $P < 0.05$, Fig. 5.10c). These results indicated the improvement in the anerobic metabolism and greater contribution of the fast-twitch fibers after the training intervention with hypothermic SM. It has been reported that divers who regularly work in cold water environments have a higher percentage of type IIx (fast glycolytic) and lower proportion of type IIa (fast oxidative glycolytic) fibers in the vastus lateralis than physically active controls [210], which might be due to the repeated recruitment of fast-twitch glycolytic fibers in cold. In animal study, shift in fiber type from type I to type IIa fibers in rat soleus muscle was observed after repeated cold exposure [211]. Thus, exercise training in cold water might selectively enhance the muscle function in the fast-twitch glycolytic fibers, which resulted in the improvement in the anaerobic power with short time to peak.

In addition to the improvement in the anaerobic power, a significant enhancement in the muscle deoxygenation during submaximal exercise was observed after the intervention of the training in cold water with hypothermic SM [170]. The deoxy [Hb+Mb] in the vastus lateralis during the LT intensity exercise was significantly greater in the final (the 12th) training session than in the first training session in cooling group (Fig. 5.11b), whereas no training effect was observed in oxy [Hb+Mb] (Fig. 5.11a). The increase in the muscle deoxygenation might indicate the improvement in the muscle oxygen extraction. In acute response with the hypothermic SM, greater muscle deoxy [Hb+Mb] [168], less muscle oxy [Hb+Mb] [169], and less StO_2 [168, 169] were observed in our previous reports, which indicated the less O_2 supply to the working muscle due to the cold-induced vasoconstriction, relative to the O_2 consumption in the muscle. The lower StO_2 in cold water was reconfirmed at the first and final intervention sessions in our training study (Fig. 5.11d), which indicated that hypoxic condition in the working muscle was repeated during the intervention. When the working SM was repeatedly in the hypoxia, there would be some adaptation in microvascular network, like an increased capillary-to-muscle fiber ratio [212, 213], as reported to be the training effect of hypoxic training. Similarly, the breath-hold divers working in cold water had greater capillary

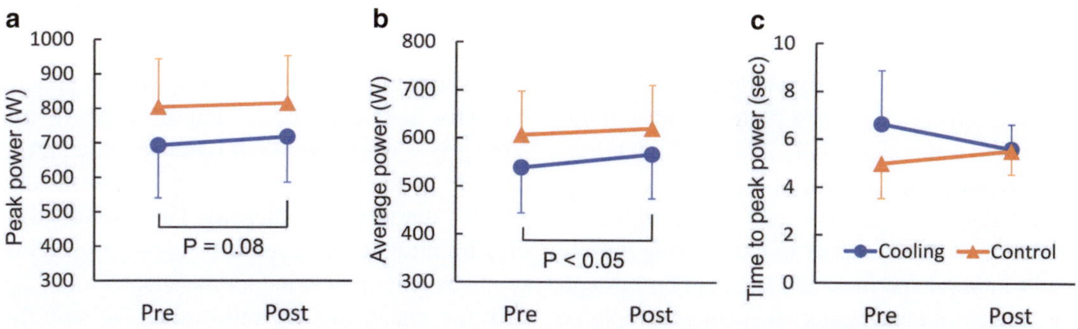

Fig. 5.10 Peak power (**a**), average power (**b**), and time to peak power (**c**) during Wingate test conducted pre- and post-intervention of exercise training in cooling and control group. Values are mean ± SD. The figures were drawn based on the data in Sakaue et al. [170]

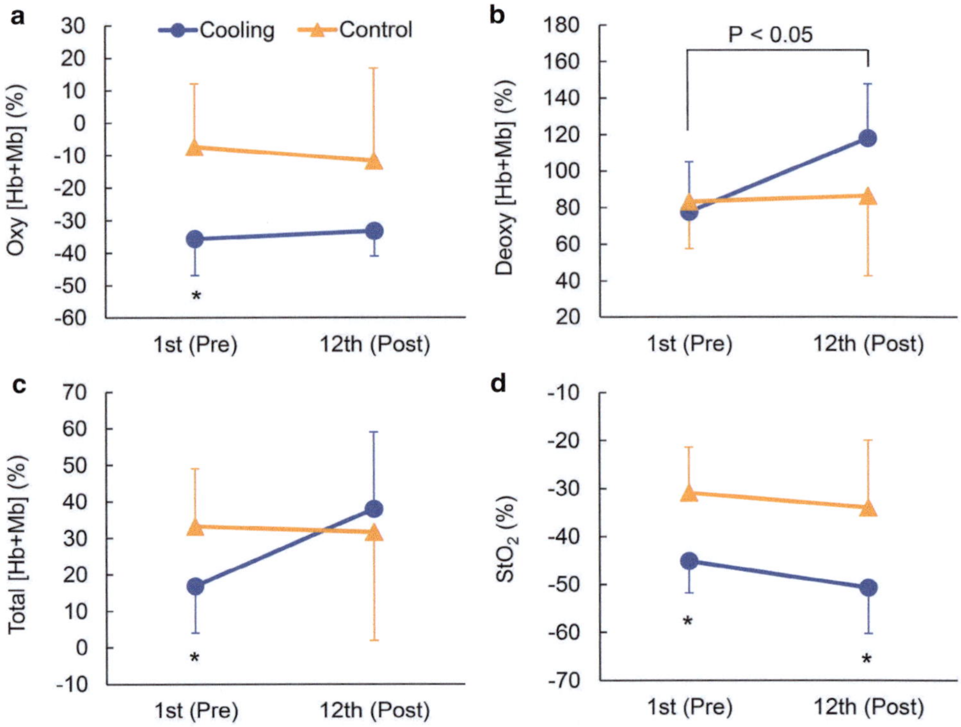

Fig. 5.11 Overall phase average values of percent change relative to the resting baseline in oxy [Hb+Mb] (**a**), deoxy [Hb+Mb] (**b**), total [Hb+Mb] (**c**), and StO_2 (**d**) in the vastus lateralis during 30-min submaximal cycling exercise at the 1st and 12th training sessions in the cooling and control groups. Values are mean ± SD. *Significant difference between groups at each training session. The figures were drawn based on the data in Sakaue et al. [170]

density and capillary-to-fiber ratio compared to the physically active controls [210]. Similarly, in animal studies, chronic cold exposure has been reported to induce a growth of capillary density and/or the capillary-to-fiber ratio [214–216]. In our study, total [Hb+Mb] during the first training session was significantly lower in the cooling group compared to the control group ($P < 0.05$), while no group difference was observed at the final session (Fig. 5.11c), which might indicate relatively enhanced muscle perfusion after training in cold water, potentially due to the microvascular remodeling [210, 214–216]. Considering these facts, the repeated exercise in cold with a limitation of the O_2 supply to the working muscle might improve the muscle O_2 extraction through the microvascular remodeling. Interestingly, a similar training effect in muscle deoxygenation [217, 218] and enhanced muscle perfusion [218] has been observed after the intervention of the hypoxic training. There could be a similarity between exercise training in a hypoxic environment and hypothermic SM in the point of O_2 restriction at the working muscle level, as shown in the less StO_2 in the vastus lateralis throughout the training sessions in the cooling group (Fig. 5.11d). However, even with the increase in the muscle deoxygenation, which might explain the enhancement in the muscle O_2 extraction with capillary angiogenesis, no improvement in the VO_{2max} was observed in our training study [170]. The discrepancy might be due to insufficient training duration and intensity. The 4-week exercise training at LT level with hypothermic SM might be sufficient to induce the adaptation in the SM for enhancing O_2 utilization, but not for inducing hematological adaptation for increasing O_2 delivery, such as increased red blood cell and hemoglobin concentrations (which were not assessed in our study). Additionally, while hypoxic training applies a systemic hypoxic load, the hypothermic SM training can only apply a localized hypoxic load, which might be insufficient to induce hematological adaptation. Thus, the improvement in the maximal aerobic capacity might not reach a significant level. Indeed, even in the previous studies on hypoxic training which found increase in the muscle deoxygenation [217, 218], a study reported a significant improvement in maximal work rate and VO_{2max} [218], while the other reported no change [217] like our study. Further research is needed for clarifying the effect of hypothermic SM training on the aerobic exercise capacity and adaptation in the SM tissue including microvascular network.

In summary, our research revealed that the hypothermic SM training in cold water improved anaerobic power, even when a moderate workload was repeated during the training session. Additionally, perhaps the O_2 extraction in the working muscle might be improved by this training. Further investigations would identify the validity, advantage, and limitations of this new training method.

5.3.2 Cognitive Performance

Cold exposure can result in various adverse consequences on cognitive performance. Cognitive performance plays an important role in the areas of orientation, safety, decision-making, work productivity, and reactions to emergency situations [219]. Several review articles have been reported on the effects of environmental stress [220], high and low temperatures [221, 222], and cold environments [219, 223] on cognitive performance.

It is known that cognitive work capacity in cold environments is strongly influenced by the level of human core temperature (T_{co}) and the complexity of cognitive performance. In this section, the relationship between T_{co} and cognitive performance is discussed at three levels of T_{co}.

The level of human consciousness in cold environments is determined by T_{co} (see Table 4.6 in Chap. 4). FitzGibbon et al. [224] evaluated the extent of alteration of electroencephalography (EEG) frequency and amplitude, and changes in visual evoked responses, in five conscious male humans who were rendered hypothermic by immersion in cold water (7 °C). Their results confirmed that for conscious, unanesthetized humans, severe hypothermia (T_{co} < 33.5 °C) was the borderline for significant alteration in the brain electrical activity [224]. Therefore, it is believed that human cognitive performance becomes impossible when the T_{co} is below 33.5 °C.

Moderate hypothermia (33.5 °C < T_{co} < 35.5 °C) is often induced by cold water exposure. Cold water immersion occurs during diving in cold water and accidental falls into cold water. As for cold water diving, cognitive performance plays a crucial role in divers' safety, decision-making, and reactions in emergencies. Coleshaw et al. [225] immersed young volunteers up to the necks in 15 °C cold water for 60 min and measured their cognitive performance when their T_{co} dropped to 34.5 °C. They found a correlation between memory registration performance and T_{co}. When T_{co} was decreased below 34.5 °C, participants were able to recall only 70% of the facts that they had learned. In the two-digit calculation test, the speed of performance was impaired by about 50% at T_{co} of 34.5 °C [225]. Giesbrecht et al. [226] investigated the effects of task complexity on the decrement in cognitive performance during cold water (8 °C) immersion up to the neck. Various tasks were measured at the following three phases: prior to cold water immersion, soon after cold water immersion before any decrease in T_{co}, after immersion when T_{co} had decreased by 2–4 °C. The results indicated that tasks requiring relatively low cognitive demands were not affected by cold water immersion, even in the phase of significant central cooling. On the other hand, more complex tasks requiring mental manipulation or mental processing and analysis showed a slight improvement immediately after cold immersion (probably due to arousal).

However, a subsequent decrease in complex tasks was observed when central cooling of 2–4 °C had occurred [226]. It was suggested that cold affected cognitive performance more if the tasks were complex or perceptually demanding and required significant concentration. Recently, Falla et al. [223] reviewed laboratory studies on cognitive performance, such as attention, executive function, memory, and processing speed, in cold water with T_{co} decreased by 0.5–2.8 °C or reached 34 °C. According to the systematic review, they concluded that the degree of impairment in cognitive performance by cold water exposure was proportional to the reduction of T_{co} [223].

Mild or non-hypothermic cold exposure ($T_{co} > 33.5$ °C) is more likely to occur during daily occupational or leisure time activities. The findings on the effects of these types of cold exposure on cognitive performance are inconsistent, with different studies reporting that performance was decreased, unaltered, or even improved in cold [227]. The lack of consistency among these studies is probably due to mild cold stress and differences in the type and duration of cold exposure and cognitive tasks. Several researchers have investigated whether the effect of moderate cold on cognitive performance differs between simple and complex tasks. Ellis et al. [228] asked eight male adults to perform 8-choice (complex) and 4-choice (simple) serial choice reaction time (SCRT) task at three times: before (pre-cold), during exposure in a cold room at −5 °C, and after rewarming (post-cold). Figure 5.12 shows errors, lags, and reaction time committed in the 8-choice and the 4-choice versions of the SCRT tasks performed at three different times, with rectal and mean skin temperature [228]. Rectal temperature decreased continuously during the experiment, and mean skin temperature rewarmed to the pre-cold levels at post-cold. They indicated that cold increased the error rate on the SCRT tasks particularly in the 8-choice version, whereas reaction times became faster in cold. In addition, lags during cold exposure were considerably reduced in the 8-choice SCRT. These data suggested that cold skin might have caused participants to alter their strategies, perhaps because of an increase in arousal, they traded off accuracy for speed. Enander [229] also conducted experiments to investigate the effects of moderate cold (5 °C) and control (20 °C) exposure on simple and complex cognitive performance. No cold effects on simple cognitive performance were observed, whereas effects on complex performance were shown as an increase in the numbers of errors and false alarms [229]. The results indicated that tasks requiring rapid and accurate responses were negatively affected by moderate cold. Two distinct theories have been proposed to explain the decline in cognitive performance due to cold exposure. The first one is the "distraction theory" proposed by Teichner [230]. Discomfort from cold distracts participants from focusing attention on a given task, leading to slower reaction

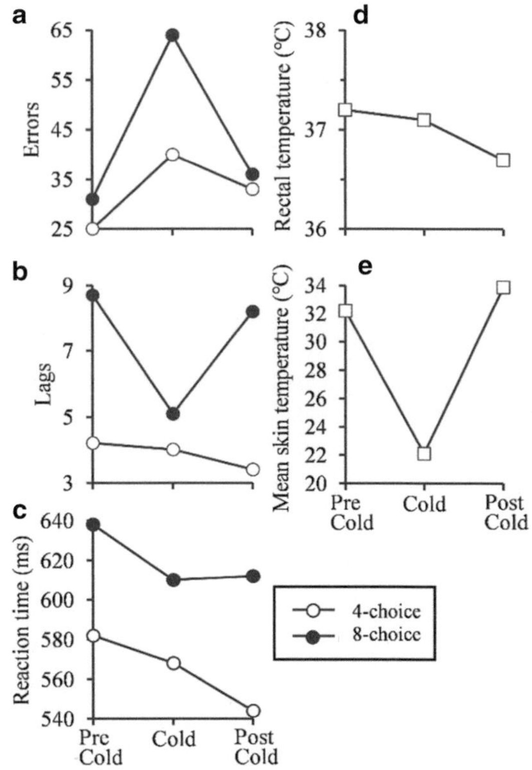

Fig. 5.12 Average values (**a**: errors, **b**: lags, **c**: reaction time) committed in the 8-choice and the 4-choice versions of the SCRT tasks performed at pre-cold, during cold, and post-cold. Average values (**d**: rectal temperature, **e**: mean skin temperature) of physiological responses at three different times. (Adapted from Ellis et al. [228])

times and reduced accuracy. The second one is the "arousal theory" proposed by several researchers [228, 229, 231]. This arousal theory can explain improvements in cognitive performance in the early stages of cold exposure. Van Orden et al. [232] supported this theory with evaluation of brain function by measuring evoked potentials of young males during 50-min exposure to air at 4 °C and 22 °C. They showed that the stress caused by moderate cold exposure (4 °C) elevated arousal levels resulting in faster central nervous system (CNS) transmission speeds, as indicted by shorter evoked potential latencies [232]. However, excessively elevated arousal levels caused by longer cold exposure or a greater decrease in T_{co} impaired attention and caused an increase in the number of missed signals and delays in reaction time.

Several investigators have reported the effects of repeated cold exposure on cognitive function. Cold acclimation, or repeated cold exposure, may be a potential strategy to attenuate cold-induced cognitive impairments for populations frequently exposed to cold environments. Thomas et al. [233] showed that the effects of moderate cold (5 °C) on complex cognitive performance were unchanged over three weekly repeated cold exposures. Mäkinen et al. [234] conducted a study of cold acclimation and cognitive performance. The effects of repeated cold exposure on cognitive performance were studied in 10 male participants who conducted 120-min cold (10 °C) exposure following 90-min control (25 °C) for 10 consecutive days. Physiological responses (rectal and skin temperatures, blood pressure, heart rate, and metabolic rate) and thermal sensation were recorded during the experiment. Only the overall thermal sensation changed from "cold" (day 1) to "cool" (day 10), indicating subjective habituation was induced by repeated cold exposure [234]. They found that cognitive performance efficiency was significantly improved, and response times were shortened over the 10-day period during both cold and control exposures, suggesting a learning effect. Since the changes in cognitive performance over the acclimation period were similar between cold and control, it was suggested that the habituation in the thermal sensation might not improve performance.

More recently, Jones et al. [235] reviewed studies on "cold acclimation and cognitive performance" including the two studies mentioned above [233, 234]. They concluded that cold acclimation, or repeated exposure to cold, did not result in any beneficial improvements in cognitive performance in laboratory or natural cold environments; in addition, they recommended that cognitive performance should be assessed not only with cognitive assessment tasks but also with measures of brain function [235].

References

1. Mekjavic IB, Eiken O. Contribution of thermal and nonthermal factors to the regulation of body temperature in humans. J Appl Physiol (1985). 2006;100(6):2065–72.
2. Werner J, Mekjavic IB, Taylor NA. Concepts in physiological regulation: a thermoregulatory perspective. In: Taylor NAS, Groeller H, editors. Physiological bases of human performance during work and exercise. Churchill Livingstone Elsevier; 2008. p. 325–40.
3. Cypess AM, Lehman S, Williams G, Tal I, Rodman D, Goldfine AB, et al. Identification and importance of brown adipose tissue in adult humans. N Engl J Med. 2009;360(15):1509–17.
4. Saito M, Okamatsu-Ogura Y, Matsushita M, Watanabe K, Yoneshiro T, Nio-Kobayashi J, et al. High incidence of metabolically active brown adipose tissue in healthy adult humans: effects of cold exposure and adiposity. Diabetes. 2009;58(7):1526–31.
5. van Marken Lichtenbelt WD, Vanhommerig JW, Smulders NM, Drossaerts JM, Kemerink GJ, Bouvy ND, et al. Cold-activated brown adipose tissue in healthy men. N Engl J Med. 2009;360(15):1500–8.
6. van Marken Lichtenbelt WD, Kingma B, van der Lans A, Schellen L. Cold exposure—an approach to increasing energy expenditure in humans. Trends Endocrinol Metab. 2014;25(4):165–7.
7. Wakabayashi H, Matsumoto K, Kobori Y, Ebara T, Matsushita M, Kameya T, et al. Multiorgan contribution to non-shivering and shivering thermogenesis and vascular responses during gradual cold exposure in humans. Eur J Appl Physiol. 2020;120(12):2737–47.
8. Yoneshiro T, Aita S, Matsushita M, Kayahara T, Kameya T, Kawai Y, et al. Recruited brown adipose tissue as an antiobesity agent in humans. J Clin Invest. 2013;123(8):3404–8.

9. Haman F, Blondin DP. Shivering thermogenesis in humans: origin, contribution and metabolic requirement. Temperature (Austin). 2017;4(3):217–26.
10. Meigal A. Gross and fine neuromuscular performance at cold shivering. Int J Circumpolar Health. 2002;61(2):163–72.
11. Muzik O, Mangner TJ, Leonard WR, Kumar A, Janisse J, Granneman JG. 15O PET measurement of blood flow and oxygen consumption in cold-activated human brown fat. J Nucl Med. 2013;54(4):523–31.
12. Din MU, Raiko J, Saari T, Kudomi N, Tolvanen T, Oikonen V, et al. Human brown adipose tissue [(15) O]O2 PET imaging in the presence and absence of cold stimulus. Eur J Nucl Med Mol Imaging. 2016;43(10):1878–86.
13. Moonen MPB, Nascimento EBM, van Marken Lichtenbelt WD. Human brown adipose tissue: underestimated target in metabolic disease? Biochim Biophys Acta Mol Cell Biol Lipids. 2019;1864(1):104–12.
14. van der Lans AA, Hoeks J, Brans B, Vijgen GH, Visser MG, Vosselman MJ, et al. Cold acclimation recruits human brown fat and increases nonshivering thermogenesis. J Clin Invest. 2013;123(8):3395–403.
15. Saito M, Wakabayashi H. Human brown adipose tissue as a regulator of systemic metabolism (in Japanese). Jpn J Physiol Anthropol. 2022;27(2):27–37.
16. Chen KY, Cypess AM, Laughlin MR, Haft CR, Hu HH, Bredella MA, et al. Brown Adipose Reporting Criteria in Imaging STudies (BARCIST 1.0): recommendations for standardized FDG-PET/CT experiments in humans. Cell Metab. 2016;24(2):210–22.
17. Saito M. Brown adipose tissue as a therapeutic target for human obesity. Obes Res Clin Pract. 2013;7(6):e432–8.
18. Yoneshiro T, Aita S, Matsushita M, Okamatsu-Ogura Y, Kameya T, Kawai Y, et al. Age-related decrease in cold-activated brown adipose tissue and accumulation of body fat in healthy humans. Obesity (Silver Spring). 2011;19(9):1755–60.
19. Bakker LEH, Boon MR, van der Linden RAD, Arias-Bouda LP, van Klinken JB, Smit F, et al. Brown adipose tissue volume in healthy lean south Asian adults compared with white Caucasians: a prospective, case-controlled observational study. Lancet Diabetes Endocrinol. 2014;2(3):210–7.
20. Nishimura T, Katsumura T, Motoi M, Oota H, Watanuki S. Experimental evidence reveals the UCP1 genotype changes the oxygen consumption attributed to non-shivering thermogenesis in humans. Sci Rep. 2017;7(1):5570.
21. Ouellet V, Routhier-Labadie A, Bellemare W, Lakhal-Chaieb L, Turcotte E, Carpentier AC, et al. Outdoor temperature, age, sex, body mass index, and diabetic status determine the prevalence, mass, and glucose-uptake activity of 18F-FDG-detected BAT in humans. J Clin Endocrinol Metab. 2011;96(1):192–9.
22. Blondin DP, Labbe SM, Noll C, Kunach M, Phoenix S, Guerin B, et al. Selective impairment of glucose but not fatty acid or oxidative metabolism in Brown adipose tissue of subjects with type 2 diabetes. Diabetes. 2015;64(7):2388–97.
23. Hanssen MJ, Hoeks J, Brans B, van der Lans AA, Schaart G, van den Driessche JJ, et al. Short-term cold acclimation improves insulin sensitivity in patients with type 2 diabetes mellitus. Nat Med. 2015;21(8):863–5.
24. Yoneshiro T, Matsushita M, Nakae S, Kameya T, Sugie H, Tanaka S, et al. Brown adipose tissue is involved in the seasonal variation of cold-induced thermogenesis in humans. Am J Physiol Regul Integr Comp Physiol. 2016;310(10):R999–R1009.
25. Blondin DP, Labbe SM, Tingelstad HC, Noll C, Kunach M, Phoenix S, et al. Increased brown adipose tissue oxidative capacity in cold-acclimated humans. J Clin Endocrinol Metab. 2014;99(3):E438–46.
26. Blondin DP, Daoud A, Taylor T, Tingelstad HC, Bezaire V, Richard D, et al. Four-week cold acclimation in adult humans shifts uncoupling thermogenesis from skeletal muscles to brown adipose tissue. J Physiol. 2017;595(6):2099–113.
27. Hanssen MJ, van der Lans AA, Brans B, Hoeks J, Jardon KM, Schaart G, et al. Short-term cold acclimation recruits brown adipose tissue in obese humans. Diabetes. 2016;65(5):1179–89.
28. Matsushita M, Nirengi S, Hibi M, Wakabayashi H, Lee SI, Domichi M, et al. Diurnal variations of brown fat thermogenesis and fat oxidation in humans. Int J Obes. 2021;45(11):2499–505.
29. Blondin DP, Labbe SM, Phoenix S, Guerin B, Turcotte EE, Richard D, et al. Contributions of white and brown adipose tissues and skeletal muscles to acute cold-induced metabolic responses in healthy men. J Physiol. 2015;593(3):701–14.
30. Haman F, Legault SR, Rakobowchuk M, Ducharme MB, Weber JM. Effects of carbohydrate availability on sustained shivering II. Relating muscle recruitment to fuel selection. J Appl Physiol. 2004;96(1):41–9.
31. Taylor NA, Mekjavic IB, Tipton MJ. The physiology of acute cold exposure, with particular reference to human performance in the cold. In: Taylor NAS, Groeller H, editors. Physiological bases of human performance during work and exercise. Churchill Livingstone Elsevier; 2008. p. 359–77.
32. Hodges GJ, Johnson JM. Adrenergic control of the human cutaneous circulation. Appl Physiol Nutr Metab. 2009;34(5):829–39.
33. Johnson JM, Minson CT, Kellogg DL. Cutaneous vasodilator and vasoconstrictor mechanisms in temperature regulation. Compr Physiol. 2014;4(1):33–89.
34. Alba BK, Castellani JW, Charkoudian N. Cold-induced cutaneous vasoconstriction in humans: function, dysfunction and the distinctly counterproductive. Exp Physiol. 2019;104(8):1202–14.

35. Rowell LB. Reflex control of the cutaneous vasculature. J Invest Dermatol. 1977;69(1):154–66.
36. Scholander PF, Krog J. Countercurrent heat exchange and vascular bundles in sloths. J Appl Physiol. 1957;10(3):405–11.
37. Walloe L. Arterio-venous anastomoses in the human skin and their role in temperature control. Temperature (Austin). 2016;3(1):92–103.
38. Clark ER. Arterio-venous anastomoses. Physiol Rev. 1938;18(2):229–47.
39. Nelms JD. Functional anatomy of skin related to temperature regulation. Fed Proc. 1963;22:933–6.
40. Donadio V, Nolano M, Provitera V, Stancanelli A, Lullo F, Liguori R, et al. Skin sympathetic adrenergic innervation: an immunofluorescence confocal study. Ann Neurol. 2006;59(2):376–81.
41. Janbu T. Blood velocities in the dorsal pedis and radial arteries during labour. Br J Obstet Gynaecol. 1989;96(1):70–9.
42. Bergersen TK, Eriksen M, Walloe L. Local constriction of arteriovenous anastomoses in the cooled finger. Am J Phys. 1997;273(3 Pt 2):R880–6.
43. Bergersen TK, Hisdal J, Walloe L. Perfusion of the human finger during cold-induced vasodilatation. Am J Physiol-Reg I. 1999;276(3):R731–7.
44. Stephens DP, Saad AR, Bennett LAT, Kosiba WA, Johnson JM. Neuropeptide Y antagonism reduces reflex cutaneous vasoconstriction in humans. Am J Physiol-Heart C. 2004;287(3):H1404–9.
45. Lang JA, Krajek AC, Smaller KA. Evidence for a functional vasoconstrictor role for ATP in the human cutaneous microvasculature. Exp Physiol. 2017;102(6):684–93.
46. Ekenvall L, Lindblad LE, Norbeck O, Etzell BM. Alpha-adrenoceptors and cold-induced vasoconstriction in human finger skin. Am J Phys. 1988;255(5 Pt 2):H1000–3.
47. Bailey SR, Mitra S, Flavahan S, Flavahan NA. Reactive oxygen species from smooth muscle mitochondria initiate cold-induced constriction of cutaneous arteries. Am J Physiol Heart Circ Physiol. 2005;289(1):H243–50.
48. Bailey SR, Eid AH, Mitra S, Flavahan S, Flavahan NA. Rho kinase mediates cold-induced constriction of cutaneous arteries: role of alpha2C-adrenoceptor translocation. Circ Res. 2004;94(10):1367–74.
49. Hodges GJ, Zhao K, Kosiba WA, Johnson JM. The involvement of nitric oxide in the cutaneous vasoconstrictor response to local cooling in humans. J Physiol. 2006;574(Pt 3):849–57.
50. Holowatz LA, Kenney WL. Local ascorbate administration augments NO- and non-NO-dependent reflex cutaneous vasodilation in hypertensive humans. Am J Physiol-Heart C. 2007;293(2):H1090–6.
51. Lewis T. Observations upon the reactions of the vessels of the human skin to cold. Heart-J Stud Circ. 1930;15(2):177–208.
52. Daanen HA. Finger cold-induced vasodilation: a review. Eur J Appl Physiol. 2003;89(5):411–26.
53. Cheung SS, Daanen HA. Dynamic adaptation of the peripheral circulation to cold exposure. Microcirculation. 2012;19(1):65–77.
54. Cheung SS. Responses of the hands and feet to cold exposure. Temperature (Austin). 2015;2(1):105–20.
55. Daanen HAM, Ducharme MB. Finger cold-induced vasodilation during mild hypothermia, hyperthermia and at thermoneutrality. Aviat Space Environ Med. 1999;70(12):1206–10.
56. Takano N, Kotani M. Influence of food intake on cold-induced vasodilatation of finger. Jpn J Physiol. 1989;39(5):755–65.
57. Hodges GJ, Mallette MM, Cheung SS. Cutaneous neural activity and endothelial involvement in cold-induced vasodilatation. Eur J Appl Physiol. 2018;118(5):971–8.
58. Wakabayashi H, Sugiyama K, Suzuki S, Sakihama Y, Hashimoto M, Barwood MJ. Influence of acute beetroot juice supplementation on cold-induced vasodilation and fingertip rewarming. Eur J Appl Physiol. 2023;123(3):495–507.
59. Daanen HA, Ducharme MB. Axon reflexes in human cold exposed fingers. Eur J Appl Physiol. 2000;81(3):240–4.
60. Gardner CA, Webb RC. Cold-induced vasodilatation in isolated, perfused rat tail artery. Am J Phys. 1986;251(1 Pt 2):H176–81.
61. Daanen HA, Van de Linde FJ, Romet TT, Ducharme MB. The effect of body temperature on the hunting response of the middle finger skin temperature. Eur J Appl Physiol Occup Physiol. 1997;76(6):538–43.
62. Flouris AD, Westwood DA, Mekjavic IB, Cheung SS. Effect of body temperature on cold induced vasodilation. Eur J Appl Physiol. 2008;104(3):491–9.
63. Fernandez N, Monge L, Garcia-Villalon AL, Garcia JL, Gomez B, Dieguez G. Cooling effects on nitric oxide production by rabbit ear and femoral arteries during cholinergic stimulation. Br J Pharmacol. 1994;113(2):550–4.
64. Sera T, Kohno T, Nakashima Y, Uesugi M, Kudo S. Low-frequency oscillations of finger skin blood flow during the initial stage of cold-induced vasodilation at different air temperatures. J Physiol Anthropol. 2020;39(1):37.
65. Yasukochi Y, Sera T, Kohno T, Nakashima Y, Uesugi M, Kudo S. Cold-induced vasodilation response in a Japanese cohort: insights from cold-water immersion and genome-wide association studies. J Physiol Anthropol. 2023;42(1):2.
66. Wickham KA, Steele SW, Cheung SS. Effects of acute dietary nitrate supplementation on cold-induced vasodilation in healthy males. Eur J Appl Physiol. 2021;121(5):1431–9.
67. Maley MJ, Eglin CM, House JR, Tipton MJ. The effect of ethnicity on the vascular responses to cold exposure of the extremities. Eur J Appl Physiol. 2014;114(11):2369–79.

68. Imray C, Grieve A, Dhillon S, Caudwell Xtreme Everest Research G. Cold damage to the extremities: frostbite and non-freezing cold injuries. Postgrad Med J. 2009;85(1007):481–8.
69. Krog J, Folkow B, Fox RH, Andersen KL. Hand circulation in the cold of Lapps and North Norwegian fisherman. J Appl Physiol. 1960;15:654–8.
70. Leblanc J, Hildes JA, Heroux O. Tolerance of Gaspe fishermen to cold water. J Appl Physiol. 1960;15:1031–4.
71. Tanaka M. Experimental studies on human reaction to cold. Vascular hunting reaction of workers to cold. Bull Tokyo Med Dent Univ. 1971;18(3):169–77.
72. Itoh S, Kuroshima A, Hiroshige T, Doi K. Finger temperature responses to local cooling in several groups of subjects in Hokkaido. Jpn J Physiol. 1970;20(4):370-+.
73. Lee JY, Park J, Koh E, Cha S. Finger cold-induced vasodilation of older Korean female divers, haenyeo: effects of chronic cold exposure and aging. Int J Biometeorol. 2017;61(7):1299–307.
74. Paik KS, Kang BS, Han DS, Rennie DW, Hong SK. Vascular responses of Korean ama to hand immersion in cold water. J Appl Physiol. 1972;32(4):446–50.
75. Lee JY, Bakri I, Matsuo A, Tochihara Y. Cold-induced vasodilation and vasoconstriction in the finger of tropical and temperate indigenes. J Therm Biol. 2013;38(2):70–8.
76. Khatun A, Ashikaga S, Nagano H, Hasib MA, Taimura A. Cold-induced vasodilation comparison between Bangladeshi and Japanese natives. J Physiol Anthropol. 2016;35
77. Geurts CL, Sleivert GG, Cheung SS. Local cold acclimation during exercise and its effect on neuromuscular function of the hand. Appl Physiol Nutr Metab. 2006;31(6):717–25.
78. Adams T, Smith RE. Effect of chronic local cold exposure on finger temperature responses. J Appl Physiol. 1962;17:317–22.
79. Geurts CL, Sleivert GG, Cheung SS. Central and peripheral factors in thermal, neuromuscular, and perceptual adaptation of the hand to repeated cold exposures. Appl Physiol Nutr Metab. 2006;31(2):110–7.
80. Livingstone SD. Changes in cold-induced vasodilation during Arctic exercises. J Appl Physiol. 1976;40(3):455–7.
81. O'Brien C, Young AJ, Lee DT, Shitzer A, Sawka MN, Pandolf KB. Role of core temperature as a stimulus for cold acclimation during repeated immersion in 20 degrees C water. J Appl Physiol (1985). 2000;89(1):242–50.
82. Geurts CL, Sleivert GG, Cheung SS. Local cold acclimation of the hand impairs thermal responses of the finger without improving hand neuromuscular function. Acta Physiol Scand. 2005;183(1):117–24.
83. Wakabayashi H, Wijayanto T, Kuroki H, Lee JY, Tochihara Y. The effect of repeated mild cold water immersions on the adaptation of the vasomotor responses. Int J Biometeorol. 2012;56(4):631–7.
84. Daanen HA, Koedam J, Cheung SS. Trainability of cold induced vasodilatation in fingers and toes. Eur J Appl Physiol. 2012;112(7):2595–601.
85. Mekjavic IB, Dobnikar U, Kounalakis SN, Musizza B, Cheung SS. The trainability and contralateral response of cold-induced vasodilatation in the fingers following repeated cold exposure. Eur J Appl Physiol. 2008;104(2):193–9.
86. Daanen HA, Van Marken Lichtenbelt WD. Human whole body cold adaptation. Temperature (Austin). 2016;3(1):104–18.
87. Keatinge WR, Evans M. The respiratory and cardiovascular response to immersion in cold and warm water. Q J Exp Physiol Cogn Med Sci. 1961;46:83–94.
88. Keatinge WR, McIlroy MB, Goldfien A. Cardiovascular responses to ice-cold showers. J Appl Physiol. 1964;19:1145–50.
89. Tipton MJ. The initial responses to cold-water immersion in man. Clin Sci. 1989;77:581–8.
90. Datta A, Tipton M. Respiratory responses to cold water immersion: neural pathways, interactions, and clinical consequences awake and asleep. J Appl Physiol (1985). 2006;100(6):2057–64.
91. Barwood MJ, Eglin C, Hills SP, Johnston N, Massey H, McMorris T, et al. Habituation of the cold shock response: a systematic review and meta-analysis. J Therm Biol. 2024;119:103775.
92. Tipton MJ, Collier N, Massey H, Corbett J, Harper M. Cold water immersion: kill or cure? Exp Physiol. 2017;102(11):1335–55.
93. Tipton MJ, Stubbs DA, Elliott DH. Human initial responses to immersion in cold water at three temperatures and after hyperventilation. J Appl Physiol (1985). 1991;70(1):317–22.
94. Cooper KE, Martin S, Riben P. Respiratory and other responses in subjects immersed in cold water. J Appl Physiol. 1976;40(6):903–10.
95. Hayward JS, French CD. Hyperventilation response to cold water immersion: reduction by staged entry. Aviat Space Environ Med. 1989;60(12):1163–5.
96. Tipton MJ, Vincent MJ. Protection provided against the initial responses to cold immersion by a partial coverage wet suit. Aviat Space Environ Med. 1989;60(8):769–73.
97. Barwood MJ, Dalzell J, Datta AK, Thelwell RC, Tipton MJ. Breath-hold performance during cold water immersion: effects of psychological skills training. Aviat Space Environ Med. 2006;77(11):1136–42.
98. Meddings DR, Scarr JP, Larson K, Vaughan J, Krug EG. Drowning prevention: turning the tide on a leading killer. Lancet Public Health. 2021;6(9):e692–5.
99. Golden FS, Tipton MJ, Scott RC. Immersion, near-drowning and drowning. Br J Anaesth. 1997;79(2):214–25.

100. Shattock MJ, Tipton MJ. 'Autonomic conflict': a different way to die during cold water immersion? J Physiol. 2012;590(14):3219–30.
101. Winter J, Tipton MJ, Shattock MJ. Autonomic conflict exacerbates long QT associated ventricular arrhythmias. J Mol Cell Cardiol. 2018;116:145–54.
102. Barwood MJ, Bates V, Long G, Tipton MJ. "Float first:" trapped air between clothing layers significantly improves buoyancy on water after immersion. Int J Aquat Res Educ. 2011;5(2)
103. Barwood MJ, Burrows H, Cessford J, Goodall S. "Float first and kick for your life": psychophysiological basis for safety behaviour on accidental short-term cold water immersion. Physiol Behav. 2016;154:83–9.
104. Mekjavic IB, Bligh J. The increased oxygen-uptake upon immersion—the raised external-pressure could be a causative factor. Eur J Appl Physiol Occup Physiol. 1989;58(5):556–62.
105. Tipton MJ, Golden FS. The influence of regional insulation on the initial responses to cold immersion. Aviat Space Environ Med. 1987;58(12):1192–6.
106. Burke WE, Mekjavic IB. Estimation of regional cutaneous cold sensitivity by analysis of the gasping response. J Appl Physiol (1985). 1991;71(5):1933–40.
107. Goode RC, Duffin J, Miller R, Romet TT, Chant RW, Ackles K. Sudden cold water immersion. Respir Physiol. 1975;23(3):301–10.
108. Johnson DG, Hayward JS, Jacobs TP, Collis ML, Eckerson JD, Williams RH. Plasma norepinephrine responses of man in cold water. J Appl Physiol Respir Environ Exerc Physiol. 1977;43(2):216–20.
109. Frey MA, Selm EA, Walther JW Jr. Reflex cardiovascular responses to cold exposure of the face or foot. Jpn Heart J. 1980;21(5):665–79.
110. Suess WM, Alexander AB, Smith DD, Sweeney HW, Marion RJ. The effects of psychological stress on respiration: a preliminary study of anxiety and hyperventilation. Psychophysiology. 1980;17(6):535–40.
111. Andersson J, Schagatay E, Gislén A, Holm B. Cardiovascular responses to cold-water immersions of the forearm and face, and their relationship to apnoea. Eur J Appl Physiol. 2000;83(6):566–72.
112. Golden FS, Tipton MJ. Human adaptation to repeated cold immersions. J Physiol. 1988;396:349–63.
113. Tipton MJ, Wakabayashi H, Barwood MJ, Eglin CM, Mekjavic IB, Taylor NA. Habituation of the metabolic and ventilatory responses to cold-water immersion in humans. J Therm Biol. 2013;38(1):24–31.
114. Tipton MJ, Eglin CM, Golden FS. Habituation of the initial responses to cold water immersion in humans: a central or peripheral mechanism? J Physiol. 1998;512(Pt 2):621–8.
115. Tipton MJ, Golden FS, Higenbottam C, Mekjavic IB, Eglin CM. Temperature dependence of habituation of the initial responses to cold-water immersion. Eur J Appl Physiol Occup Physiol. 1998;78(3):253–7.
116. Tipton MJ, Mekjavic IB, Eglin CM. Permanence of the habituation of the initial responses to cold-water immersion in humans. Eur J Appl Physiol. 2000;83(1):17–21.
117. Barwood MJ, Datta AK, Thelwell RC, Tipton MJ. Breath-hold time during cold water immersion: effects of habituation with psychological training. Aviat Space Environ Med. 2007;78(11):1029–34.
118. Eglin CM, Tipton MJ. Repeated cold showers as a method of habituating humans to the initial responses to cold water immersion. Eur J Appl Physiol. 2005;93(5-6):624–9.
119. Barwood MJ, Corbett J, Tipton M, Wagstaff C, Massey H. Habituation of the cold shock response is inhibited by repeated anxiety: implications for safety behaviour on accidental cold water immersions. Physiol Behav. 2017;174:10–7.
120. Hensel H, Banet M. Thermoreceptor activity, insulative and metabolic changes in cold and warm adapted cats. In: Houdas Y, Guideu J, editors. New trends in thermal physiology. Paris: Masson; 1978. p. 53–5.
121. Hensel H, Schafer K. Static and dynamic activity of cold receptors in cats after long-term exposure to various temperatures. Pflug Arch Eur J Phy. 1982;392(3):291–4.
122. Kozyreva TV. Adaptive changes in temperature sensitivity in humans under the conditions of cold, heat, and prolonged exercise. Hum Physiol. 2006;32(6):721–5.
123. Keatinge WR, Nadel JA. Immediate respiratory response to sudden cooling of the skin. J Appl Physiol. 1965;20:65–9.
124. Glaser EM, Griffin JP. Influence of the cerebral cortex on habituation. J Physiol. 1962;160(3):429–45.
125. Griffin JP. The role of the frontal areas of the cortex upon habituation in man. Clin Sci. 1963;24:127–34.
126. Fechir M, Klega A, Buchholz HG, Pfeifer N, Balon S, Schlereth T, et al. Cortical control of thermoregulatory sympathetic activation. Eur J Neurosci. 2010;31(11):2101–11.
127. IUPS TC. Glossary of terms for thermal physiology, Third edition. Jpn J Physiol. 2001;51(2):245–85.
128. Hammel HT. Terrestrial animals in cold: recent studies of primitive man. In: Dill DB, Adolph EF, editors. Handbook of physiology. Washington, DC: American Physiological Society; 1964. p. 413–34.
129. LeBlanc J. Adaptation of man to cold. In: Wang LCH, Hudson JW, editors. Strategies in cold. New York: Academic; 1978. p. 695–715.
130. Leblanc J. Factors affecting cold acclimation and thermogenesis in man. Med Sci Sports Exerc. 1988;20(5 Suppl):S193–6.
131. Young AJ. Homeostatic responses to prolonged cold exposure: human cold acclimatization. In: Fregly MJ, Blatteis CM, editors. Handbook of physiology, section 4: environmental physiology. Oxford University Press; 1996. p. 419–38.
132. Tipton MJ, Pandolf KB, Sawka MN, Werner J, Taylor NA. Physiological adaptation to hot and cold

environments. In: Taylor NAS, Groeller H, editors. Physiological bases of human performance during work and exercise. Churchill Livingstone Elsevier; 2008. p. 379–400.
133. Castellani JW, Young AJ. Human physiological responses to cold exposure: acute responses and acclimatization to prolonged exposure. Auton Neurosci. 2016;196:63–74.
134. Hammel HT, Elsner RW, Lemessurier DH, Andersen HT, Milan FA. Thermal and metabolic responses of the Australian aborigine exposed to moderate cold in summer. J Appl Physiol. 1959;14(4):605–15.
135. Andersen KL, Loyning Y, Nelms JD, Wilson O, Fox RH, Bolstad A. Metabolic and thermal response to a moderate cold exposure in nomadic Lapps. J Appl Physiol. 1960;15:649–53.
136. Rennie DW, Covino BG, Howell BJ, Song SH, Kang BS, Hong SK. Physical insulation of Korean diving women. J Appl Physiol. 1962;17:961–6.
137. Hong SK, Rennie DW, Park YS. Cold acclimatization and deacclimatization of Korean women divers. Exerc Sport Sci Rev. 1986;14:231–68.
138. Hanna JM, Hong SK. Critical water temperature and effective insulation in scuba divers in Hawaii. J Appl Physiol. 1972;33(6):770–3.
139. Park YS, Rennie DW, Lee IS, Park YD, Paik KS, Kang DH, et al. Time course of deacclimatization to cold water immersion in Korean women divers. J Appl Physiol Respir Environ Exerc Physiol. 1983;54(6):1708–16.
140. Veicsteinas A, Rennie DW. Thermal insulation and shivering threshold in Greek sponge divers. J Appl Physiol Respir Environ Exerc Physiol. 1982;52(4):845–50.
141. Hart JS, Sabean HB, Hildes JA, Depocas F, Hammel HT, Andersen KL, et al. Thermal and metabolic responses of coastal Eskimos during a cold night. J Appl Physiol. 1962;17:953–60.
142. Leonard WR, Sorensen MV, Galloway VA, Spencer GJ, Mosher MJ, Osipova L, et al. Climatic influences on basal metabolic rates among circumpolar populations. Am J Hum Biol. 2002;14(5):609–20.
143. Kang BS, Song SH, Suh CS, Hong SK. Changes in body temperature and basal metabolic rate of the Ama. Wilderness Environ Med. 2003;14(3):193.
144. Cannon B, Nedergaard J. Brown adipose tissue: function and physiological significance. Physiol Rev. 2004;84(1):277–359.
145. Hemingway A, Price WM, Stuart D. The calorigenic action of catecholamines in warm acclimated and cold acclimated non-shivering cats. Int J Neuropharmacol. 1964;3(5):495–503.
146. LeBlanc J, Robinson D, Sharman DF, Tousignant P. Catecholamines and short-term adaptation to cold in mice. Am J Phys. 1967;213(6):1419–22.
147. Kang BS, Han DS, Paik KS, Park YS, Kim JK, Kim CS, et al. Calorigenic action of norepinephrine in the Korean women divers. J Appl Physiol. 1970;29(1):6–9.
148. Itoh S, Doi K, Kuroshima A. Enhanced sensitivity to noradrenaline of the Ainu. Int J Biometeorol. 1970;14(2):195-200.
149. Levy SB, Klimova TM, Zakharova RN, Fedorov AI, Fedorova VI, Baltakhinova ME, et al. Brown adipose tissue thermogenesis among young adults in northeastern Siberia and Midwest United States and its relationship with other biological adaptations to cold climates. Am J Hum Biol. 2022;34(6)
150. Scholander PF, Hammel HT, Hart JS, Lemessurier DH, Steen J. Cold adaptation in Australian aborigines. J Appl Physiol. 1958;13(2):211–8.
151. Hong SK. Pattern of cold adaptation in women divers of Korea (ama). Fed Proc. 1973;32(5):1614–22.
152. Hong SK, Lee CK, Kim JK, Song SH, Kennie DW. Peripheral blood flow and heat flux of Korean women divers. Fed Proc. 1969;28(3):1143–8.
153. Budd GM, Brotherhood JR, Beasley FA, Hendrie AL, Jeffery SE, Lincoln GJ, et al. Effects of acclimatization to cold baths on men's responses to whole-body cooling in air. Eur J Appl Physiol Occup Physiol. 1993;67(5):438–49.
154. Jansky L, Janakova H, Ulicny B, Sramek P, Hosek V, Heller J, et al. Changes in thermal homeostasis in humans due to repeated cold water immersions. Pflug Arch Eur J Phy. 1996;432(3):368–72.
155. Bittel JH. Heat debt as an index for cold adaptation in men. J Appl Physiol. 1987;62(4):1627–34.
156. Scholander PF, Hammel HT, Andersen KL, Loyning Y. Metabolic acclimation to cold in man. J Appl Physiol. 1958;12(1):1–8.
157. Budd GM, Brotherhood JR, Thomas DW, Beasley FA, Hendrie AL, Jeffery SE, et al. Cardiovascular and metabolic responses to noradrenaline in men acclimatized to cold baths. Eur J Appl Physiol Occup Physiol. 1993;67(5):450–6.
158. Jansky L, Sramek P, Savlikova J, Ulicny B, Janakova H, Horky K. Change in sympathetic activity, cardiovascular functions and plasma hormone concentrations due to cold water immersion in men. Eur J Appl Physiol. 1996;74(1-2):148–52.
159. Young AJ, Muza SR, Sawka MN, Gonzalez RR, Pandolf KB. Human thermoregulatory responses to cold air are altered by repeated cold water immersion. J Appl Physiol (1985). 1986;60(5):1542–8.
160. Heaton JM. The distribution of brown adipose tissue in the human. J Anat. 1972;112(Pt 1):35–9.
161. Adams T, Covino BG. Racial variations to a standardized cold stress. J Appl Physiol. 1958;12(1):9–12.
162. Leblanc J, Pouliot M, Rheaume S. Thermal balance and biogenic amine excretion in Gasp'e fishermen exposed to cold. J Appl Physiol. 1964;19:9–12.
163. Castellani JW, Tipton MJ. Cold stress effects on exposure tolerance and exercise performance. Compr Physiol. 2015;6(1):443–69.
164. Heus R, Daanen HA, Havenith G. Physiological criteria for functioning of hands in the cold: a review. Appl Ergon. 1995;26(1):5–13.

165. Racinais S, Oksa J. Temperature and neuromuscular function. Scand J Med Sci Sports. 2010;20(Suppl 3):1–18.
166. Wakabayashi H, Oksa J, Tipton MJ. Exercise performance in acute and chronic cold exposure. J Phys Fitness Sports Med. 2015;4(2):177–85.
167. Wakabayashi H, Sakaue H, Sengoku Y, Takagi H. Physiological mechanism of performance impairment during exercise with hypothermic skeletal muscle and application for exercise training (in Japanese). Jpn J Exerc Sports Physiol. 2020;27(1):9–13.
168. Wakabayashi H, Osawa M, Koga S, Li K, Sakaue H, Sengoku Y, et al. Effects of muscle cooling on kinetics of pulmonary oxygen uptake and muscle deoxygenation at the onset of exercise. Physiol Rep. 2018;6(21):e13910.
169. Sakaue H, Wakabayashi H, Sengoku Y, Takagi H, Nabekura Y. Enhanced glycolytic metabolism and skeletal muscle hypoxia during sub-maximum exercise in cold water environment (in Japanese). Jpn J Physiol Anthropol. 2023;28(3):57–69.
170. Sakaue H, Sengoku Y, Takagi H, Nabekura Y, Wakabayashi H. Improvement of glycolysis metabolic capacity by exercise training under local muscle hypoxia in a cold water environment (in Japanese). Jpn J Phys Fitness Sports Med. 2024;73(4):157–68.
171. Clarke RS, Hellon RF, Lind AR. The duration of sustained contractions of the human forearm at different muscle temperatures. J Physiol. 1958;143(3):454–73.
172. Davies CT, Mecrow IK, White MJ. Contractile properties of the human triceps surae with some observations on the effects of temperature and exercise. Eur J Appl Physiol Occup Physiol. 1982;49(2):255–69.
173. Bergh U, Ekblom B. Influence of muscle temperature on maximal muscle strength and power output in human skeletal muscles. Acta Physiol Scand. 1979;107(1):33–7.
174. Oksa J. Neuromuscular performance limitations in cold. Int J Circumpolar Health. 2002;61(2):154–62.
175. Oksa J, Rintamaki H, Rissanen S. Muscle performance and electromyogram activity of the lower leg muscles with different levels of cold exposure. Eur J Appl Physiol Occup Physiol. 1997;75(6):484–90.
176. Oksa J, Rintamaki H, Makinen T, Hassi J, Rusko H. Cooling-induced changes in muscular performance and EMG activity of agonist and antagonist muscles. Aviat Space Environ Med. 1995;66(1):26–31.
177. Crowley GC, Garg A, Lohn MS, Van Someren N, Wade AJ. Effects of cooling the legs on performance in a standard Wingate anaerobic power test. Br J Sports Med. 1991;25(4):200–3.
178. Nadel ER, Holmer I, Bergh U, Astrand PO, Stolwijk JA. Energy exchanges of swimming man. J Appl Physiol. 1974;36(4):465–71.
179. Fujimoto T, Sasaki Y, Wakabayashi H, Sengoku Y, Tsubakimoto S, Nishiyasu T. Maximal workload but not peak oxygen uptake is decreased during immersed incremental exercise at cooler temperatures. Eur J Appl Physiol. 2016;116(9):1819–27.
180. Meigal AY, Oksa J, Hohtola E, Lupandin YV, Rintamaki H. Influence of cold shivering on fine motor control in the upper limb. Acta Physiol Scand. 1998;163(1):41–7.
181. Holmer I, Bergh U. Metabolic and thermal response to swimming in water at varying temperatures. J Appl Physiol. 1974;37(5):702–5.
182. Tipton MJ, Eglin C, Gennser M, Golden F. Immersion deaths and deterioration in swimming performance in cold water. Lancet. 1999;354(9179):626–9.
183. Shiojiri T, Shibasaki M, Aoki K, Kondo N, Koga S. Effects of reduced muscle temperature on the oxygen uptake kinetics at the start of exercise. Acta Physiol Scand. 1997;159(4):327–33.
184. Koga S, Kano Y, Barstow TJ, Ferreira LF, Ohmae E, Sudo M, et al. Kinetics of muscle deoxygenation and microvascular PO(2) during contractions in rat: comparison of optical spectroscopy and phosphorescence-quenching techniques. J Appl Physiol (1985). 2012;112(1):26–32.
185. Koga S, Rossiter HB, Heinonen I, Musch TI, Poole DC. Dynamic heterogeneity of exercising muscle blood flow and O2 utilization. Med Sci Sports Exerc. 2014;46(5):860–76.
186. Rennie D, Park Y, Veicsteinas A, Pendergast D. Metabolic and circulatory adaptation to cold water stress. In: Cerretelli P, Whipp BJ, editors. Exercise bioenergetics and gas exchange. Amsterdam: Elsevier/North-Holland Biomedical Press; 1980. p. 315–21.
187. Thorsson O, Lilja B, Ahlgren L, Hemdal B, Westlin N. The effect of local cold application on intramuscular blood flow at rest and after running. Med Sci Sports Exerc. 1985;17(6):710–3.
188. Gregson W, Black MA, Jones H, Milson J, Morton J, Dawson B, et al. Influence of cold water immersion on limb and cutaneous blood flow at rest. Am J Sports Med. 2011;39(6):1316–23.
189. Enander A. Performance and sensory aspects of work in cold environments: a review. Ergonomics. 1984;27(4):365–78.
190. Tanaka M, Tochihara Y, Yamazaki S, Ohnaka T, Yoshida K. Thermal-reaction and manual performance during cold-exposure while wearing cold-protective clothing. Ergonomics. 1983;26(2):141–9.
191. De Jong RH, Hershey WN, Wagman IH. Nerve conduction velocity during hypothermia in man. Anesthesiology. 1966;27(6):805–10.
192. Morton R, Provins KA. Finger numbness after acute local exposure to cold. J Appl Physiol. 1960;15:149–54.
193. Oosterveld FG, Rasker JJ, Jacobs JW, Overmars HJ. The effect of local heat and cold therapy on the intraarticular and skin surface temperature of the knee. Arthritis Rheum. 1992;35(2):146–51.

194. Hunter J, Kerr EH, Whillans MG. The relation between joint stiffness upon exposure to cold and the characteristics of synovial fluid. Can J Med Sci. 1952;30(5):367.
195. Lockhart JM, Kiess HO, Clegg TJ. Effect of rate and level of lowered finger surface temperature on manual performance. J Appl Psychol. 1975;60(1):106–13.
196. Ozaki H, Nagai Y, Tochibara Y. Physiological responses and manual performance in humans following repeated exposure to severe cold at night. Eur J Appl Physiol. 2001;84(4):343–9.
197. Kim TG, Tochihara Y, Fujita M, Hashiguchi N. Physiological responses and performance of loading work in a severely cold environment. Int J Ind Ergon. 2007;37(9-10):725–32.
198. Koga S, Barstow TJ, Okushima D, Rossiter HB, Kondo N, Ohmae E, et al. Validation of a high-power, time-resolved, near-infrared spectroscopy system for measurement of superficial and deep muscle deoxygenation during exercise. J Appl Physiol (1985). 2015;118(11):1435–42.
199. Wakabayashi H, Nishimura T, Wijayanto T, Watanuki S, Tochihara Y. Effect of repeated forearm muscle cooling on the adaptation of skeletal muscle metabolism in humans. Int J Biometeorol. 2017;61(7):1261–7.
200. Severinghaus JW. Oxyhemoglobin dissociation curve correction for temperature and pH variation in human blood. J Appl Physiol. 1958;12(3):485–6.
201. Astrup P, Engel K, Severinghaus JW, Munson E. The influence of temperature and pH on the dissociation curve of oxyhemoglobin of human blood. Scand J Clin Lab Invest. 1965;17(6):515–23.
202. Bigland-Ritchie B, Donovan EF, Roussos CS. Conduction velocity and EMG power spectrum changes in fatigue of sustained maximal efforts. J Appl Physiol Respir Environ Exerc Physiol. 1981;51(5):1300–5.
203. Mucke R, Heuer D. Behaviour of EMG-parameters and conduction velocity in contractions with different muscle temperatures. Biomed Biochim Acta. 1989;48(5-6):S459–64.
204. Petrofsky J, Laymon M. Muscle temperature and EMG amplitude and frequency during isometric exercise. Aviat Space Environ Med. 2005;76(11):1024–30.
205. Wakabayashi H, Wijayanto T, Tochihara Y. Neuromuscular function during knee extension exercise after cold water immersion. J Physiol Anthropol. 2017;36(1):28.
206. Mallette MM, Green LA, Gabriel DA, Cheung SS. The effects of local forearm muscle cooling on motor unit properties. Eur J Appl Physiol. 2018;118(2):401–10.
207. Ranatunga KW. The force-velocity relation of rat fast- and slow-twitch muscles examined at different temperatures. J Physiol. 1984;351(1):517–29.
208. Rome LC, Loughna PT, Goldspink G. Muscle fiber activity in carp as a function of swimming speed and muscle temperature. Am J Phys. 1984;247(2 Pt 2):R272–9.
209. Rome LC, Choi IH, Lutz G, Sosnicki A. The influence of temperature on muscle function in the fast swimming scup. I. Shortening velocity and muscle recruitment during swimming. J Exp Biol. 1992;163(1):259–79.
210. Bae KA, An NY, Kwon YW, Kim C, Yoon CS, Park SC, et al. Muscle fibre size and capillarity in Korean diving women. Acta Physiol Scand. 2003;179(2):167–72.
211. Walters TJ, Constable SH. Intermittent cold exposure causes a muscle-specific shift in the fiber type composition in rats. J Appl Physiol (1985). 1993;75(1):264–7.
212. Desplanches D, Hoppeler H, Linossier MT, Denis C, Claassen H, Dormois D, et al. Effects of training in normoxia and normobaric hypoxia on human muscle ultrastructure. Pflugers Arch. 1993;425(3–4):263–7.
213. Mizuno M, Juel C, Brorasmussen T, Mygind E, Schibye B, Rasmussen B, et al. Limb skeletal-muscle adaptation in athletes after training at altitude. J Appl Physiol. 1990;68(2):496–502.
214. Suzuki J, Gao M, Ohinata H, Kuroshima A, Koyama T. Chronic cold exposure stimulates microvascular remodeling preferentially in oxidative muscles in rats. Jpn J Physiol. 1997;47(6):513–20.
215. Deveci D, Egginton S. Differing mechanisms of cold-induced changes in capillary supply in m. tibialis anterior of rats and hamsters. J Exp Biol. 2002;205(Pt 6):829–40.
216. Sillau AH, Aquin L, Lechner AJ, Bui MV, Banchero N. Increased capillary supply in skeletal muscle of Guinea pigs acclimated to cold. Respir Physiol. 1980;42(3):233–45.
217. Kime R, Karlsen T, Nioka S, Lech G, Madsen O, Saeterdal R, et al. Discrepancy between cardiorespiratory system and skeletal muscle in elite cyclists after hypoxic training. Dyn Med. 2003;2(1):4.
218. Wang JS, Wu MH, Mao TY, Fu TC, Hsu CC. Effects of normoxic and hypoxic exercise regimens on cardiac, muscular, and cerebral hemodynamics suppressed by severe hypoxia in humans. J Appl Physiol (1985). 2010;109(1):219–29.
219. Makinen TM. Human cold exposure, adaptation, and performance in high latitude environments. Am J Hum Biol. 2007;19(2):155–64.
220. Martin K, McLeod E, Periard J, Rattray B, Keegan R, Pyne DB. The impact of environmental stress on cognitive performance: a systematic review. Hum Factors. 2019;61(8):1205–46.
221. Enander AE. Effects of thermal stress on human performance. Scand J Work Environ Health. 1989;15(Suppl 1):27–33.
222. Pilcher JJ, Nadler E, Busch C. Effects of hot and cold temperature exposure on performance: a meta-analytic review. Ergonomics. 2002;45(10):682–98.

223. Falla M, Micarelli A, Hüfner K, Strapazzon G. The effect of cold exposure on cognitive performance in healthy adults: a systematic review. Int J Environ Res Public Health. 2021;18(18)
224. FitzGibbon T, Hayward JS, Walker D. EEG and visual evoked potentials of conscious man during moderate hypothermia. Electroencephalogr Clin Neurophysiol. 1984;58(1):48–54.
225. Coleshaw SR, Van Someren RN, Wolff AH, Davis HM, Keatinge WR. Impaired memory registration and speed of reasoning caused by low body temperature. J Appl Physiol Respir Environ Exerc Physiol. 1983;55(1 Pt 1):27–31.
226. Giesbrecht GG, Arnett JL, Vela E, Bristow GK. Effect of task complexity on mental performance during immersion hypothermia. Aviat Space Environ Med. 1993;64(3 Pt 1):206–11.
227. Palinkas LA. Mental and cognitive performance in the cold. Int J Circumpolar Health. 2001;60(3):430–9.
228. Ellis HD, Wilcock SE, Zaman SA. Cold and performance: the effects of information load, analgesics, and the rate of cooling. Aviat Space Environ Med. 1985;56(3):233–7.
229. Enander A. Effects of moderate cold on performance of psychomotor and cognitive tasks. Ergonomics. 1987;30(10):1431–45.
230. Teichner WH. Reaction-time in the cold. J Appl Psychol. 1958;42(1):54–9.
231. Ellis HD. The effects of cold on the performance of serial choice reaction-time and various discrete tasks. Hum Factors. 1982;24(5):589–98.
232. Van Orden KF, Ahlers ST, Thomas JR, House JF, Schrot J. Moderate cold exposure shortens evoked potential latencies in humans. Aviat Space Environ Med. 1990;61(7):636–9.
233. Thomas JR, Ahlers ST, House JF, Schrot J. Repeated exposure to moderate cold impairs matching-to-sample performance. Aviat Space Environ Med. 1989;60(11):1063–7.
234. Mäkinen TM, Palinkas LA, Reeves DL, Pääkkönen T, Rintamäki H, Leppäluoto J, et al. Effect of repeated exposures to cold on cognitive performance in humans. Physiol Behav. 2006;87(1):166–76.
235. Jones DM, Bailey SP, Roelands B, Buono MJ, Meeusen R. Cold acclimation and cognitive performance: a review. Auton Neurosci. 2017;208:36–42.

6 Physiological Responses and Performance in Hot Environments

Titis Wijayanto

6.1 Physiological Responses to Hot Environments

In a thermoneutral environment and at rest, the human body maintains a temperature of approximately 37 °C by balancing heat production with heat exchange with the environment. This temperature setpoint is not a steady state or constant but a dynamic equilibrium subject to continuous adjustment [1]. The body temperature will increase when exposed to heat stress from environmental factors such as temperature, humidity, solar radiation, physical exertion, and clothing insulation. In response, the human body initiates a process known as thermoregulation to regulate its internal body temperature by adjusting the heat transfer between the body and the environment to achieve heat balance and maintain a relatively constant body temperature of 37 °C.

The thermoregulation system can be broadly divided into behavioral and physiological thermoregulation. Behavioral thermoregulation involves voluntary behavioral adjustments, such as turning on an air conditioner or wearing light clothing, to reduce heat gain or increase heat loss. This form of thermoregulation is essential to our everyday lives as we continuously adjust our behavior to adapt to changing environmental conditions.

Physiological thermoregulation, on the other hand, is an involuntary process that operates autonomously and independently of voluntary behavior. Physiological thermoregulation is a complex system that maintains core body temperature within a narrow range. During heat exposure, the body activates heat loss mechanisms like sweating and vasodilation to prevent overheating. The body's heat balance depends on the heat it produces and loses to the environment. When the heat balance is disrupted due to metabolic or environmental factors, the autonomic nervous system triggers the sudomotor function (i.e., activation of sweat gland) and vasomotor function (i.e., redistribution of heat flow from the body core to the skin) to adjust body heat transfer from the body to the environment [2]. To dissipate heat, the body increases blood flow to the skin, which carries heat from the core to the skin surface [3]. Further, it activates the sweating mechanism, releasing moisture that evaporates and cools the skin. The explanation of these two processes will be discussed in the following subchapters.

6.1.1 Vasomotor Responses

Exposure to environmental heat stress impairs the body's ability to dissipate heat, causing an increase in both skin and core body temperature.

T. Wijayanto (✉)
Department of Mechanical and Industrial Engineering, Faculty of Engineering, Universitas Gadjah Mada, Yogyakarta, Indonesia
e-mail: twijaya@ugm.ac.id

Y. Tochihara (ed.), *The Thermal Environment*, https://doi.org/10.1007/978-981-96-3009-7_6

This rise in skin temperature triggers the withdrawal of vasoconstriction and a slight increase in blood flow [4]. For example, warming reduces vasoconstrictor nerve activity when the body starts from a cooler state, whereas when the body is more comfortable, vasoconstrictor activity is minimal or absent, allowing an initial rise in blood flow [4, 5]. As heat stress continues, these early vasomotor adjustments give way to more pronounced responses aimed at preventing overheating.

As skin temperature further increases, the elevated body temperature activates vasodilation significantly increasing skin blood flow. This process is mediated by local responses involving the endothelium, adrenergic system, sensory mechanisms, and thermoregulatory pathways [6, 7]. Increased skin blood flow transfers metabolic heat from the core to the skin, facilitating dry heat dissipation when thermal gradients are favorable [8]. Once sweating starts, the amplified skin blood flow transfers heat to the skin. This heat is then released through radiation, convection, and evaporation [8]. The transition from dry heat loss to evaporative cooling represents the body's escalating efforts to cope with rising temperatures. The efficiency of these mechanisms depends largely on the core-to-skin temperature gradient; as this gradient narrows, skin blood flow increases reflexively [9, 10]. However, skin temperature often rises more than core temperature in hot environments or during exercise. This makes changes in skin temperature a significant factor in the body's drive to dilate blood vessels in the skin and produce sweat.

In extreme heat, where evaporative cooling is less effective due to environmental conditions, the core-to-skin temperature gradient may reverse, with skin temperature exceeding core temperature. Under these conditions, increased skin blood flow can raise core temperature through convective heat gain [11], undermining the ability of the body to regulate its temperature [12]. This transition from beneficial to detrimental heat transfer underscores the complexity of thermoregulatory responses in extreme environments and highlights the limitations of the body's compensatory mechanisms.

Aging significantly impacts the effectiveness of these thermoregulatory responses, particularly vasomotor control [13]. In older adults, maximum skin blood flow is reduced, even when smooth muscle tone in the skin is minimized through prolonged heating [14, 15]. During passive heat stress, the elderly exhibit a diminished capacity for cutaneous vasodilation [16]. This reduced vasodilatory response is partly due to a weaker cardiac output during thermal stress and a decreased ability to redirect blood flow from internal organs to the skin, despite similar increases in core and skin temperatures compared to younger individuals [17]. As the body ages, these weakened responses become even more apparent during physical exertion in hot conditions, with older adults showing smaller increases in skin blood flow and cardiac output than their younger counterparts [18]. The cumulative effects of aging on heat dissipation mechanisms make older adults particularly vulnerable to heat-related illnesses, further emphasizing the importance of understanding how physiological responses to heat change across the lifespan.

6.1.2 Sweating Responses

Heat dissipation from the body to the environment heavily relies on sweat evaporation, especially during exposure to hot environments or intense physical activity. Sweat evaporation is the main avenue of heat loss when a human is exposed to a hot environment [19]. When sweat production is compromised due to excessive physical exertion or prolonged exposure to heat, the body's ability to regulate its temperature declines. This cooling mechanism involves converting water on the skin surface into water vapor, removing 2.4 KJ of heat for each gram of evaporated sweat [20, 21]. The effectiveness of this process depends not only on the rate of sweat secretion by eccrine glands but also on the environment's capacity to absorb moisture from the evaporated sweat [2]. However, as an individual approaches their maximum potential for sweat evaporation, the cooling process becomes less efficient [22]. This sets the stage for exploring how sweat production and control mechanisms contribute to maintaining thermal balance in challenging conditions.

Sweating control involves two key parameters: (1) the central control, which sets the threshold temperature for the onset of sweating, and (2) the peripheral control, which determines the sensitivity of the response, indicated by the relationship between increases in internal or core temperature and sweating [23, 24]. This framework was first established in a study by Nadel et al. [25], which directly correlated the rising sweat rates with the changes in internal temperature in humans. These findings led to the understanding that sweating is regulated by both the internal or mean body temperature at which sweating begins and the degree to which sweating increases as body temperature rises. This detailed regulation highlights how both central and peripheral mechanisms work together to manage heat loss under different conditions.

While changes in mean skin temperature affect sweating through central control, the sweat rate can also be affected by the local temperature of the sweat gland through peripheral control [26]. It is possible for an increase in sweating to be caused by either an increase in the density of activated sweat glands, an increase in the sweat output per gland, or a combination of both these factors [19, 23]. According to Kondo et al. [27], the early rise in sweat rate observed during heat stress is mainly due to the activation of more sweat glands. As heat stress continues, further increases in sweat production are driven by greater output per gland. This nuanced control of sweat production ensures that the body can adapt to increasing heat loads, gradually escalating its efforts to dissipate heat through both the number of glands activated and the amount of sweat produced.

Humans possess two types of sweat glands: eccrine sweat glands and apocrine sweat glands [28, 29]. Eccrine sweat glands are distributed across the body surface. They are primarily responsible for body temperature regulation during exposure to hot environments or physical exercise. In contrast, the apocrine glands, found in specific areas like the axillae and genital region, contribute minimally to cooling [28, 30]. The human body contains between 2 and 4 million eccrine sweat glands, with varying densities across different regions [19, 30]. The density of human eccrine sweat glands is not uniform across the body but varies by region, with the greatest concentrations found on the fingers and toes. According to a review by Taylor and Machado-Moreira [31], the forehead averages about 200 glands/cm^2, while the back, torso, and limbs have lower densities, ranging from 70 to 140 glands/cm^2. The variation in sweat gland density across different body parts may influence, or be affected by, these regions' varying heat dissipation capabilities. Additionally, the overall density of eccrine glands among individuals has been correlated with body surface area, suggesting that the differences in gland density between different body segments are largely a result of how these segments grow during an individual's development [19]. Understanding these variations provides insight into how different body regions contribute to overall heat loss.

Sudomotor function, which includes gland densities, secretion rates, thermal sensitivity, and the timing of sweating onset, also differs among body segments at rest and during exercise [19, 32]. Sweat production can vary substantially even within the same body segment. Studies have identified that sweat distribution is not uniform within the foot, head, and torso [32, 33], and these disparities in sweat output tend to be more pronounced with an increase in core body temperature [32]. These disparities highlight the body's complex and region-specific approach to managing heat stress, where some areas contribute more significantly to cooling than others.

Similar to the decline in vasomotor responses with age, sudomotor function also diminishes as people get older [34, 35]. Inoue and Shibasaki [35] reported that decrements in the active sweat gland density and sweat gland output associated with age may occur after cutaneous vasodilation decreases. The decline in heat dissipation functions tends to start in the lower limbs, advancing next to the back of the upper body, then to the front of the upper body, and finally affecting the head [35] (Fig. 6.1). This pattern indicates that

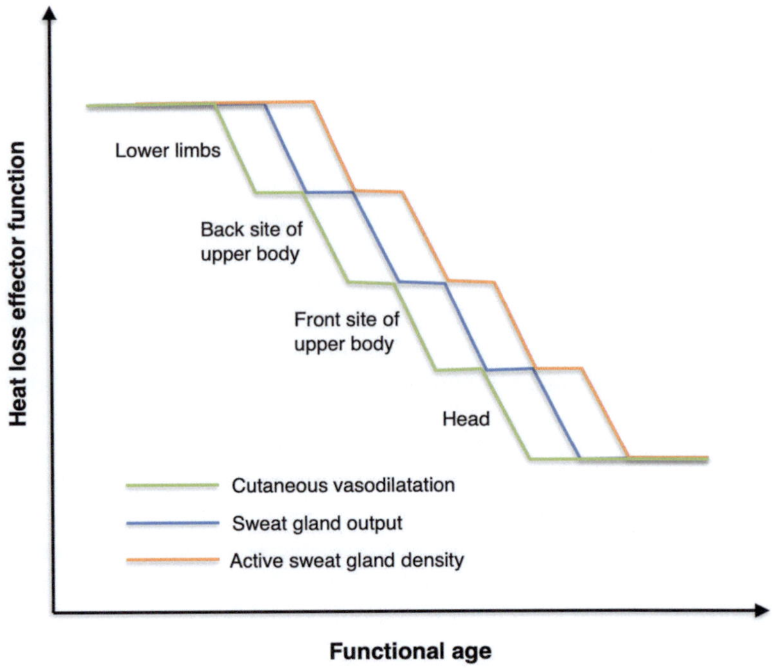

Fig. 6.1 Diagram of the effect of increasing age on sweating and skin blood flow responses. (Adapted from Inoue and Shibasaki [35])

the ability of the body to regulate its temperature through these mechanisms becomes less efficient with age, and the decrement proceeds in a somewhat systematic manner across different body regions.

6.2 Heat Adaptation

Heat adaptation is a major research area within physiological anthropology, focusing on the physiological changes that occur in response to repeated exposure to conditions that elevate skin and core temperatures and induce profuse sweating, whether due to exercise, hot weather, or a combination of both [36, 37]. Repeated exposure to heat, whether in a controlled setting or a natural environment, enhances an individual's ability to tolerate heat stress over time [38, 39]. These adaptive responses can occur in two primary ways: through artificial heat exposure in laboratory settings, known as heat acclimation, or through natural exposure to hot environments, referred to as heat acclimatization. While both terms stem from the same root, they are attributed with different meanings based on the context of exposure [40].

Heat acclimation refers to adaptations resulting from repeated exposure to heat in controlled, experimental settings [40]. This process leads to a range of physiological adjustments that enhance the thermoregulatory capabilities and cardiovascular function, ultimately improving physical performance and reducing the risk of severe heat-related illnesses [37]. These adaptations encompass improved sweating responses [41, 42], enhanced cutaneous blood flow [42, 43], lower core and skin temperatures [44], decreased cardiovascular strain, and improved fluid balance [45]. By improving these physiological responses, heat acclimation enables individuals to perform better in hot conditions while reducing the likelihood of heat stress.

Heat acclimatization, on the other hand, occurs through prolonged exposure to naturally hot environments, leading to adaptations that may differ from those seen with heat acclimation [40]. This process is particularly relevant to individuals who live in tropical regions or have long-term exposure to hot climates. While the general physiological outcomes are similar, such as enhanced thermoregulation, the mechanisms may vary due to the uncontrolled and dynamic nature of natural environments. Understanding these distinctions between heat acclimation and heat acclimatization is essential for appreciating the nuances of how the human body adapts to heat in different contexts. This section will explore heat acclimation and heat acclimatization separately to better understand the differences between the two processes.

6.2.1 Heat Acclimation

Heat acclimation, achieved through controlled environmental conditions, enhances human thermoregulation and tolerance during repeated heat exposures [40]. Various methods exist to induce heat acclimation, including a short-term, high-intensity exercise protocol [46] or a passive approach involving sauna or hot water immersion after exercise [47, 48]. The exercise-induced heat exposure typically involves 90–120 min of treadmill walking in a warm environment over 10–14 days [39, 49].

This adaptation involves complex mechanisms across thermoregulatory and cardiovascular systems. Heat acclimation brings about several significant alterations. Resulting in significant physiological benefits. These adaptations include reductions in core body temperature thresholds for initiating heat loss [38, 50], increased sweat rate and sensitivity [41, 42], and improved cutaneous vascular function [42, 51]. These changes enable the body to manage heat more efficiently, enhancing thermal comfort and physical performance. Early in the acclimation process, alterations in sweat rate and composition are among the first observed responses, indicating the body's immediate effort to optimize cooling [39, 49]. As a result, heat acclimation provides a strong foundation for improved thermoregulation, as summarized in Table 6.1.

Moreover, heat acclimation increases plasma volume [52, 53], stroke volume, and cardiac output [53], improving heat dissipation. These outlined adaptations primarily maintain the core body temperature within an optimal range, enhancing heat dissipation and temperature regulation. The cardiovascular adjustments brought about by heat acclimation enable increased work output, resulting in heightened heat tolerance, thermal comfort, and improved performance in hot conditions for the acclimated individual [51, 54]. Together, these thermoregulatory and cardiovascular adaptations contribute to better heat tolerance, heightened performance, and reduced risk of heat-related illness.

Table 6.1 Physiological adaptation and functional consequences associated with heat acclimation

Adaptation	Consequences	Adaptation	Consequences
Core temperature	Reduced	Cardiovascular stability	Improved
Rest (temperature)—Decreased		Heart rate—Lowered	
Exercise—Decreased		Stroke volume—Increased	
Sweating	Improved	Cardiac output—Better sustained	
Onset threshold—Decreased		Blood pressure—Better sustained	
Rate—Increase		Myocardial compliance—Increased	
Sensitivity—Increased		Myocardial efficiency—Increased	
Skin temperature	Reduced	Cardioprotection—Improved	
Skin blood flow	Improved	Skeletal muscle metabolism	Improved
Onset threshold—Decreased		Muscle glycogen—Spared	
Rate—Increase		Lactate threshold—Increased	
Sensitivity—Increased		Muscle and plasma lactate—Lowered	
Fluid balance	Improved	Muscle force production—Increased	
Thirst—Improved		Whole-body metabolic rate	Lowered
Electrolyte losses—Reduced		Acquired thermal tolerance	Improved
Total body water—Increased		Heat shock proteins expression—Increased	
Plasma volume—Increased		Cytoprotection—Improved	

Adapted from Periard et al. [37]

During the initial phases of heat acclimation, improvements in heart rate, skin temperature, core temperature, and sweat rate are typically observed within the first week [38, 55] (Fig. 6.2). Notably, heart rate adaptation is often the earliest physiological change reported, reflecting the body's quick cardiovascular adjustments to improve stability. A review by Periard et al. [56] suggested that heart rate adaptation occurs within the first 4–5 days of heat acclimation as cardiovascular stability improves. In addition, Hodge et al. [57] also reported an 18% decrease in heart rate over 8 days of acclimation, with similar findings from other studies lasting 10–12 days [58, 59]. While these changes are rapid initially, they continue slowly for approximately one more week. Most improvement in skin and core body temperature also occurs within 8 days of heat acclimation [39], with the thermoregulatory benefit from heat acclimation being approximately 75% developed after the first week and is generally thought to be completed after 10–14 days of acclimation [60]. Pandolf [38] suggests that nearly complete heat acclimation for hot, dry, or hot humid conditions occurs after 7–10 days of exposure.

Acclimation also brings about significant improvements in sweat rate, with increases typically occurring by the second day of exposure [41]. Armstrong and Maresh [39] note that this increase generally develops over 5–14 days of acclimation, indicating improved efficiency in the body cooling processes. This augmentation arises from central and peripheral adaptation [25]. At the central level, heat acclimation shifts the threshold body temperature for sweating, meaning that sweat begins at a lower internal body temperature than before acclimation [37]. At the peripheral level, there is an increase in the volume of sweat produced for each degree of body temperature rise [61]. Alongside these sudomotor adaptations, there's also a shift in cutaneous blood flow regulation. The threshold for the start of skin vasodilation decreases while the responsiveness of blood vessels to heat increases, allowing for better blood flow to the skin for heat dissipation [61] (Fig. 6.3).

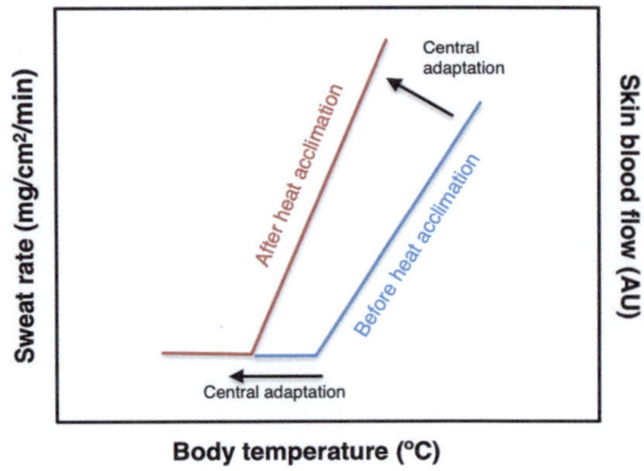

Fig. 6.3 Schematic representation of the central and peripheral adaptations that occur in response to heat acclimation. (Adapted from Periard et al. [37])

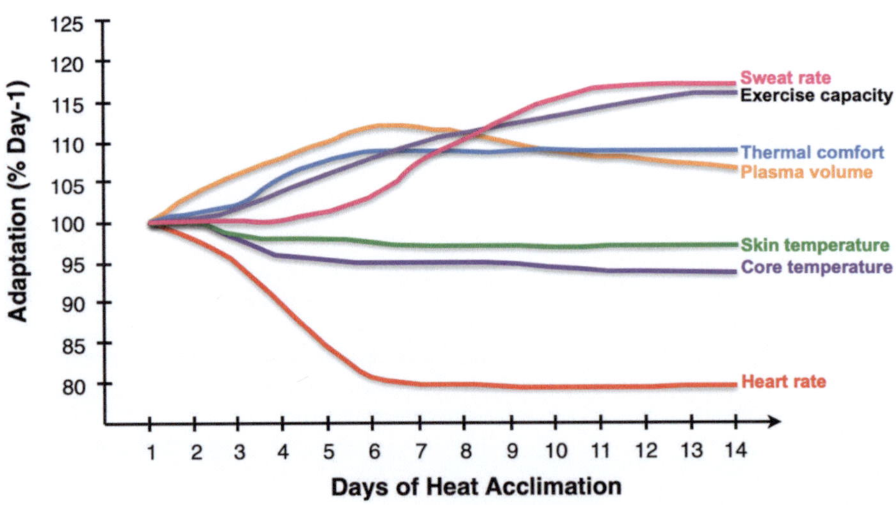

Fig. 6.2 Time course of induction in heat acclimation. (Adapted from Periard et al. [37])

There is some evidence that heat acclimation may also lead to shifts in sweat distribution. A classical heat acclimation study documented shifts in regional sweat redistribution toward the limbs [62]. However, more literature is needed regarding the presence of such redistribution. Some studies propose a change in sweat distribution pattern [63, 64], while others observe an overall increase in sweat rates without altering the distribution pattern [65, 66]. The variation in sweating rate findings across studies can be attributed, in part, to challenges in directly comparing results due to differences in environmental conditions (such as hot and humid versus dry), diverse acclimation protocols and durations, and varying exercise modes and intensities.

Finally, heat acclimation induces several positive alterations in fluid balance management. These comprise an enhanced alignment of fluid intake with the body's water requirements [54], a rise in overall body water content, and an augmentation of blood volume [67, 68]. Individuals who have undergone heat acclimation demonstrate the ability to uphold their state of hydration during exercise in hot conditions. This adaptation is crucial, as heat acclimation increases sweating, necessitating a higher fluid intake to maintain proper hydration levels.

6.2.2 Heat Acclimatization

Heat acclimatization is an adaptive process that occurs when individuals are exposed to a natural environment for an extended period. This process is particularly relevant to individuals from tropical regions exposed to hot climates for a prolonged time. Studies have shown tropical individuals have better physiological functions to tolerate heat exposure than those from the temperate areas [69, 70]. Understanding the differences in physiological adaptations between tropical and nontropical natives is key to recognizing how heat stress is managed across different populations. To explore these differences, our research compared heat-acclimatized Malaysians to Japanese subjects under various heating conditions, offering valuable insights into how tropical natives adapt to heat stress [71–74] (Table 6.2).

One notable finding was that Malaysians showed a higher core (rectal) temperature than Japanese individuals while resting in a thermoneutral environment [74] (see Fig. 6.4a). This characteristic aligns with previous studies that reported higher resting core temperatures in heat-acclimatized individuals from Singapore [75], India [76], Vietnam [77, 78], and Malaysia [79] compared to those from cooler climates. This elevated resting core temperature is thought to be a result of vascular adjustments that improve heat release [74], possibly through increasing the temperature differential between the core and skin, thereby enhancing convective heat loss [80]. Interestingly, while Malaysians had higher resting core temperatures, there were no significant ethnic differences in core temperature at the end of passive or active heating sessions [72, 73] (Fig. 6.4a). However, Malaysians showed a smaller rise in core temperature during thermal stress, indicating superior heat tolerance compared to Japanese individuals [72, 73, 81] (see Fig. 6.4b). This stability in core temperature among tropical natives is consistent with Wilder's law of initial value, suggesting that tropical natives may have more stable core temperatures and extremities better acclimatized to heat [74] (see Fig. 6.5).

In addition to core temperature differences, Malaysians exhibited lower hand and foot temperatures [74]. This was coupled with a marked increase in hand–skin temperature during both passive [72] and active heat exposure [73], suggesting an adaptive blood flow pattern in tropical individuals. Such a mechanism supports countercurrent heat exchange, where cooler venous blood helps cool warmer arterial blood, reducing the increase in core body temperature during heat stress in passive [72] and active heat exposure [73], suggesting an adaptive blood flow pattern in tropical individuals. This physiological trait may reflect an evolved efficiency in body temperature regulation, further supporting the ability of tropical natives to manage heat exposure effectively.

Table 6.2 Adaptation traits of tropical individuals from Malaysia and temperate individual from Japan during passive heat exposure [72] and active heat exposure [73]

Adaptation traits	Passive heating	Active heating
Initial rectal temperature (°C)	JP < MY*	JP < MY*
Rectal temperature at the last stage (°C)	NS	NS
Change in rectal temperature (°C)	JP > MY***	JP > MY***
Mean skin temperature at the last stage (°C)	NS	JP > MY*
Change in mean skin temperature (°C)	NS	JP > MY*
Initial hand/foot skin temperature (°C)	JP < MY*	JP < MY*
Hand/foot skin temperature at the last stage (°C)	NS (hand)–(foot)	NS (hand) JP > MY* (foot)
Change in hand/foot skin temperature (°C)	JP < MY*	JP < MY*
Total sweat loss ($g\ h\ m^{-2}$)	NS	NS
Local sweat rate at the last stage ($mg\ cm^{-2}\ min^{-1}$)	JP > MY* (forehead and thigh)	NS (forearm and back)
Sweat onset time (min)	JP < MY* (thigh)	–
Active sweat gland density at the 40th min ($gland\ cm^{-2}$)	JP > MY*(forehead, back and thigh)	–
Sweat gland output at the 40th min ($mg\ gland^{-1}\ min^{-1}$)	NS	–
Skin blood flow at the last stage (% SkBF)	JP > MY* (forehead and forearm)	JP > MY* (forearm)
Mean arterial blood pressure (mmHg)	NS	NS
Thermal sensation at the last stage	NS	NS
Thermal comfort at the last stage	NS	NS
Sweat sensation at the last stage	NS	NS
Thirst sensation at the last stage	JP > MY*	JP > MY**

*indicates $p < 0.05$
**indicate $p < 0.01$
***indicate $p < 0.001$

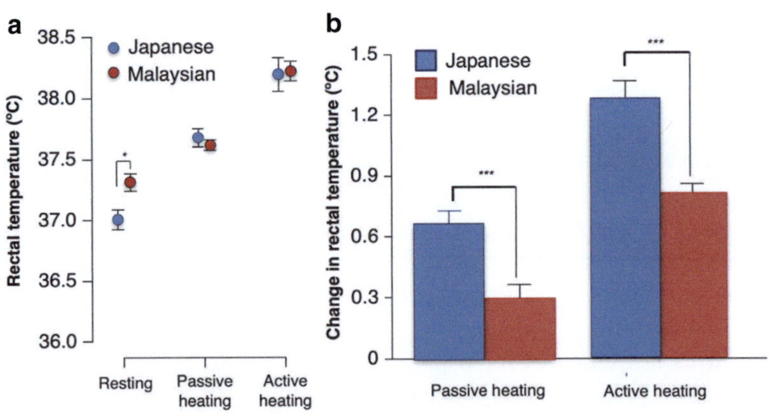

Fig. 6.4 (**a**) Rectal temperature of tropical individuals from Malaysia and temperate individuals from Japan during rest [74], passive heating [72], and active heat exposure [73]. (**b**) Change in rectal temperature tropical individuals from Malaysia and temperate individual from Japan passive heating [72] and active heat exposure [73]. Values are mean and SE. The asterisk indicates a significant difference at $p < 0.05$. Triple asterisks indicate a significant difference at $p < 0.05$

In terms of sudomotor function, individuals native to tropical regions have been shown to exhibit lower sweat rates compared to nontropical populations, a feature often attributed to long-term heat acclimatization. This pattern has been observed in a variety of tropical populations, including Nigerians [83], New Guineans [84], Thai [85, 86], and Malaysians [87, 88]. It is considered that in hot climates, where sweating is frequent, more sweat glands are recruited

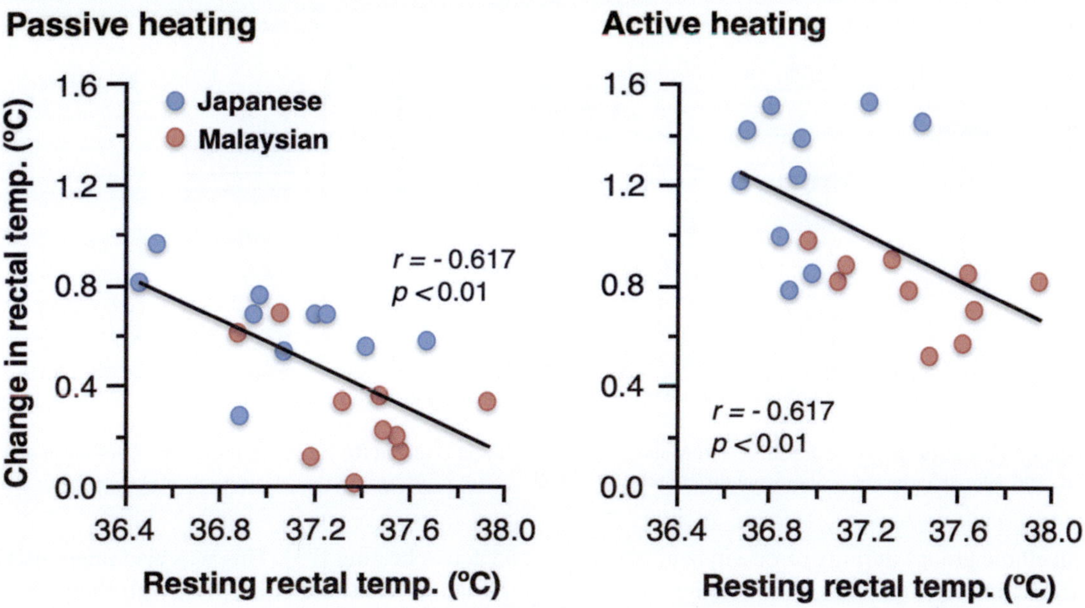

Fig. 6.5 Law of initial value in rectal temperatures during passive and active heating. (Adapted from Lee et al. [74])

and remain active. Conversely, in cooler climates, where sweating is less common, there is a lower density of active sweat glands [19]. This hypothesis is illustrated by comparing sweat gland densities between individuals from cooler and hotter climates (Fig. 6.6). These findings suggest that tropical natives may develop different sweating responses to heat as part of their long-term acclimatization to hot, often humid environments.

In the series of our heat acclimatization studies, Wijayanto et al. [72] further supported this finding, particularly in the context of passive heat exposure (specifically, hot water leg immersion), where Malaysian tropical natives displayed lower local sweat rates on the forehead and thigh than Japanese. The Malaysians also showed a lower density of active sweat glands on the forehead, upper back, forearm, and thigh than the Japanese, with no significant differences in sweat gland output between groups in any body region [72]. Interestingly, no significant differences between Malaysians and Japanese in local sweat rates were observed on the upper back and forearm (Fig. 6.7a) [72]. It is plausible that tropical natives from Malaysia may exhibit a more uniform sweating pattern across the body or limbs, providing an advantage in evaporative sweat loss within hot and humid environments. This uniform sweating pattern is likely an adaptive response to the prevailing climate conditions [72, 82]. In contrast to passive heat exposure, Wakabayashi et al. [73] did not find any discernible difference in local sweat rates across differ-

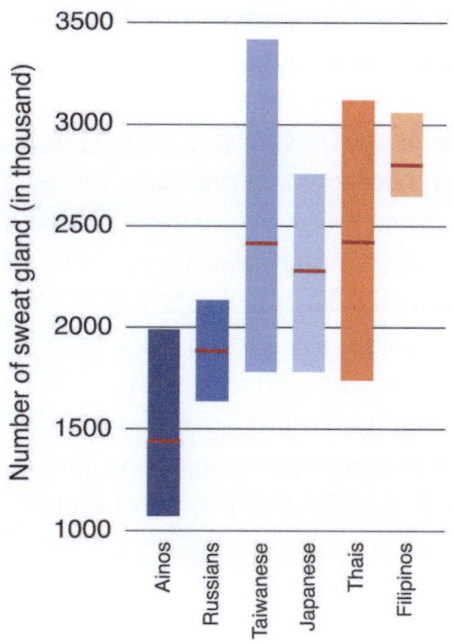

Fig. 6.6 Active sweat glands density in Ainos, Russians, Taiwanese, Japanese, Thais, and Filipinos. Values are max, mean, min. (Data obtained from Kuno [19])

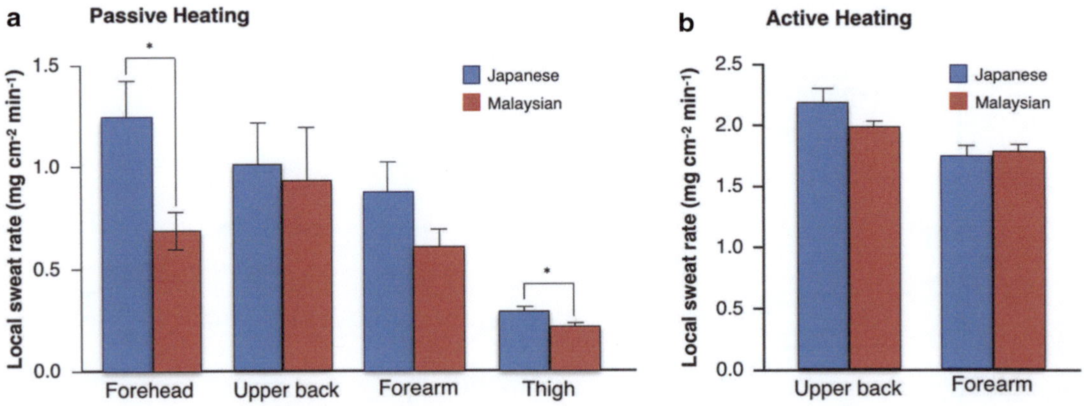

Fig. 6.7 Local sweat rate in Japanese and Malaysians during (**a**) passive and (**b**) active heating. Values are mean and SE. The asterisk indicates a significant difference at $p < 0.05$

ent ethnic groups during exercise in heat or active heat exposure (Fig. 6.7b), contradicting an earlier study that reported lower sweat rates in tropical natives during exercise in a neutral-temperature environment [85].

Regarding the comparison of total sweat loss between Japanese and Malaysians, no significant differences were found, whether during passive or active heat exposure [72, 73]. This is consistent with other research that hasn't found notable differences in sweat loss across various populations under similar conditions [79, 89, 90]. Despite this similarity in sweat output, Malaysians exhibited a less pronounced thirst sensation after both passive and active heating [71–73]. This could imply that tropical natives are less inclined to feel thirsty in response to equivalent body fluid loss than their nontropical counterparts. The reduced thirst could be linked to a delayed central drive for sweating, which may correspond with the later onset of sweating in Malaysians [72]. Furthermore, the diminished sensation of thirst in Malaysians may indicate an acclimatization strategy focused on conserving body fluid by minimizing urine output rather than increasing water intake [71].

Heat acclimatization also impacts cutaneous vasomotor function differently across ethnicities, affecting skin blood flow. Studies comparing Malaysians and Japanese revealed that Malaysians have lower forearm skin blood flow in both passive and active heat conditions [72] and active heating [73]. This was also previously reported by Katsuura et al. [89], who found that Japanese long-term residents in a tropical environment (Brazil) also exhibited lower forearm blood flow and higher skin temperatures than those living in Japan, suggesting that higher skin temperatures may counteract environmental heat and promote evaporative cooling [91]. Similar adaptive responses were reported in Negroid relative to Caucasians [83] and in Malaysians compared to Japanese during exercise in hot-humid climates [88]. Additionally, Fox et al. [84] observed reduced hand blood flow in indigenous New Guineans from hot-humid and cool-dry areas during controlled hyperthermia, in contrast to heat-acclimated and nonacclimated Caucasians. This suggests that, along with developing efficient sweating mechanisms, indigenous populations may also experience a reduction in cutaneous blood flow as part of their acclimatization to high temperatures [84].

In our study examining the long-term heat acclimatization impact on blood volume control, Wakabayashi et al. [71] examined hydration's effects on body fluid and temperature regulation during exercise in hot and humid conditions (32 °C, 70% RH), with or without hydration (12 mL/kg body weight). The results revealed that the hydrated tropical group exhibited notably lower postexercise core body temperatures than their nonhydrated counterparts, a distinction not observed in the temperate group. Additionally, tropical individuals demonstrated superior body fluid regulation to Japanese individuals, especially when adequately hydrated. This was evidenced by reduced dehydration, diminished inclination toward excessive sweat production, and unnecessary urinary loss. Consequently, they could sustain higher post-exercise plasma volumes, facilitating more efficient heat dissipation through enhanced skin blood flow. This adept fluid conservation may grant tropical natives an edge in maintaining lower core body temperatures when adequately hydrated, optimizing their heat dissipation responses without undue fluid loss.

6.2.3 Decay of Heat Adaptation

6.2.3.1 Decay of Heat Acclimation

Heat acclimation benefits can diminish if not maintained through regular heat exposure. Among the physiological adaptations achieved during acclimation, heart rate improvements are typically the first to regress, deteriorating more quickly than thermoregulatory adaptations such as core temperature regulation [38, 92]. Although it is well established that heat acclimation decays over time, the exact rate at which it is lost varies across studies [38, 93]. One contributing factor to the rate of decline is the frequency of heat exposure, with individuals who experience heat more frequently able to maintain their adaptations for longer [49].

In general, it is suggested that heat acclimation enhances heat tolerance for about 2 weeks, but the benefits begin to decline without regular exposure to heat [94]. The rate of heat acclimation decay is relatively rapid [38, 95], with Givoni et al. [96] proposing that for every 2 days without heat exposure, 1 day of heat acclimation is lost. Pandolf [38] further suggests that the benefits can dissipate entirely within 6–21 days if heat exposure ceases. The first adaptations to regress are typically those related to cardiovascular function, such as increased plasma volume and reduced exercise heart rate [38, 39]. This suggests that to retain the advantages of heat acclimation, regular exposure to heat is necessary, and even short breaks from heat exposure can lead to significant reductions in acclimation levels.

Research on the loss of heat acclimation effects over time has revealed decreased benefits, particularly in heart rate response. Saat et al. [97] demonstrated that 2 weeks after the final heat exposure, there was a 35% loss in heart rate adaptation during exercise at 60% VO_2max in temperate conditions (18 °C). Similarly, Ashley et al. [98] found that heart rate improvements were retained for about a week following heat acclimation but were largely gone by the third week. Garrett et al. [99] reported a 31% decline in heart rate adaptation after 2 weeks without heat exposure. These studies collectively illustrate that cardiovascular responses to heat are among the first to decline when heat exposure is interrupted.

While decay in heart rate adaptation was evident, changes in core temperature appeared to decline at a slower pace. Ashley et al. [98] observed no significant decay in core temperature immediately following heat acclimation, and Weller et al. [75] reported a modest 5% decay in postexercise core temperature after 26 days. Similarly, Garrett et al. [99] noted an 8% mean decay in core temperature, equivalent to a change of just 0.02 °C, after 1 week without heat exposure. On average, core temperature adaptations decline gradually, with a rate of around 2.6% per day [93]. This slower decay suggests that while the heart rate response to heat quickly returns to preacclimated levels after a period without heat exposure, core temperature adaptations are more resilient and take longer to regress.

6.2.3.2 Decay in Heat Acclimatization

Studies examining the decay of heat acclimation, particularly in short-term and artificially induced conditions, have received more attention, likely due to the controlled nature and ease of conducting such research. In contrast, studies on the decay of heat acclimatization, which occurs in natural environments over longer periods, have been less extensive. Heat acclimatization is often considered a partially phenotypic adaptation, meaning it can be influenced by both genetic predisposition and environmental factors. This phenotypic adaptation is evident in tropical individuals who spend extended periods in temperate or colder climates, where reversible alterations in their sweating responses, such as sweat onset time and sweat rate, have been documented [100]. For example, a study by Saat et al. [101] found that individuals from Malaysia who had resided in temperate climates like Japan for over 27 months developed an earlier sweat onset than those who had been there for less than 15 months. Similarly, Lee et al. [100] observed that Malaysians who had spent 2–72 months in Japan gradually lost their heat acclimatization, as measured by changes in their sweat responses to acetylcholine iontophoresis.

In our previous study, which focused on passive heat exposure responses in tropical indigenes from various Southeast Asian countries residing in Japan [102], we observed a gradual decay in heat acclimatization, particularly in sweating responses. Specifically, the longer individuals had lived in Japan, the shorter their sweating onset time and the greater their total sweat rate (Fig. 6.8). Over time, the sweat rate of tropical individuals residing in colder climates began to resemble that of individuals already acclimatized to cold conditions. These findings align with previous studies documenting the alteration in sweating responses following extended stays in temperate climates [100, 101]. Despite the observed changes in sweating responses, other physiological markers, such as resting rectal temperature and the increase in rectal temperature during heat exposure, remained unchanged. This suggests that while there may be a partial decay in heat acclimatization, the core heat tolerance developed from living in a tropical climate may persist, even after residing in temperate or colder environments for an extended period [102].

6.3 Performance in a Hot Environment

6.3.1 Physical Performance

Physical performance in hot environments places significant demands on physiological mechanisms, particularly the cardiovascular system. Environmental factors such as solar radiation, wind speed and water vapor pressure further intensify the environmental stress experienced by athletes [103, 104]. These stressors collectively contribute to a well-documented decline in endurance and high-intensity exercise performance under heat stress [105, 106]. Numerous studies conducted in controlled environments have shown that endurance exercise capacity decreases in hot environments, as indicated by earlier time to fatigue [107, 108] and overall performance declines [109]. Similar performance decrements

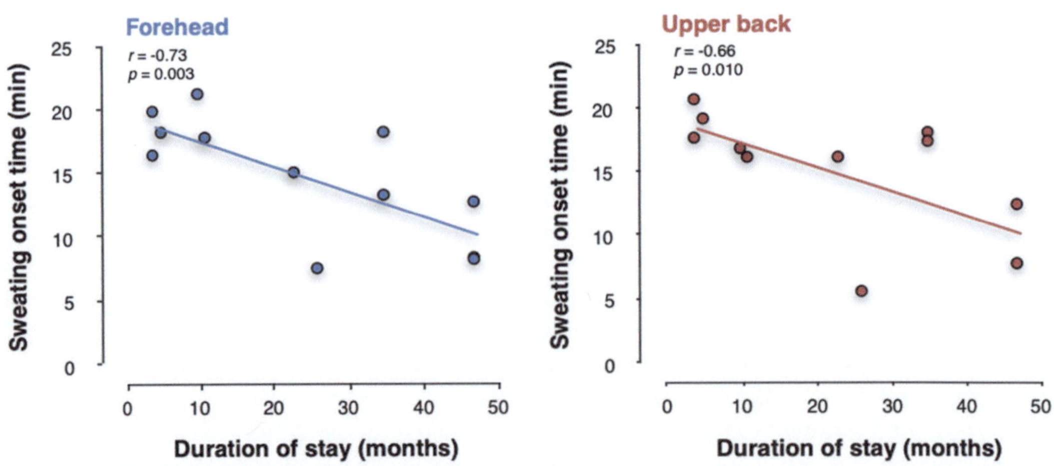

Fig. 6.8 Relationships between duration of stay in Japan and sweating onset time on the forehead and upper back. (Data obtained from Wijayanto et al. [102])

have been observed in high-intensity sprints [110, 111] and marathon running [112, 113]. For example, Galloway and Maughan [107] reported that exercising duration at 70% maximum oxygen uptake was significantly shorter in a hot environment (Tair = 31 °C, 51.6 ± 3.7 min) compared to a cooler environment (Tair = 11 °C, 93.5 ± 6.2 min), with participants reporting a higher rating of perceived exertion (RPE) in the heat.

One of the primary causes of this performance impairment is hyperthermia, defined as a core temperature exceeding 38.5 °C, which has a direct impact on the central nervous system [111]. Hyperthermia disrupts the brain's role in regulating exercise performance, particularly in maintaining cardiovascular function. As body temperature rises, the individual's motivation to exercise declines, influenced by sensory feedback and the direct effect of elevated body temperature on muscle activation. Brück and Olschewski [114] proposed that hyperthermia disrupts the balance between circulation, heat, and muscle discomfort, reducing the willingness to persist in exercise. Nielsen et al. [53] highlighted that hyperthermia, rather than circulatory limitations, is the primary cause of fatigue during exercise in hot environments. Their findings show that once core temperature reaches a certain threshold, central nervous system function is compromised, decreasing the drive for physical activity.

Interestingly, despite the onset of exhaustion, cardiac output and muscle blood flow remain relatively stable. Immediately after reaching exhaustion, individuals can still generate force, but over time, this force diminishes more rapidly in hot environments compared to cooler conditions [115]. This reduction is linked to central fatigue, specifically a decline in voluntary muscle activation caused by overheating. Some theories suggest that as brain temperature rises, the hypothalamus may send inhibitory signals that reduce motor activity [116]. Thus, hyperthermia-induced central fatigue plays a crucial role in limiting performance, even when muscle capacity is maintained.

In addition to the central nervous system's role, the cardiovascular system faces increased demands due to the increased need for skin blood flow to regulate temperature. This thermoregulatory demand competes with the oxygen supply needed by exercising muscles and vital organs, further complicating performance [117]. During exercise in the heat, the cardiovascular system must balance the distribution of blood flow between the skin and muscles [118]. Studies show that in temperatures as high as 43 °C, stroke volume and central blood volume both decline, suggesting a reduction in cardiac output during high-intensity activities [121]. This decrease in cardiac output correlates with reduced maximal oxygen uptake during intense heat stress [120]. Notably, the cardiac output in hot conditions does not exceed that achieved at equivalent effort levels in cooler environments [121]. Moreover, if blood flow to the active muscles is maintained throughout prolonged exercise sessions, it compromises skin blood flow and impairs internal temperature regulation, further reducing physical performance [119].

In summary, physical activity in hot environments presents a range of challenges to both physiological and cognitive systems, with the cardiovascular and central nervous systems most affected. The interaction between these internal systems and external environmental stressors, such as high temperatures and humidity, is key to determining performance outcomes. Managing the balance between thermoregulation and the cardiovascular demands of exercise is crucial for optimizing performance and reducing the risk of heat-related performance decrements.

6.3.2 Cognitive Performance

Despite a well-understood relationship between heat and physical performance, the connection between hot environments and cognitive functions remains more ambiguous. While some studies report a negative impact on cognitive functions in the heat [122, 123], others find no evident effects [124, 125], and a few even suggest enhancements in performance under certain conditions [126, 127]. Pilcher et al. [128] conducted a meta-analysis revealing that extreme temperatures affect cognitive performance according to an inverted U-curve, where moderate heat may enhance performance while extreme heat impairs it. The complexity of these findings highlights the multifaceted nature of heat's impact on cognitive performance, where variables such as the intensity and duration of heat stress, core temperature changes, and task complexity all play important roles [129, 130].

One of the critical factors influencing cognitive performance in hot environments is core body temperature [130]. When heat stress is insufficient to significantly elevate core temperature, cognitive functions such as sustained attention often remain stable [131] or, in some cases, are even improved [132]. When core temperature increases by 1.5 °C or more, cognitive processes such as memory recall and pattern recognition are negatively affected [133, 134]. However, smaller changes in core temperature are less likely to impair performance. For example, Wijayanto et al. [124] found that minor fluctuations in core temperature during passive heat stress had minimal effects on short-term memory, emphasizing the threshold at which cognitive performance starts to decline.

The effect of heat on cognitive tasks involving executive function is further complicated by the interaction between task complexity and the extremity of temperature exposure. For example, exposure to extreme heat (44 °C for 5 h) did not impair performance on a simple planning task but did affect performance on more complex tasks [133]. Meanwhile, an increase of 1 °C in core body temperature had no effect on either simple or complex tasks [135], suggesting that the interplay between task difficulty or complexity and the degree of heat stress is crucial in predicting performance outcomes. In one study, performance accuracy and reaction time were reduced by roughly 6% after 4.5 h of exposure at 30 °C, compared to 22 °C [136, 137]. Interestingly, grammatical reasoning remained unaffected by 30 min of exposure to 40 °C [138], but it declined by 25% after 4.5 h at 30 °C [137], indicating that both the duration of exposure and task complexity are key factors in cognitive performance under heat stress.

Mixed findings regarding cognitive performance in hot environments may be explained by the brain's ability to maintain function through compensatory mechanisms. One potential explanation is the brain's capacity to reallocate neural resources to sustain performance during heat stress [139]. This compensatory effort involves neural resource reallocation, as evidenced by changes in the brain's electrical activity, measured through electroencephalography [140], or cerebral oxygenation [124, 125]. However, when heat stress leads to a reduction in cerebral blood flow and oxygenation, cognitive performance may suffer. This decrease in blood flow and oxygenation limits the brain's ability to manage both the task at hand and hyperthermia, leading to cognitive declines under severe heat exposure [140, 141].

In addition to neural resource allocation, an individual's adaptation level can significantly influence cognitive performance in the heat [130]. Adapted individuals, who have undergone physiological adaptations to better tolerate heat stress, are more resistant to performance declines than their nonacclimatized counterparts. For example, Radakovic et al. [142] found that a 10-day heat acclimation regimen helped preserve attention during heat exposure. Similarly, Wijayanto et al. [125] observed that nonadapted individuals showed declines in mental arithmetic tasks under heat stress, while the acclimatized individuals maintained their performance (Fig. 6.9a). Furthermore, during task performance in heated conditions, the nonacclimatized group exhibited higher oxyhemoglobin (oxyHb) levels (Fig. 6.9b), possibly leading to overarousal and subsequent declines in performance.

In summary, the effects of heat on cognitive functions are complex and influenced by multiple factors, including temperature thresholds, duration of exposure, task complexity, core body temperature, and the individual's adaptation level. Neural resource allocation and physiological adaptations, such as heat acclimatization, play critical roles in determining how well cognitive performance is maintained under heat stress. These findings emphasize the nuanced and multifactorial nature of the relationship between heat and cognitive function, as summarized in Fig. 6.10.

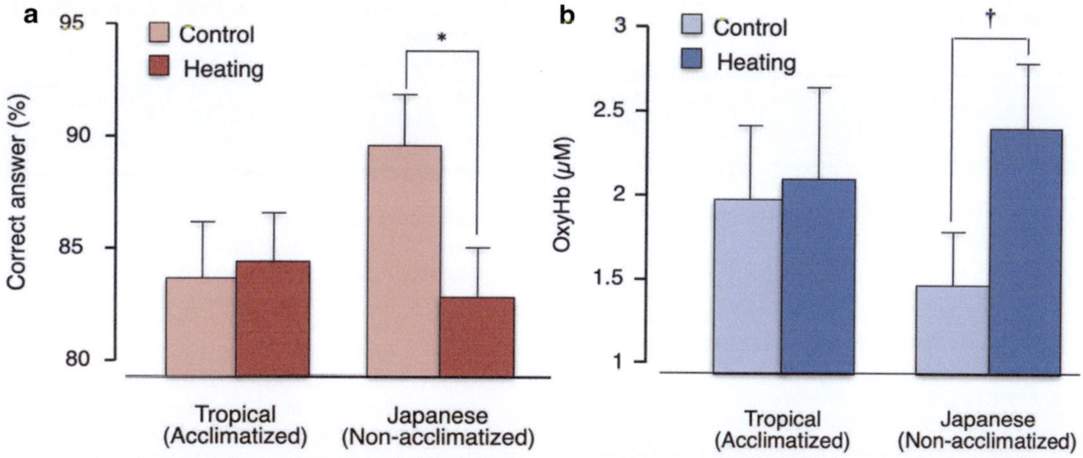

Fig. 6.9 (**a**) Cognitive task performance of the mental arithmetic task and (**b**) Concentration changes of oxyhemoglobin (OxyHb) during control and passive heating conditions. Values are mean and SE. The asterisk indicates a significant difference at $p < 0.05$. A dagger indicates a tendency toward $p < 0.1$. (Data obtained from Wijayanto et al. [125])

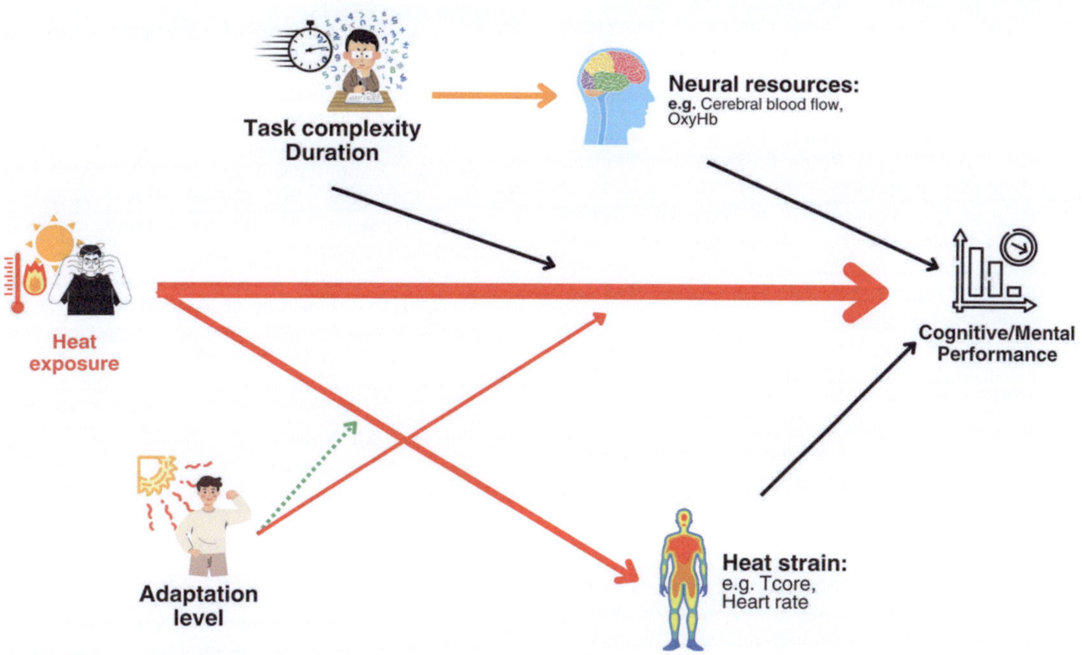

Fig. 6.10 Factors affecting cognitive/mental performance during heat exposure

References

1. Parsons K. Human thermal environments: the effects of hot, moderate, and cold environments on human health, comfort, and performance. 3rd ed; 2014.
2. Kenny GP, Flouris AD. The human thermoregulatory system and its response to thermal stress. In: Protective clothing: managing thermal stress. Woodhead Publishing; 2014.
3. Nielsen TA, da Silva LB, Arendt-Nielsen L, Gazerani P. The effect of topical capsaicin-induced sensitization on heat-evoked cutaneous vasomotor responses. Int J Physiol Pathophysiol Pharmacol. 2013;5(3):148-160.
4. Fox RH, Edholm OG. Nervous control of the cutaneous circulation. Br Med Bull. 1963;19:110-114.
5. Edholm OG, Fox RH, Macpherson RK. Vasomotor control of the cutaneous blood vessels in the human forearm. J Physiol. 1957;139(3):455-65.
6. Johnson JM, Kellogg DL. Local thermal control of the human cutaneous circulation. J Appl Physiol. 2010;109(4):1229-1238.
7. Johnson JM, Kellogg DL. Thermoregulatory and thermal control in the human cutaneous circulation. Front Biosci–(Schol ed). 2010:2(3):825-853.
8. Kenney WL, Johnson JM. Control of skin blood flow during exercise. Med Sci Sports Exerc. 1992;24(3):303-312.
9. Johnson JM. Park MK. Reflex control of skin blood flow by skin temperature: role of core temperature: J Appl Physiol Respir Environ Exerc Physiol; 1979;47(6):1188-1193.
10. Pérgola PE, Johnson JM, Kellogg DL, Kosiba WA. Control of skin blood flow by whole body and local skin cooling in exercising humans. Am J Physiol Heart Circ Physiol. 1996;270(1):H208-H215.
11. Rowell LB. Human cardiovascular adjustments to exercise and thermal stress. Physiol Rev. 1974;54:75–159.
12. Kenney WL, Stanhewicz AE, Bruning RS, Alexander LM. Blood pressure regulation III: what happens when one system must serve two masters: temperature and pressure regulation? Eur J Appl Physiol. 2014;114(3):467-479.
13. Holowatz LA, Kenney WL. Acute localized administration of tetrahydrobiopterin and chronic systemic atorvastatin treatment restore cutaneous microvascular function in hypercholesterolaemic humans. J Physiol. 2011;589(19):4787–4797.
14. Rooke GA, Savage MV, Brengelmann GL. Maximal skin blood flow is decreased in elderly men. J Appl Physiol. 1994;77(1):11-14.
15. Martin HL, Loomis JL, Kenney WL. Maximal skin vascular conductance in subjects aged 5-85 yr. J Appl Physiol. 1995;79(1):297-301.
16. Kenney WL, Tankersley CG, Newswanger DL, Hyde DE, Puhl SM, Turner NL. Age and hypohydration independently influence the peripheral vascular response to heat stress. J Appl Physiol. 1990;68(5):1902-1908.
17. Holowatz LA, Thompson-Torgerson C, Kenney WL. Aging and the control of human skin blood flow. Front Biosci. 2010;15:718–739.
18. Ho CW, Beard JL, Farrell PA, Minson CT, Kenney WL. Age, fitness, and regional blood flow during exercise in the heat. J Appl Physiol. 1997;82(4):1126-1135.
19. Kuno Y. Human perspiration: Springfield. Illinois: Charles C Thomas Publisher; 1956.
20. Wenger CB. Heat of evaporation of sweat: thermodynamic considerations. J Appl Physiol. 1972;32(4):456-459.
21. Sato K, Dobson RL. Regional and individual variations in the function of the human eccrine sweat gland. J Invest Dermatol.1970;54(6):443-449.
22. Candas V, Libert JP, Vogt JJ. Human skin wettedness and evaporative efficiency of sweating. J Appl Physiol Respir Environ Exerc Physiol. 1979;46(3):522-528.
23. Kondo N, Takano S, Aoki K, Shibasaki M, Tominaga H, Inoue Y. Regional differences in the effect of exercise intensity on thermoregulatory sweating and cutaneous vasodilation. Acta Physiol Scand. 1998;164(1):71-78.
24. Nadel ER, Bullard RW, Stolwijk JA. Importance of skin temperature in the regulation of sweating. J Appl Physiol. 1971;31(1):80-87.
25. Nadel ER, Mitchell JW, Saltin B, Stolwijk JA. Peripheral modifications to the central drive for sweating. J Appl Physiol. 1971;31(6):828-33.
26. Van Beaumont W, Bullard RW. Sweating: direct influence of skin temperature. Science. 1965;147(3664):1465-7.
27. Kondo N, Shibasaki M, Aoki K, Koga S, Inoue Y, Crandall CG. Function of human eccrine sweat glands during dynamic exercise and passive heat stress. J Appl Physiol. 2001;90:1877–81.
28. Sato K. Update on pharmacology of the eccrine sweat gland. Trends Pharmacol Sci. 1984;5: 391-393
29. Shibasaki M, Wilson TE, Crandall CG. Neural control and mechanisms of eccrine sweating during heat stress and exercise. J Appl Physiol. 2006;100:1692–701.
30. Baker LB. Physiology of sweat gland function: The roles of sweating and sweat composition in human health. Temperature (Austin). 2019;6(3):211–59.
31. Taylor NAS, Machado-Moreira CA. Regional variations in transepidermal water loss, eccrine sweat gland density, sweat secretion rates and electrolyte composition in resting and exercising humans. Extrem Physiol Med. 2013;2(1):4
32. Machado-Moreira CA, Wilmink F, Meijer A, Mekjavic IB, Taylor NAS. Local differences in sweat secretion from the head during rest and exercise in the heat. Eur J Appl Physiol. 2008;104:257–64.
33. Smith CJ, Havenith G. Body mapping of sweating patterns in male athletes in mild exercise-

induced hyperthermia. Eur J Appl Physiol. 2011;111:1391–404.
34. Inoue Y, Shibasaki M, Ueda H, Ishizashi H. Mechanisms underlying the age-related decrement in the human sweating response. Eur J Appl Physiol Occup Physiol. 1999;79:121–9.
35. Inoue Y. Regional differences in age related decrements of the cutaneous vascular and sweating responses to passive heating. Eur J Appl Physiol Occup Physiol. 1996;74(1-2):78-84.
36. Taylor NAS. Ethnic differences in thermoregulation: genotypic versus phenotypic heat adaptation. J Therm Biol. 2006;31(1-2):90-104.
37. Périard JD, Racinais S, Sawka MN. Adaptations and mechanisms of human heat acclimation: applications for competitive athletes and sports. Scand J Med Sci Sports. 2015;25:20–38.
38. Pandolf KB. Time course of heat acclimation and its decay. Int J Sports Med. 1998;19:S157–60.
39. Armstrong LE, Maresh C. The induction and decay of heat acclimatisation in trained athletes. Sports Med. 1991;12(5):302-12
40. IUPS Thermal Commission I. Glossary of terms for thermal physiology. Third edition. Revised by The Commission for Thermal Physiology of the International Union of Physiological Sciences. Jpn J Physiol. 2001;51(2):245-80
41. Nielsen B, Strange S, Christensen NJ, Warberg J, Saltin B. Acute and adaptive responses in humans to exercise in a warm, humid environment. Pflugers Arch. 1997;434:49–56.
42. Yamazaki F, Hamasaki K. Heat acclimation increases skin vasodilation and sweating but not cardiac baroreflex responses in heat-stressed humans. J Appl Physiol. 2003;95(4):1567-74.
43. Roberts MF, Wenger CB, Stolwijk JAJ, Nadel ER. Skin blood flow and sweating changes following exercise training and heat acclimation. J Appl Physiol Respir Environ Exerc Physiol. 1977:43(1):133-7.
44. Buono MJ, Heaney JH, Canine KM. Acclimation to humid heat lowers resting core temperature. Am J Phys. 1998;274:R1295–9.
45. Patterson MJ, Stocks JM, Taylor NA. Sustained and generalized extracellular fluid expansion following heat acclimation. J Physiol. 2004;559:327–34.
46. Garrett AT, Creasy R, Rehrer NJ, Patterson MJ, Cotter JD. Effectiveness of short-term heat acclimation for highly trained athletes. Eur J Appl Physiol. 2012;112(5):1827-37
47. Casadio JR, Kilding AE, Cotter JD, Laursen PB. From lab to real world: heat acclimation considerations for elite athletes. Sports Med. 2017:47(8):1467-1476
48. Taylor NAS, Cotter JD. Heat adaptation: guidelines for the optimisation of human performance. Int SportMed J. 2006;7(1):33-57
49. Armstrong L. Heat acclimatization. In: Fahey T, editor. Encyclopedia of sports medicine and science. Internet Society for Sport Science; 1998.
50. Horowitz M. Heat acclimation, epigenetics, and cytoprotection memory. Compr Physiol. 2014:4(1):199-230
51. Lorenzo S, Minson CT. Heat acclimation improves cutaneous vascular function and sweating in trained cyclists. J Appl Physiol. 2010;109(6):1736-43
52. Andres T, Hexamer M, Werner J. Heat acclimation of humans: hot environment versus physical exercise. J Therm Biol. 2000;25(1-2): 139-142
53. Nielsen BYB, Hales JRS, Strange S, Christensen Juel N, Warberg J, Saltin B, et al. Human circulatory and thermoregulatory adaptations with heat acclimation and exercise in a hot, dry environment. J Physiol. 1993;460:467–85.
54. Racinais S, Alonso JM, Coutts AJ, Flouris AD, Girard O, González-Alonso J, et al. Consensus recommendations on training and competing in the heat. Br J Sports Med. 2015;49(7):925-38
55. Shapiro Y, Pandolf KB, Avellini BA, Pimental NA, Goldman RF. Physiological responses of men and women to humid and dry heat. J Appl Physiol. 1980;49(1):1-8
56. Périard JD, Travers GJS, Racinais S, Sawka MN. Cardiovascular adaptations supporting human exercise-heat acclimation. Auton Neurosci. 2016;196:52–62.
57. Hodge D, Jones D, Martinez R, Buono MJ. Time course of the attenuation of sympathetic nervous activity during active heat acclimation. Auton Neurosci. 2013;177(2):101-3
58. Frank A, Belokopytov M, Moran D, Shapiro Y, Epstein Y. Changes in heart rate variability following acclimation to heat. J Basic Clin Physiol Pharmacol. 2001;12(1):19-32.
59. Epstein Y, Moran DS, Kobo R, Lewkowicz M, Levitan J, Heled Y. Acclimation to heat interpreted from the analysis of heart-rate variability by the multipole method. J Basic Clin Physiol Pharmacol. 2010;21(4):315-23
60. Sawka MN, Wenger CB, Pandolf KB. Thermoregulatory responses to acute exercise-heat stress and heat acclimation. Compr Physiol. 1996;157-185.
61. Nadel ER, Pandolf KB, Roberts MF, Stolwijk AJ, Stolwijk JA. Mechanisms of thermal acclimation to exercise and heat. J Appl Physiol. 1974;37(4):515-20
62. Höfler W. Changes in regional distribution of sweating during acclimatization to heat. J Appl Physiol. 1968;25(5):503-6
63. Poirier MP, Gagnon D, Kenny GP. Local versus whole-body sweating adaptations following 14 days of traditional heat acclimation. Appl Physiol Nutr Metab. 2016:41.(8):816-24
64. Smith CJ, Havenith G. Upper body sweat mapping provides evidence of relative sweat redistribution towards the periphery following hot-dry heat acclimation. Temperature. 2019:6(1):50-65
65. Patterson MJ, Stocks JM, Taylor NAS. Humid heat acclimation does not elicit a preferential sweat

redistribution toward the limbs. Am J Physiol Regul Integr Comp Physiol. 2004;286(3):R512-8
66. Cotter JD, Patterson MJ, Taylor NAS. Sweat distribution before and after repeated heat exposure. Eur J Appl Physiol Occup Physiol. 1997;76:181–6.
67. Mack GW, Nadel ER. Body fluid balance during heat stress in humans. In: Fregly MJ, Blatteis CM, editors. Handbook of physiology section 4: environmental physiology. New York: Oxford University Press; 1996.
68. Sawka MN, Coyle EF. Influence of body water and blood volume on thermoregulation and exercise performance in the heat. Exerc Sport Sci Rev. 1999;27:167-218.
69. Duncan MT, Horvath SM. Physiological adaptations to thermal stress in tropical Asians. Eur J Appl Physiol Occup Physiol. 1988;57(5):540–4.
70. Lee JKW, Nio AQX, Fun DCY, Teo YS, Von CE, Lim CL. Effects of heat acclimatisation on work tolerance and thermoregulation in trained tropical natives. J Therm Biol. 2012;37(5):366-373
71. Wakabayashi H, Wijayanto T, Lee JY, Hashiguchi N, Saat M, Tochihara Y. A comparison of hydration effect on body fluid and temperature regulation between Malaysian and Japanese males exercising at mild dehydration in humid heat. J Physiol Anthropol. 2014;33:1–11.
72. Wijayanto T, Wakabayashi H, Lee JY, Hashiguchi N, Saat M, Tochihara Y. Comparison of thermoregulatory responses to heat between Malaysian and Japanese males during leg immersion. Int J Biometeorol. 2011;55:491–500.
73. Wakabayashi H, Wijayanto T, Lee JY, Hashiguchi N, Saat M, Tochihara Y. Comparison of heat dissipation response between Malaysian and Japanese males during exercise in humid heat stress. Int J Biometeorol. 2011;55:509–17.
74. Lee JY, Wakabayashi H, Wijayanto T, Hashiguchi N, Saat M, Tochihara Y. Ethnic differences in thermoregulatory responses during resting, passive and active heating: application of Werner's adaptation model. Eur J Appl Physiol. 2011;111:2895–905.
75. Adam JM, Ferres HM. Observations on oral and rectal temperatures in the humid tropics and in a temperate climate. J Physiol. 1954;125:21P.
76. Raynaud J, Martineaud JP, Bhatnagar OP, Vieillefond H, Durand J. Body temperatures during rest and exercise in residents and sojourners in hot climate. Int J Biometeorol. 1976;20:309–17.
77. Nguyen M, Tokura H. Observations on normal body temperatures in Vietnamese and Japanese in Vietnam. J Physiol Anthropol. 2002;21(1):59-65
78. Nguyen MH, Tokura H. Sweating and tympanic temperature during warm water immersion compared between Vietnamese and Japanese living in Hanoi. J Hum Ergol (Tokyo). 2003;32:9–16.
79. Saat M, Sirisinghe RG, Singh R, Tochihara Y. Effects of short-term exercise in the heat on thermoregulation, blood parameters, sweat secretion and sweat composition of tropic-dwelling subjects. J Physiol Anthropol Appl Hum Sci. 2005;24:541–9.
80. Taylor N, Kondo N, Kenney W. The physiology of acute heat exposure, with implications for human performance in the heat. In: Taylor N, Groeller H, editors. Physiological bases of human performance during work and exercise. London: Churcill Livingstone; 2008. p. 379–400.
81. Wakabayashi H, Wijayanto T, Lee J-Y, Hashiguchi N, Saat M, Tochihara Y. A comparison of hydration effect on body fluid and temperature regulation between Malaysian and Japanese males exercising at mild dehydration in humid heat. J Phisiol Anthropol. 2014;33:1–11.
82. Tochihara Y, Wakabayashi H, Lee JY, Wijayanto T, Hashiguchi N, Saat M. How humans adapt to hot climates learned from the recent research on tropical indigenes. J Physiol Anthropol. 2022;41:1–16.
83. Thomson ML. A comparison between the number and distribution of functioning eccrine sweat glands in Europeans and Africans. J Physiol. 1954;123(2):225-33
84. Fox RH, Budd GM, Woodward PM, Hackett AJ, Hendrie AL. A study of temperature regulation in New Guinea people. Philos Trans R Soc Lond Ser B Biol Sci. 1974;268(893):375-91
85. Inoue Y, Ueda H, Kubota T, Kondo N, Matsumoto T, Sutabhaha T, et al. Sweat gland function in Thai and Japanese males in relation to physical training. In: Proceedings of the 13th international conference of environmental ergonomics; 2009. p. 491–6.
86. Matsumoto T, Kosaka M, Yamauchi M, Tshuchiya K, Ohwatari N, Motomura M, et al. Study on mechanisms of heat acclimatization due to thermal sweating—comparison of heat-tolerance between Japanese and Thai subjects. Tropical Med. 1993;35(1):23-34
87. Lee JB, Bae JS, Matsumoto T, Yang HM, Min YK. Tropical Malaysians and temperate Koreans exhibit significant differences in sweating sensitivity in response to iontophoretically administered acetylcholine. Int J Biometeorol. 2009;53(2):149-57.
88. Saat M, Tochihara Y, Hashiguchi N, Sirisinghe RG, Fujita M, Chou CM. Effects of exercise in the heat on thermoregulation of Japanese and Malaysian males. J Physiol Anthropol Appl Hum Sci. 2005;24:267–75.
89. Katsuura T, Tachibana ME, Okada A, Kikuchi Y. Comparison of thermoregulatory responses to heat between Japanese Brazilians and Japanese. J Therm Biol. 1993;18(5-6):299-302
90. Martineaud JP, Raynaud J, Duhaze P, Vieillefond H. Hand blood flow at rest and during submaximal exercise in acute and chronic heat stress. J Physiol Paris. 1976;72:233–47.
91. Eichna LW, Park CR, Nelson N, Horvath SM, Palmes ED. Thermal regulation during acclimatization in a hot, dry (desert type) environment. Am J Phys. 1950;163(3):585-97.

92. Pandolf KB, Burse RL, Goldman RF. Role of physical fitness in heat acclimatisation, decay and reinduction. Ergonomics. 1977;20:399–408.
93. Daanen HAM, Racinais S, Périard JD. Heat acclimation decay and re-induction: a systematic review and meta-analysis. Sports Med 2017 48:2. 201748:409–430.
94. Lind AR. Physiologic responses to heat. In: Licht S, editor. Medical climatology. New Haven: Elizabeth Licht; 1964.
95. Williams CG, Wyndham CH, Morrison JF. Rate of loss of acclimatization in summer and winter. J Appl Physiol. 1967;22:21–6.
96. Givoni B, Goldman RF. Predicting effects of heat acclimatization on heart rate and rectal temperature. J Appl Physiol. 1973;35:875–9.
97. Saat M, Sirisinghe RG, Singh R, Tochihara Y. Decay of heat acclimation during exercise in cold and exposure to cold environment. Eur J Appl Physiol. 2005;95:313–20.
98. Ashley CD, Ferron J, Bernard TE. Loss of heat acclimation and time to re-establish acclimation. J Occup Environ Hyg. 2015;12(5):302-8.
99. Garrett AT, Rehrer NJ, Patterson MJ. Induction and decay of short-term heat acclimation in moderately and highly trained athletes. Sports Med. 2011;41(9):757-71
100. Lee JB, Bae JS, Lee MY, Yang HM, Min YK, Song HY, et al. The change in peripheral sweating mechanisms of the tropical Malaysian who stays in Japan. J Therm Biol. 2004;29(7-8):743-747
101. Saat MI, Lee JB, Matsumoto T, Kosaka M, Ohwatari N, Motomura K, et al. Relationship between the duration of stay in Japan of Malaysian subjects and the suppression of sweat gland sensitivity by iontophoretically applied acetylcholine. Acta Med Nagasaki. 1999;44:9-53
102. Wijayanto T, Toramoto S, Wakabayashi H, Tochihara Y. Effects of duration of stay in temperate area on thermoregulatory responses to passive heat exposure in tropical southeast Asian males residing in Japan. J Physiol Anthropol. 2012;31(1):25
103. González-Alonso J. Human thermoregulation and the cardiovascular system. Exp Physiol. 2012;97(3):340–6.
104. Brocherie F, Girard O, Pezzoli A, Millet GP. Outdoor exercise performance in ambient heat: time to overcome challenging factors? Int J Hyperth. 2014;30(8):547-9.
105. Ely BR, Cheuvront SN, Kenefick RW, Sawka MN. Aerobic performance is degraded, despite modest hyperthermia, in hot environments. Med Sci Sports Exerc. 2010;42(1):135-41.
106. Tucker R, Rauch L, Harley YXR, Noakes TD. Impaired exercise performance in the heat is associated with an anticipatoiy reduction in skeletal muscle recruitment. Pflugers Arch. 2004;448(4):422-30
107. Galloway SDR, Maughan RJ. Effects of ambient temperature on the capacity to perform prolonged cycle exercise in man. Med Sci Sports Exerc. 1997;29(9):1240-9.
108. Parkin JM, Carey MF, Zhao S, Febbraio MA. Effect of ambient temperature on human skeletal muscle metabolism during fatiguing submaximal exercise. J Appl Physiol. 1999;86(3):902-8.
109. Tatterson AJ, Hahn AG, Martini DT, Febbraio MA. Effects of heat stress on physiological responses and exercise performance in elite cyclists. J Sci Med Sport. 2000;3(2):186-93.
110. Morris JG, Nevill ME, Boobis LH, Macdonald IA, Williams C. Muscle metabolism, temperature, and function during prolonged, intermittent, high-intensity running in air temperatures of 33° and 17°C. Int J Sports Med. 2005;26(10):805-14.
111. Drust B, Rasmussen P, Mohr M, Nielsen B, Nybo L. Elevations in core and muscle temperature impairs repeated sprint performance. Acta Physiol Scand. 2005;183(2):181-90.
112. Trapasso LM, Cooper JD. Record performances at the Boston Marathon: biometeorological factors. Int J Biometeorol. 1989;33(4):233-7.
113. El Helou N, Tafflet M, Berthelot G, Tolaini J, Marc A, Guillaume M, et al. Impact of environmental parameters on Marathon running performance. PLoS One. 2012;7(5):e37407.
114. Bruck K, Olschewski H. Body temperature related factors diminishing the drive to exercise. Can J Physiol Pharmacol. 1987;65(6):1274-80.
115. Nybo L, Nielsen B. Hyperthermia and central fatigue during prolonged exercise in humans. J Appl Physiol. 2001;91(3):1055-60.
116. Nybo L, Rasmussen P, Sawka MN. Performance in the heat-physiological factors of importance for hyperthermia-induced fatigue. Compr Physiol. 2014:4(2):657-89.
117. González-Alonso J, Crandall CG, Johnson JM. The cardiovascular challenge of exercising in the heat. J Physiol. 2008;586(1):45-53.
118. Johnson JM. Exercise in a hot environment: the skin circulation. Scand J Med Sci Sports. 2010;20 Suppl 3:29-39.
119. Johnson JM. Regulation of skin circulation during prolonged exercise. Ann N Y Acad Sci. 1977;301:195-212.
120. Rowell LB. Competition between skin and muscle for blood flow during exercise. In: Problems with temperature regulation during exercise. New York: Academic Press; 1977.
121. Rowell LB, Marx HJ, Bruce RA, Conn RD, Kusumi F. Reductions in cardiac output, central blood volume, and stroke volume with thermal stress in normal men during exercise. J Clin Invest. 1966;45(11):1801–1816
122. Lenzuni P, Capone P, Freda D, Del Gaudio M. Is driving in a hot vehicle safe? Int J Hyperth. 2014;30(3):338-342
123. Jiang Q, Yang X, Liu K, Li B, Li L, Li M, et al. Hyperthermia impaired human visual short-

term memory: an fMRI study. Int J Hyperth. 2013;29(3):219-24.
124. Wijayanto T, Toramoto S, Tochihara Y. Passive heat exposure induced by hot water leg immersion increased oxyhemoglobin in pre-frontal cortex to preserve oxygenation and did not contribute to impaired cognitive functioning. Int J Biometeorol. 2013;57:557–67.
125. Wijayanto T, Toramoto S, Maeda Y, Son S-Y, Umezaki S, Tochihara Y. Cognitive performance during passive heat exposure in Japanese males and tropical Asian males from Southeast Asian living in Japan. J Physiol Anthropol. 2017;36:1–11.
126. Colquhoun WP, Goldman RF. Vigilance under induced hyperthermia. Ergonomics. 1972;15(6):621-32
127. Nunneley SA, Dowd PJ, Myhre LG, Stribley RF, McNee RC. Tracking-task performance during heat stress simulating cockpit conditions in high-performance aircraft. Ergonomics. 1979;22(5):549-55
128. Pilcher JJ, Nadler E, Busch C. Effects of hot and cold temperature exposure on performance: a meta-analytic review. Ergonomics. 2002:45(10):682-98.
129. Hancock P, Vasmatzidis I. Effects of heat stress on cognitive performance: the current state of knowledge. Int J Hyperth. 2003;19(3):355-72.
130. Ramsey JD. Task performance in heat: a review. Ergonomics. 1995;38(1):154-65.
131. Hancock PA. Sustained attention under thermal stress. Psychol Bull. 1986;99(2):263-81.
132. Schmit C, Hausswirth C, Le Meur Y, Duffield R. Cognitive functioning and heat strain: performance responses and protective strategies. Sports Med. 2017;47(7):1289-1302.
133. Gaoua N, Racinais S, Grantham J, El Massioui F. Alterations in cognitive performance during passive hyperthermia are task dependent. Int J Hyperth. 2011;27(1):1-9.
134. Racinais S, Gaoua N, Grantham J. Hyperthermia impairs short-term memory and peripheral motor drive transmission. J Physiol. 2008;586(19):4751-62.
135. Schlader ZJ, Gagnon D, Adams A, Rivas E, Cullum CM, Crandall CG. Cognitive and perceptual responses during passive heat stress in younger and older adults. Am J Physiol Regul Integr Comp Physiol. 2015;308(10):R847-54
136. Lan L, Wargocki P, Lian Z. Thermal effects on human performance in office environment measured by integrating task speed and accuracy. Appl Ergon. 2014;45(3):490-5
137. Lan L, Wargocki P, Wyon DP, Lian Z. Effects of thermal discomfort in an office on perceived air quality, SBS symptoms, physiological responses, and human performance. Indoor Air. 2011;21(5):376-90
138. Ely BR, Sollanek KJ, Cheuvront SN, Lieberman HR, Kenefick RW. Hypohydration and acute thermal stress affect mood state but not cognition or dynamic postural balance. Eur J Appl Physiol. 2013;113(4):1027-1034
139. Razmjou S, Kjellberg A. Sustained attention and serial responding in heat: mental effort in the control of performance. Aviat Space Envir Md. 1992;63:594–601.
140. Hocking C, Silberstein RB, Lau WM, Stough C, Roberts W. Evaluation of cognitive performance in the heat by functional brain imaging and psychometric testing. Comp Biochem Physiol A Mol Integr Physiol. 2001;128:719–34.
141. Gaoua N, Grantham J, Racinais S, El Massioui F. Sensory displeasure reduces complex cognitive performance in the heat. J Environ Psychol. 2012;32:158–63.
142. Radakovic SS, Maric J, Surbatovic M, Radjen S, Stefanova E, Stankovic N, et al. Effects of acclimation on cognitive performance in soldiers during exertional heat stress. Mil Med. 2007;172:133–6.

7 Protective Clothing in Thermal Environments

Joo-Young Lee

Clothing is the second skin of the human body and a representative behavioral thermoregulatory tool in thermal environments. Personal protective clothing is worn to protect workers from noxious materials and harmful environments. However, a dilemma arises when wearing protective clothing, as it creates a serious thermal burden, especially in hot environments. This is due to the fact that typical protective clothing consists of several layers, which can be semipermeable or impermeable to air and water vapor. Heat transfer between the human body and its environment occurs through conduction, convection, radiation, and evaporation. Protective clothing alters the properties of heat exchange between the human skin and its environment. Among the four avenues, convective and radiative heat loss are more crucial than conduction in thermoneutral environments. However, conductive heat transfer becomes more significant when wearing cooling or heating garments with protective clothing in hot or cold environments. In this chapter, we delve into the thermal strain experienced by workers who wear protective clothing in hot or cold environments, and explore test methods to evaluate the thermal performance of protective clothing.

J.-Y. Lee (✉)
Department of Fashion and Textiles, College of Human Ecology, Seoul National University, Seoul, South Korea
e-mail: leex3140@snu.ac.kr

7.1 A Pyramid of Protective Clothing Research

Researchers who focus on evaluating and improving the performance of protective clothing often grapple with a balancing act between protection and comfort. While protective clothing should afford adequate protection for workers, greater protection can result in increased discomfort, posing an additional physical burden to workers (Fig. 7.1).

There has been a discussion about "What is an appropriate protection?" It would be ideal to protect workers safely without deteriorating comfort. In order to develop and evaluate protective clothing, researchers typically follow six steps: survey, textile performance tests, manikin tests, human wear trials in controlled environments, field tests, and standardization (Fig. 7.2). As the first step, conducting surveys with workers and monitoring their work environments can provide insights and requirements to improve the existing

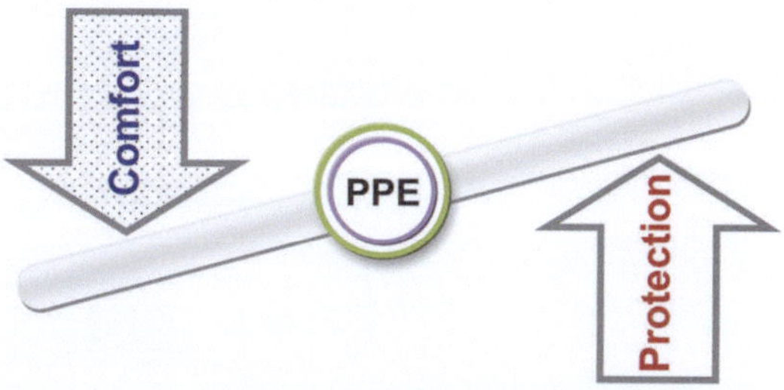

Fig. 7.1 A balance between protective and comfort functions of personal protective equipment (PPE)

Y. Tochihara (ed.), *The Thermal Environment*, https://doi.org/10.1007/978-981-96-3009-7_7

protective clothing. For example, manufacturers have developed real-time biomonitoring undershirts for elderly healthcare, but survey results can indicate that most older workers seldom wear tight clothing. Physical performance tests of textiles reflect the fundamental requirements for protective functions and are a necessity for creating protective clothing. However, textile tests assess two-dimensional properties without accounting for the influences of the microclimate inside the clothing. Clothing is worn around the human body as a three-dimensional entity, which is affected by factors such as air layers, openings, closures, body movements, and sweating. Clothing serves as protection against external hazards while also acting as a barrier that can disrupt the heat transfer from the human body to the environment. Quantitative analyses of heat exchange between the skin, clothing, and the surrounding environment can be simulated using a thermal manikin. Physiological burdens and wearing comfort are assessed through human wear trials conducted in laboratories or field settings. Test methods and standards for protective clothing are then developed based on results from these five steps (Fig. 7.2).

7.2 [Step 1] Survey on Protective Clothing

Personal protective clothing is categorized based on its application into various types, including chemical protective clothing, biological protective clothing, nuclear protective clothing, ballistic/puncture-resistant/cut-resistant protective clothing, and thermal protective clothing (e.g., firefighting, cooling, or cold-protective clothing). Over the past several decades, numerous surveys focusing on diverse types of protective clothing have been documented. We also conducted surveys with professionals from various sectors such as firefighters, healthcare workers, agricultural workers, delivery workers, and military soldiers regarding their protective clothing in thermal environments. This chapter, among the array of surveys conducted, presents the survey results specifically related to firefighting protective clothing.

An international survey pertaining to the development of the next generation of personal protective equipment (PPE) for firefighters spanned across Australia, Japan, Korea, and the United States [1]. The survey results revealed that

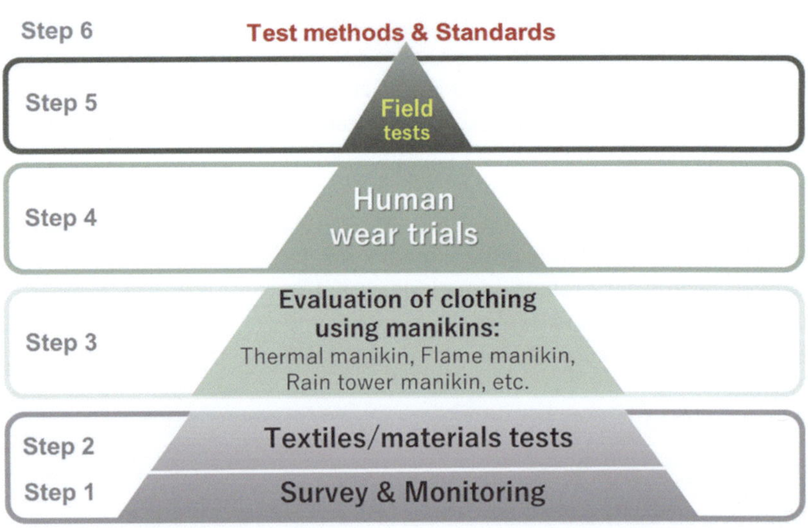

Fig. 7.2 A pyramid of protective clothing research

the most preferred smart feature among the firefighters was a location monitoring system and an automatic body cooling system. This indicates that firefighters' primary concern was the mitigation of heat stress, rather than merely enhancing protective functions. Kim et al. [2] investigated instances of heat-related illnesses among Korean firefighters during their duty hours, considering factors such as resting/working hours and behaviors related to removing PPE during rest periods. Among the 674 Korean firefighters surveyed, a 75% reported experiencing symptoms of heat-related illnesses. Notably, 36% of these firefighters consistently removed their helmets during rest, whereas only 3% removed their protective boots. Despite the efficacy of cooling the feet as one of the most efficient interventions to alleviate heat strain, firefighters seldom remove their boots during breaks due to their cumbersomeness. Similarly, as reported by Son et al. [3], 58% of firefighters in Japan experienced heat strain while on duty. Moreover, almost 90% of these Japanese firefighters rated the ventilation function of their PPE as inadequate.

Firefighters encounter varying levels of fire intensities, including flashovers, backdrafts, flame-overs, and rollovers, which are categorized as extremely hazardous fire scenarios. Lee et al. [4] conducted a survey involving 794 firefighters to assess their experiences and the performance of their PPE during flashover or backdraft situations. Firefighters expressed a strong desire for a reduction in the overall weight of the PPE, accompanied by improved mobility, enabling them to swiftly navigate high-risk environments. Their priority was the ability to escape swiftly, rather than enhancing the protective functionalities of the PPE materials or design. Naturally, the primary consideration remains the protective efficacy of the PPE. Presently, however, the balance between protection and comfort has taken precedence, as the existing protective capabilities of PPE are generally sufficient to safeguard firefighters in typical fire scenarios. According to Baek et al. [5], who surveyed 106 firefighters, the dissatisfaction rate regarding the comfort aspect of their PPE stood at 80%, whereas the dissatisfaction rate concerning the protective function was 55%. Furthermore, engaging in firefighting activities during flashovers or backdrafts is not advisable for firefighters due to the extreme hazards associated with such scenarios.

We deliberated on a balance between enhancing the protective and comfort functions when advancing firefighting PPE. Despite the emphasis on the balance, there remains a persistent need for enhancing protective functions, particularly in protective gloves, visors, and protective hoods. Kim et al. [6] conducted a comprehensive investigation into burn injuries sustained by 536 firefighters during firefighting operations. Their findings revealed that 22% of the firefighters had experienced burns, encompassing first-degree burns. Notably, the most frequently affected body regions were the hands (37%) and the head (including the face and neck) (34%). Firefighters cited their gloves as the most vulnerable component. Furthermore, Kim et al. [7] conducted a survey involving 270 firefighters, focused on fire protective gloves. The firefighters were queried about the significance that they attributed to various performance factors of these gloves. It emerged that hand dexterity ranked as the most vital factor, followed by fire/flame protection and cut resistance. Supporting these results, another survey by Baek et al. [5] reported that actively serving firefighters identified fire-protective gloves as the item most in need of improvement among various PPE components.

As the initial stage in the evaluation and enhancement of PPE (Fig. 7.2), it is imperative to undertake surveys involving stakeholders and workers. Researchers must adopt a balanced standpoint that considers both protection and comfort while evaluating PPE. Designers and manufacturers ought to acknowledge that the protective attributes of PPE must harmoniously coexist with comfort, thereby delivering enhanced safeguards to workers, particularly in exceptionally challenging thermal environments.

7.3 [Step 2] Textile Performance Tests

Testing and standards for protective clothing predominantly focus on material properties and prerequisites. These aspects have been meticulously documented in the following classic reference texts: Fourt and Hollies [8], Hollies and Goldman [9], and Renbourn and Rees [10]. Notably, the factors exerting influence over the thermal attributes of textiles encompass several dimensions. These include structural factors such as the fabric type, threads, yarns, and the specific heat of fibers. Additionally, surface characteristics encompassing texture, hairiness, coatings, and applied films play a role, along with factors tied to dry and latent heat transfer, as well as environmental parameters.

Typically, thermal resistance, thermal absorptivity, thermal conductivity, breathability, and air/water vapor permeability are scrutinized as key thermal properties of textiles. However, these physical properties only partially encapsulate the thermal performance of clothing. Textiles envelop the human body, engendering a distinct microclimate detached from the ambient surroundings. Heat transfer through clothing is markedly influenced by the air layers presented within textiles and the established clothing microclimate, rather than being solely determined by the intrinsic properties of the fibers themselves. The microclimate created by clothing constitutes a barrier to the transfer of heat and water vapor from the skin to the external environment. According to NIOSH [11], research focusing on clothing materials have culminated in the determination that the insulating capacity of clothing generally adheres to a linear relationship with its thickness, depending upon the extent of the enclosed still air layer. However, differences in fibers or fabric weave yield only negligible impacts on insulation, unless these differences tangibly influence the fabric's thickness or its vapor and air permeability. Notably, fibers exhibit higher conductivity than air, and augmenting fiber density can paradoxically lead to a reduction in their thermal insulation. It is important to note that assessing "clothing comfort" cannot be feasibly achieved through mere textile testing. This stems from the fact that comfort is fundamentally a subjective notion, denoting the psychological state of individuals expressing satisfaction with their thermal environment [12–14].

The conundrum between ensuring protection and optimizing comfort presents a noteworthy challenge. A case in point is evident in the imperative requirements of protective clothing for firefighters, where attributes of thermal stability and insulation become paramount when facing extreme heat and flames. In this scenario, the selection of textiles boasting heightened thermal and flame resistance emerges as a judicious choice for crafting firefighting protective clothing. However, firefighters must effectively facilitate the transmission of metabolic heat and evaporation of sweat vapor emanating from firefighters' bodies. Moreover, the pursuit of heightened protection can inadvertently introduce a counterpoint of escalated stress. While integrating bulkier thermal liners and water-impermeable, partially breathable moisture barriers can indeed enhance defense against heat and flames, their deployment may simultaneously impose excessive thermal stress on firefighters. This juxtaposition underlines the complexity of harmonizing protection and comfort, necessitating a judicious synthesis of both imperatives.

7.4 [Step 3] Thermal Manikin Tests

In the robust and multilayered protective clothing, the dynamics of heat transfer between the body and the environment assume a complex nature. A thermal manikin has been used as an analogous tool of the human body with human-like functions such as sweating, walking, breathing, and multilayered structures. Employing the thermal manikin offers quantified data of gauging the thermal and water vapor resistance of clothing. The development of the first thermal manikin dates back to the early 1940s, as a single-segment copper manikin was devised for the US Army. In 1942, Belding used the first heated manikin to delineate the comfort parameters for sleeping bags and Arctic uniforms. Subsequent advancements were found from Dr. Goldman simulating sweating by draping his copper manikin with a closely fitting cotton clothing that he saturated with water before being garbed [15]. Modern thermal manikins provide more detailed information about regional variation, having 15–34 separately controlled heaters. Furthermore, comprehensive thermal manikins encompass a spectrum spanning from infants to adult males and females. To accurately simulate body movement and walking dynamics, motion systems have been incorporated, accommodating walking speeds.

Thermal insulation signifies the resistance against heat transfer between the human body and the environment, and stands as the reciprocal of thermal conductance. Given that thermal conductance is typically expressed in W/(m²·°C), thermal resistance is quantified in °C·m²/W. Gagge et al. [16] introduced the "clo" unit as a measure to depict the thermal resistance of clothing, particularly in the context of military uniforms worn by soldiers in cold climates. The clo unit provides a practical means of conveying the relative thermal insulation value. Specifically, 1 clo designates the thermal insulation conferred by a standard business suit at an air temperature of 21 °C with the relative humidity of 50%, which is defined as 0.155 °C·m²/W. Standard clothing for summer work, comprising half-sleeve shirts and long pants, yields a thermal insulation value of approximately 0.6 clo, while winter clothing ensemble registers around 2.0 clo. Examples of clothing thermal insulation are provided in ISO 9920 [17]. In ISO 9920 [17], "a garment" is defined as an individual constituent of a clothing ensemble, and a clothing ensemble is defined as the combination of garments covering the body.

With the progressive refinement and applications of thermal manikins in various thermal settings, several issues concerning test methodologies and influential factors have come under discussion. These encompass aspects such as partial and overall clothing insulation, water vapor resistance, or the impacts of air velocity. The issue surrounding partial and overall thermal insulation has stemmed from technological advancements of thermal manikins. In the present landscape, commercially available thermal manikins can be categorized into two groups: full-body manikins and body part manikins that simulate specific body sections. Among the latter category, manikins emulating the head, one hand, or one foot are prevalent. Before our discussion moving forward, it is essential to clarify that in the unit of thermal resistance, the square meter represents the surface area of the human body (e.g., 1.8 m² for an adult male; 1 clo = 0.155 °C·m²/W signifies thermal insulation per "m²." Let's remember that the clo unit is a relative unit). Consider, for instance, the scenario of evaluating the thermal insulation of a protective boot. This could be carried out using a full-body manikin or a thermal foot manikin. The full-body male manikin typically has a body surface area ranging from 1.8 to 2.0 m², while the surface area of a foot thermal manikin corresponds to the area of one side of the foot and half of the calf. The insulation value yielded by the full-body manikin signifies the thermal insulation of a pair of boots, devoid of any additional clothing layers (e.g., wearing boots only over the nude body, about 90%BSA unclothed). Conversely, the insulation value derived from the foot thermal manikin reflects the thermal insulation of an oversized boot enveloping the entire body (e.g., wearing a huge boot over the entire body, most skin clothed). As addressed, both insulation values are normalized to the unit body surface area (°C·m²/W). Hence, in this context, a direct comparison between these two values is inadvisable. Researchers should explicitly specify the employed manikin type in their measurements.

The second concern may indeed be intertwined with the first. While the foundational copper manikin was essentially a single entity, contemporary thermal manikins are composed of up to 20 or more body zones. Though limited in number, a few studies have reported thermal insulations specific to certain body regions that are encompassed by clothing. This arises due to the capability of modern manikins to control over distinct body zones. However, it is crucial to acknowledge that the clo value extrapolated from a full-body thermal manikin encapsulates the insulation afforded to the entire body, comprising both clothed and unclothed portions.

The third concern aligns closely with the first and second issues related to segmented body zones. With the evolution of thermal manikins from the singular-segment model to those over 20 individually heated segments, distinct methodologies have arisen for computing thermal insulation. These methodologies encompass the global, parallel, and serial methods [17]. The global method involves deriving thermal insulation by calculating the overall mean surface temperature and heat flux using all regional body heat fluxes and surface temperatures, with each being weighted by area [17]. On the other hand, the serial method entails summing the local thermal insulation for each body segment, assuming that each segment possesses its own heat production and loss dynamics [17]. Empirical observations tend to converge on the conclusion that values obtained through the serial method tend to exceed those values through the global method [18, 19]. It is worth noting if the thermal manikin is covered with exactly the same level of insulation over all sections, results from the two methods would be almost the same [20]. Lee et al. [21] validated clothing insulation derived from the global and serial methods in comparison with human wear trials. They encompassed the assessment of 150 individual garments and 38 clothing ensembles using a thermal manikin, alongside 26 seasonal clothing ensembles that underwent human trials (Fig. 7.3).

The serial values were found to exceed the global values, exhibiting increments of 39% (0.18 clo) for spring/fall wear, 63% (0.15 clo) for summer wear, and 65% (0.69 clo) for winter wear (Fig. 7.3b). Notably, the global method produced systematically lower clothing insulation values across all 26 seasonal ensembles compared to those attained through human wear trials (Fig. 7.3a). This result suggests that the values generated by the global calculation can be more accurately aligned with data from human wear trials through suitable corrections.

The aforementioned three concerns collectively account for the holistic thermal dynamics inherent in the human body. This underlines the necessity for a judicious and circumspect interpretation of data extracted from analyses involving body regions, as facilitated by advanced thermal manikins.

The fourth issue is around the methodology employed for controlling sweating when estimat-

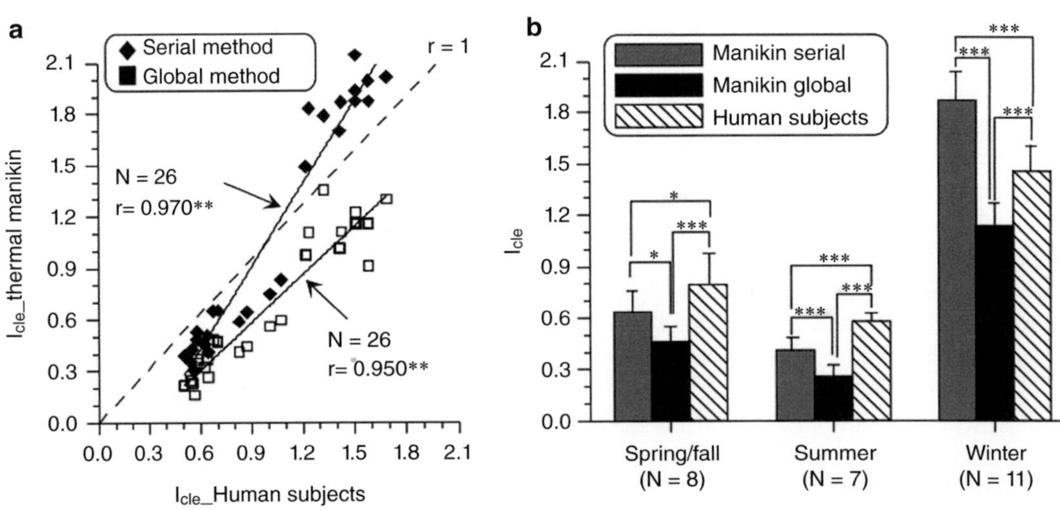

Note: Significance at: *P<0.05, **P<0.01 and ***P<0.001

Fig. 7.3 Relationships between thermal insulation of clothing ensembles estimated with a thermal manikin and human subjects (**a**) and seasonal clothing ensembles (**b**). (Redrawn from Lee et al. [21])

ing evaporative resistance. The evaporative resistance of the business suit with 1 clo is 0.033 kPa·m²/W [22], and the evaporative resistance of the still air layer around the nude body is approximately 0.014 kPa·m²/W. An inherent flaw arises in obtaining the values of the evaporative resistance measured by the sweating thermal manikin because the calculation involves sweating mass loss (dripping). The problem becomes particularly pronounced when attempting to measure the evaporative resistance of (semi) impermeable protective clothing, relating to the regulation of the sweating so that there is no sweating dripping. Consequently, the rate of sweating must be judiciously regulated, and this can be diversified based on distinct body segments as well as the specific types of protective clothing being evaluated.

Last issue is the effect of air flow and still air layer on thermal insulation. For a nude body in air, the thermal insulation of the air layer around the body is given the symbol $\mathbf{I_a}$. It is reported that the still air layer formed around the surface of the clothing is around 6 mm thick [23]. Air layer thickness is affected by the clothing's fit, body posture, movements, and the wind around the body. The intrinsic insulation of the clothing does not include this air layer and is given the symbol $\mathbf{I_{cl}}$. As the insulation of the air around clothing decreases with the increase in surface area and hence the thickness of clothing, the symbol used $\mathbf{f_{cl}}$ (clothing area factor), which is the ratio of the clothed surface area to the nude surface area. The total insulation is then given as $I_T = I_{cl} + I_a/f_{cl}$. Values greater than 2.5 clo are difficult to achieve as adding clothing layers provides diminishing return on thermal insulation partly because the surface area for heat exchange with the environment becomes large. Thermal insulation is remarkably decreased by air flow because the thermal insulation of air around the body decreases in an exponential way as wind increases. We investigated the thermal insulation of clothing ensembles, including padded jackets with two different filling types in the following four conditions (two air temperatures × two air velocities) [24]. The polyester filling ensemble (I_{cl} = 1.25 clo at the air temperature of 21 °C) was warmer than a goose down ensemble (1.21 clo) below 0.15 m·s^{-1}, while the goose down ensemble (0.67 clo) had greater thermal insulation than the polyester filling ·ensemble (0.61 clo) at an air velocity of 1.5 m·s^{-1}. Thermal insulation was reduced 40–50% when air velocity reached 1.5 m·s^{-1} for both ensembles.

Clothing ensemble of 1.0 clo in still air yields 0.4 clo for wind at around 1.0 m/s, and 0.1 clo at around 20 m/s. Wind velocity up to 1 m/s does little to disturb the intrinsic clothing insulation, due to the inability to penetrate through the materials and to disturb the inner air layers inside clothing [25]. At the wind velocity of 1.5 m/s, the thermal insulation of the still air layer becomes abruptly decreased [23, 26, 27]. Nielsen et al. [28] found a 30%, 50%, 36% reduction in the I_{cl} and a 20% reduction in I_a during cycling. The combined effect of a wind and movement reduced the clothing insulation up to 60% [27]. Also, the body motion affects the enclosed and surrounding air layers, while the wind mainly affects the surrounding air layer and the layer under the outer garment. The combination of air velocity and body movements reduces the intrinsic clothing insulation and the insulation of the surface air.

7.5 [Step 4] Human Wear Trials in Hot and Cold Environments

7.5.1 Heat Strain of Workers and Protective Clothing

Heat stress is an any change in the thermal relationship between a temperature regulator and its environment, which, if uncompensated by temperature regulation, would result in hyperthermia [29]. Heat strain is defined as a deviation of body temperature induced by sustained heat stress that cannot be fully compensated by temperature regulation [29]. There are four heat-related disorders: heat cramps, heat exhaustion, heat syncope, and heat stroke. It is known that the most vulnerable workers to heat-related disorders are construction workers, military soldiers, or agricultural workers during summer, but workers who should wear heavy PPE can experience heat-related disorders even in thermoneutral environments. In addition to resistance to dry heat transfer (conduction, convection, and radiation), heavy PPE provides resistance to the evaporation of sweat. The evaporative resistance depends upon the vapor permeability properties of the clothing, but most PPE is heavy, multilayered, encapsulated, or (semi-) impermeable. Level A or Type 1 PPE, such as chemical and biological protective clothing, protects the respiratory tract and the whole body against vapor and liquid. These types of clothing place a great deal of thermal burden on the wearers, as they are heavy, bulky, and stiff, preventing microclimate heat and vapor transfer. Occupational Safety and Health Administration (OSHA) classifies the PPE levels from Level A to D according to the protective level. However, the heat stress experienced when wearing these PPE is not always proportional to the protective level. We evaluated the heat strain of workers who wear firefighting PPE, CBRN PPE, healthcare workers' PPE, or military combat uniforms in hot environments.

7.5.1.1 Firefighters' Protective Clothing and Heat Strain

Firefighting is one of the most physically demanding occupations, with a considerably high rate of on-duty fatalities worldwide. Physiological strains in firefighting activities result from the combination of strenuous physical work, extreme thermal environments, and heavy protective clothing [30–36]. Holmér and Gavhed [37] reported that male firefighters sustain a mean metabolic rate of 474 W·m^{-2} for periods of 20 min and longer. In particular, firefighters' protective clothing exacerbates heat strain by increasing the metabolic rate due to the addition of clothing weight load and inhibiting body heat dissipation from the skin to the air. A typical firefighting PPE, including self-contained breathing apparatus (SCBA), weighs approximately ~30 kg [32], and its thermal insulation is about ~2.44 clo [38].

We extracted results on the core body temperatures of firefighters wearing various combination of PPE during exercise at air temperatures of 22, 28, 30, or 32 °C (Fig. 7.4). In Fig. 7.4a, four types of firefighters' jackets (Type A only is aluminized coated, while Type B, C, and D are typical turnout jackets) were worn along with Japanese stationary uniform (CON), and firefighters exercised at 30%, 45%, and 60% of VO_{2max} for the three cycling sessions [39]. Figure 7.4b was from Baek et al. [40], which compared Japanese, Korean, and US firefighters' helmets. Firefighters conducted a 30-min exercise at 5 km·h^{-1} on a treadmill. Figure 7.4c shows the increases in rectal temperature when wearing PPE without SCBA, with the current SCBA, mass-lightening SCBA, and mass-lightening and improved harness strap SCBA. They conducted a 30-min running at 6 km·h^{-1} on a treadmill [30]. Roh et al. [41] evaluated the effect of firefighters' boot weight from 3.2 to 5.3 kg on rectal temperature during 30-min exercise at 5.5 km·h^{-1} on a treadmill (Fig. 7.4d). In Fig. 7.4e, eight types of PPE conditions, excluding each PPE component, were compared, and subjects conducted a 30-min exercise at 5.5 km·h^{-1} on a treadmill. Figure 7.4f shows the effects of wearing body cooling vest with an ice drinking function on rectal temperature during 30-min exercise at 5.5 km·h^{-1} on a treadmill [42]. Most experimental protocols consisted of 30-min exercise with a 10-min rest and a 20-min recovery, as published in Tochihara et al. [43]. Taken together, we can consider that core body temperature of firefighters on duty increases up to around 38.5 °C with a rise of 1.5–2.0 °C in hot environments, and the increases continued even during recovery session. In particular, under heat radiation from fire and flames, core body temperature can increase up to 39 °C when wearing a heat-reflective aluminized jacket (Fig. 7.4a). Improving the design of the SCBA harness strap (Fig. 7.4c) and helmets with a protective hood (Fig. 7.4b), as well as reducing the weight of SCBA (Fig. 7.4c), can alleviate the heat strain of firefighters in hot environments. Furthermore, it is worth noting that rectal tem-

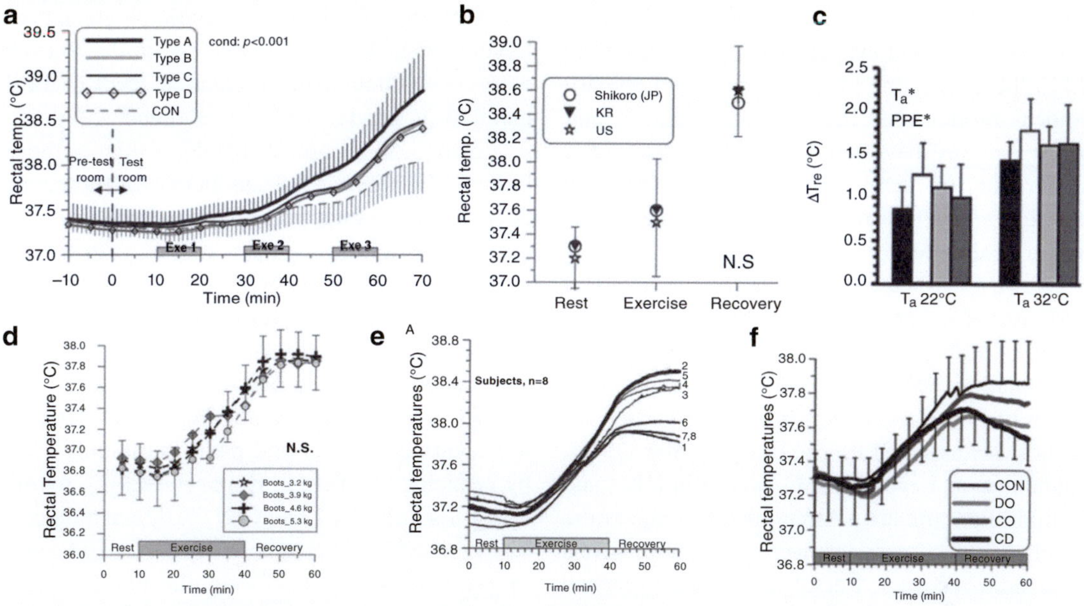

Fig. 7.4 Rectal temperatures when wearing firefighting personal protective equipment in thermoneutral and hot environments: (**a**) air temperature of 30 °C with relative air humidity of 50% with an additional infrared heat radiation (1.1 kW·m^{-2}) from Chou et al. [39], (**b**) 32 °C with 70%RH from Baek et al. [40], (**c**) 22 and 32 °C with 50%RH from Bakri et al. [30], (**d**) 28 °C with 50%RH from Roh et al. [41], (**e**) Lee et al. [35], and (**f**) 30 °C with 50%RH from Kim et al. [42]

perature was lower for the condition without protective boots (Condition 6 in Fig. 7.4e) compared to No SCBA (#3), No helmet (#4), No Gloves (#5), and full PPE (#2) (Fig. 4e), which could be applied to passive cooling strategies during breaks. Wearing a cooling vest with an ice drinking function (Condition CD in Fig. 7.4f) effectively reduced rectal temperature, especially during the recovery period (Fig. 7.4f).

Figure 7.4 summarizes the rectal temperature only when wearing various combinations of firefighters' protective clothing in thermoneutral and hot environments from the six studies. However, the impacts of wearing PPE on cardiovascular responses, sweat responses, or psychological responses should be considered comprehensively to improve firefighters' protective clothing. Most of all, we should consider the balance between protection and comfort. In this regard, effective body cooling strategies can be developed, encompassing both passive and active cooling technologies. This will be introduced in Sect. 7.5.2 of this chapter.

7.5.1.2 Chemical Protective Clothing and Heat Strain

In terms of the chemical protective functions of PPE, ISO 16602 [44] specifies six types of PPE: gas-tight PPE with internal or external breathing air supply (Type 1a/1b/1c), non-gas-tight PPE with breathable air supply (Type 2), liquid-tight PPE (Type 3), spray-tight PPE (Type 4), PPE against airborne solid chemicals (Type 5), and PPE with limited protective performance against liquid chemicals (Type 6). In the United States, four classification levels are used by the Occupational Safety and Health Administration (OSHA). Level A provides the highest protection, including skin, respiratory, and eye protection. Level B provides the highest respiratory protection but a less skin protection. Level C protects from airborne substance along with liquid splashes and Level D is a work uniform affording minimal protection [45]. Thermal insulation and vapor resistance of the various types/levels of PPE have been reported [46–50].

These studies suggest that the greater the thermal insulation and evaporation resistance of the PPE, the greater the metabolic cost and physiological burden. As Hanson [51] pointed out, many studies have generally demonstrated a continuum of increasing heat stress running through a vapor-permeable to liquid and vapor-impermeable fabrics.

However, whether or not protection levels of PPE are indeed proportional to the physiological strain of PPE is a question which needs further examination. We evaluated whether the heat strain of PPE wearers is proportional to the protection levels of PPE, classified by the OSHA guidelines as Level A, B, C, and D in a hot and humid environment. After rectal temperature reached 39.0 °C, subjects were required to have a 5–20 min recovery. Rectal temperature reached 39 °C latest for Level D and the earliest for Level A, while the tolerance time was not significantly different between Level B and C, but 6-min shorter with Level C than Level B. The speed of increase in rectal temperature was faster for Level C than that for Level B (Fig. 7.5a). The results indicate that there is no continuous relationship between the protection level and physiological strain, especially Level B and C. The results suggest that physiological strain could be greater for Level C than B, or they might be similar to each other, even though the protection level of PPE materials is greater for Level B than C.

Figure 7.5b introduces results similar to those in Fig. 7.5a, wherein we conducted a comparative analysis of the thermal burden of three different types of pesticide protective clothing with similar protective levels (Level C): (1) PC, representing traditional work wear (Level C1, polyester and cotton, 782 g, 1.26 clo, and 42.0 Pa·m^{2}·W^{-1}), (2) TV, denoting Tyvek 600 coverall (Level C1, 172 g, 1.16 clo, 47.5 Pa·m^{2}·W^{-1}), and (3) CU, representing a jacket and pants commonly used by pesticide handlers in Korea (Level C2, nylon, 764 g, 1.42 clo, 54.1 Pa·m^{2}·W^{-1}). Rectal temperature reached 38.3 °C in both the CU and TV conditions with no statistically significant differences (Fig. 7.5b). Notably, despite the Tyvek condition exhibiting lower clothing mass and thermal insulation compared to the CU condition (nylon work wear), the physiological burden during human wear trials was frequently more pronounced for the nylon work wear than for the Tyvek condition. Small design factors (coverall vs. separated pieces) could surpass the difference in thermal properties of the clothing.

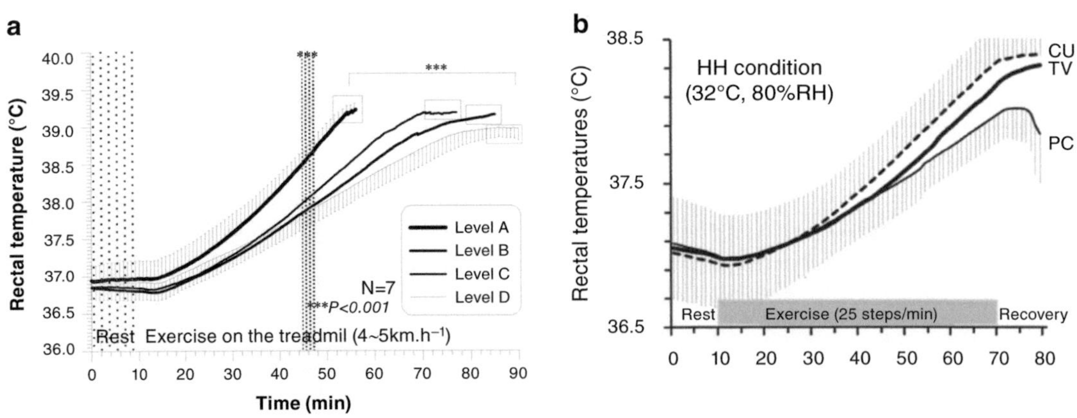

Fig. 7.5 Time courses in rectal temperature during exercise and recovery while wearing personal protective equipment at an air temperature of 33 °C with 70%RH (Left) [52] and at an air temperature of 32 °C with 80%RH (Right) [53]. Where, PC, TV, and CU represent three types of pesticide protective clothing at Level C1, Level C1, and Level C2, respectively

7.5.2 Skin Wetting and Humidity Inside Clothing

While one might assume that the high temperature inside the PPE microclimate is the most critical factor in heat-related injuries of PPE workers, a high level of humidity within the PPE is even riskier for workers, especially when wearing semipermeable or impermeable PPE (e.g., Level A and B). High humidity inside the PPE is strongly related to high skin wettedness of workers. Skin wettedness has been defined as the fraction of the total body surface area (BSA) covered by sweat (the wetted area) [29]. It is generally agreed that unpleasantness during heavy sweating is not only due to thermal sensation, but also due to the size of the total wetted area over the skin. We newly developed a method for quantifying perceived skin wettedness of workers wearing personal protective clothing [21]. The perceived skin wettedness finally reached 51.6% of the whole body surface area when wearing encapsulated PPE at rest in 25 °C, while the value rose to 89.8% at rest in 32 °C (Fig. 7.6). For exercise conditions, the perceived skin wettedness of the encapsulated protective clothing reached around 90% of BSA within 30 min at both thermal environments. The perceived skin wettedness had significant relationships with rectal temperature, heart rate, total sweat rate, thermal

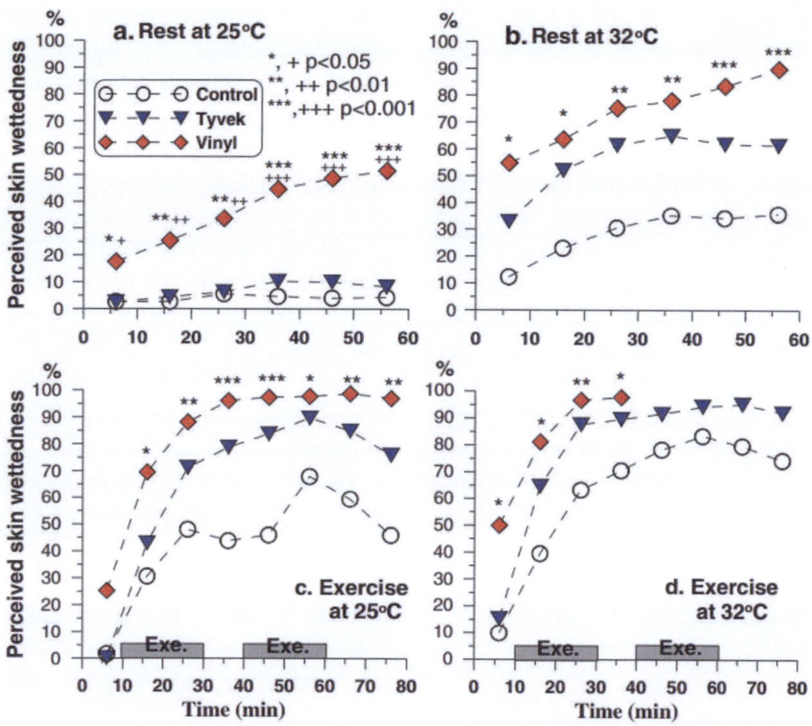

Fig. 7.6 Time courses of perceived skin wettedness (%BSA) for the three clothing conditions at rest and during exercise (Control: half-sleeve shirt and half-pants; Tyvek: coverall, Vinyl: encapsulated protective clothing without SCBA) [21]

comfort, and humidity sensation [21]. Sweat efficiency, defined as the ratio between evaporative rate and sweat rate, decreased when skin wettedness exceeded 0.74 [54].

Recently, we monitored clothing microclimate humidity of workers wearing chemical and flame protective clothing while doing a simulated mobility protocol at an air temperature of 19 °C with air humidity of 57%RH [55]. The clothing microclimate humidity on the chest exceeded 80%RH and reached as high as 98%RH in an hour, which caused fogging inside the visor and impaired visibility. These results indicate that wearing encapsulated personal protective clothing causes high microclimate humidity inside the PPE because of the high evaporative resistance of PPE, which induces greater skin wettedness of workers. According to NIOSH [11], at the extent of skin wetted with sweat approaches 20%BSA, the discomfort begins to be noted, and performance decrements can appear with between 20% and 40% wetting of BSA. Performance decrements become increasingly noted wetted skin area approaches 60%BSA. Sweat begins to be wasted dripping at 70%BSA, and physiological strain becomes marked between 60% and 80%BSA. Increases of the sweated area above 80%BSA result in limited tolerance, even for physically fit young workers. Remember that PPE-workers' skin becomes wet greater than 80%BSA, and the microclimate becomes humid over 80%RH in a half-hour, especially during exercise in hot environments. In this regard, more studies on the dehumidification inside clothing are required for PPE-workers.

7.5.3 Body Cooling Strategies in Hot Environments

Various actions can be implemented to mitigate heat stress and strain among workers in hot environments. Engineering controls, exemplified by the utilization of air conditioning, aim to diminish environmental heat stress. Administrative controls can be employed to regulate factors such as heat acclimation, work schedule, hydration, rest/recovery, or medical evaluations. However, when it comes to workers clad in protective clothing, the implementation of both engineering and administrative controls becomes technically unfeasible or impractical. Full personal protective equipment, such as firefighters' protective clothing or NBC protective clothing, comprises protective over garments (jacket/pants, coverall), boots, a mask, a hood, and gloves. The wearing of such gear imposes an augmented metabolic burden due to its bulkiness and weight. Additionally, the high evaporative resistance of the outer garments hinders the dissipation of metabolic heat, exacerbating the challenges faced by wearers.

Indeed, protective clothing in environments above 24 °C decreases tolerance times for continuous, moderately heavy physical work to only 30 min [56]. Microclimate cooling inside the clothing is the most efficient intervention to prevent heat-related to injuries of workers. Microclimate cooling systems vary from simple approaches, such as applying frozen materials beneath clothing, to more intricate systems. Four auxiliary cooling techniques have been evaluated: (1) liquid-cooled garments, (2) air-cooled garments, (3) ice-cooled vest, and (4) wetted over garments. Enhancing cooling efficacy necessitates careful consideration of the coverage extent and specific body areas addressed by the cooling garment. This is crucial because cutaneous vasomotor adjustments and skin blood flow are not uniformly distributed [57] (Table 7.1).

Speckman et al. [57] conducted a comprehensive review of studies on the effectiveness of the three different body cooling systems—liquid, ice, and air—carried out at the US Army Research Institute of Environmental Medicine (USARIEM). While various microclimate cooling systems, encompassing ice-, air-, and liquid-cooled garments, showed effectiveness in mitigating heat stress during light exercise while wearing chemical protective clothing in hot-wet and/or hot-dry environments, liquid cooling systems emerged as more efficient in cooling the body compared to ice- or air-cooled counterparts. For the liquid-cooling systems, key factors con-

Table 7.1 Studies on the influence of cooling garments on alleviating heat strain duringe exercise and recovery in heat [42]

Author (year)	Subjects	Experimental condition	Cooling method (total clothing mass)	T_a (°C) H_a (%RH)	Experimental clothing	Exercise mode	Main outcomes by cooling
Nishihara et al. (2002)	4 males	① No cooling ② Cooling vest A ③ Cooling vest B	[Cooling vest A] Ice vest with 6 bags [120 g/bag] on chest and 10 bags [100 g/each] on back (2.1 kg)	30 °C, 37%	Long shirt and pants (0.88 clo)	10-min rest, 90-min walking, 10-min rcv (110 min)	• Lower chest temp. • TS, SS, TC improved
Sigurbjo¨rn et al. (2004)	17 runners	① No cooling ② Cooling vest during warm-up (precooling)	Ice vest with 8 ice packs (450–500 ml/pack); 2 on chest, 2 on stomach, 2 on shoulder, and 2 on lower back (4.5 kg)	30 °C, 50%	T-shirts & short pants	38-min warm up, 5-km run	• Blunting increases of Tre & HR • TC improved • Run time shortened
Smolander et al. (2004)	4 firefighters	① No cooling ② Ice cooling vest	Ice vests with 5 packs (1.0 ~ 1.1 kg)	45 °C, 30%	Firefighter's PPE (21 ~ 23 kg)	30-min waking ×4 times with 5-min rest for each	• 10 bpm lower HR & 13% smaller TSR (approx. 50 g) • Lower trunk temp; • No effect on extremity temp, VO_2, & Tre. • RPE improved (score 1) & TS improved (score 1)
Webster et al. (2005)	16 atheletes	① No cooling ② Ice vest A ③ Ice vest B ④ Ice vest C (precooling)	(2.8 ~ 3.0 kg)	37 °C, 50%		30-min run (70%VO_{2max})	• Lowered Tre, Tsk • TSR (10 ~ 23%) • No effect on HR • Tolerance time, TS, & SS improved

(continued)

Table 7.1 (continued)

Author (year)	Subjects	Experimental condition	Cooling method (total clothing mass)	T_a (°C) H_a (%RH)	Experimental clothing	Exercise mode	Main outcomes by cooling
Yoshida et al. (2005)	6 males	No cooling, water perfused suit and vest with water temp of 14, 20, and 26 °C (7 conditions)	Water suits (86% CBSA) Water vest (25% CBSA)_	WBGT 28 °C	Fencing uniforms	20-min cycling ×3 times (250 W/m^2)	• Lowered Tre, Tsk, HR, & TSR • TS improved
Hamada et al. (2006)	7 males	① Fan cooling (2.5 m/s) ② Ice cooling (last 20-min exe cooling)	Ice pack on the bilateral carotid (500 g)	30 °C, 40%	Trunks	40-min bicycle	• Lowered Tty, SR, and HR • No effect on Tre • TS improved
Choi et al. (2008)	12 males	① No cooling ②③ 2 Cooling scarfs ④ Cooling hat ⑤ Cooling vest ⑥⑦ Combinations	Scarf A (0.07 kg, 0.4%BSA cooled) Scarf B (0.2 kg, 0.8%) Hat (0.5 kg, 0.8%) Vest (0.8 kg)	33 °C, 65%	Long sleeved shirts & pants	120-min simulated red pepper harvest work	• Vest: lowered Tre, Tsk, and HR • Lowered PSI • TS improved
Kenny et al. (2011)	10 males	① Seminude with no cooling ② NBC suit with cooling vest ③ NBC suit without cooling	Ice packs (4.1 kg)	35 °C, 65%	NBC protective suit	120-min walking	• Lowered in Tes (0.3 °C) and HR (10 bpm) • TS, RPE (2 scores), & tolerance time improved
Stannard et al. (2011)	8 male runners	① No cooling ② Cooling vest (precooing)	4 pockets (2 on chest, 2 on back)	24 ~ 26 °C, 29 ~ 33%	T-shirts & short pants	10-km running (about 42 min)	• No effect on Tre, HR, RPE, & TS
Luomala et al. (2012)	7 male cyclists	① No cooling ② Ice cooling vest	4 packs (1 kg)	30 °C, 40%	Cycle wear	Cycling at 80%VO_{2max}	• No effect on Tre & Tsk • Lowered trunk temp • 21.5% tolerance time improved
House et al. (2013)	10 males	① No cooling ②③④⑤ PCM cooling vests with 0, 10, 20, 30 °C melting points	2 packs (1 kg/pack)	40 °C, 46%	Firefighting clothing	45-min stepping and 45-min recivery	• Lowered in Tre, Tsk, HR, & TSR

Teunissen et al. (2014)	9 males	① No cooling ②③ 2 cooling vests	4 cool pads (1 kg) Water perfusion (4 kg)	30 °C, 50%	EU firefighters' coverall	30-min walking and 10-min recovery	• Lowered Tsk & TSR • No effect on Tre, HR • TS improved
Butts et al. (2017)	16 males	① No cooling ② PCM cooling vest	8 pockets (chest, abdomen, back, thighs, and hamstrings) with melting point at 10 °C (~3.6 kg)	35 °C, 53%		55-min exercise	• Lowered Tre & Tsk • TS improved
Bartkowiak et al. (2017)	6 adults	① No cooling ② Cooling vest	Liquid cooling (cooling capacity 300 W)	30 °C, 40%	Aluminized PPC	45-min walking	• Lowered Tsk • TS improved
Chan et al. (2017)	140 construction workers	① No cooling ② Cooling vest (break-cooling)	8 PCM packs cooling vest with 2 ventilation fans; 80 g/pack. Melting point 28 °C. 2 on the chest, 2 on the abdoment, 4 on the back (1.26 kg)	WBGT 29-31 °C	Work wear (field study)	Construction work	• RPE 0.93 ~ 1.34 improved
Butts et al. (2017)	20 males	① No cooling ② Phase change material (PCM) cooling vest	Set to change phase at 10 °C (~3.6 kg)	34 °C, 55%	Coverall suit	20-min simulated work×2 times	• Lowered HR, Tsk, Tre, PSI, PeSI, & HS • RPE, thirst, TS improved
Mejuto et al. (2018)	7 cyclists	① Fluid inghestion ② Ice slurry ingestion (pre) ③ Ice slurry ingestion (pre + mid)	-1 °C ice slurry	32 °C, 50%	Cycle wear	45-min cycling×3 times at 70%VO_{2max}	• Lowered in Tre • No effect on tolerance time • RPE improved

tributing to effectiveness include the proportion of the total body surface area covered by the cooling garments, the specific body regions targeted for cooling, and the appropriateness of the garment fit. Nevertheless, liquid-cooled garments necessitate an external apparatus for operation, comprising a battery, circulating pump, heat exchanger, fluid container, and control pad. The weight and volume of the operational device may curtail the wearer's mobility and impose an additional weight burden. On the other hand, air-cooled garments distribute cooling air in close proximity to the skin. A vortex tube, which is the source of the cooled air, requires a continuous supply of compressed air delivered via an air hose. Unless the worker's mobility is restricted, the vortex tube can be perceived as a straightforward source of cooled air. Consequently, air-cooled garments prove efficacious for workers engaged in activities lower metabolic rate, such as light exercise. Nevertheless, it is imperative to avoid using air-cooling garments in air-polluted environments, as exemplified by situations like the "COVID-19 pandemic."

Cooling vests consist of multiple cooling packs, either made of ice or phase-change materials. The effectiveness of the cooling varies over time and depends on factors such as the contact pressure with the body surface and the ambient temperature. Unlike other cooling methods, cooling vests can be worn noiselessly and are independent of any energy source or umbilical cord, preserving the wearer's mobility. However, the practicality of ice-cooled garments diminishes once the ice melts, rendering them an additional burden for wearers [58]. To sustain the cooling effects, workers must replace melted ice packs with new ones, which proves impractical for workers mandated to wear full PPE. Additionally, while ice-cooled vests are more economical than their liquid and air-cooled counterparts, they lack the ability to control the cooling temperature.

From a practical standpoint, we investigated the effects of various types of cooled garments on physiological and psychological responses of workers in hot environments. In a study by Choi et al. [59], the effects of frozen-gel cooling systems applied to the head, neck, or trunk were examined using neck scarves, a hat, a vest, and the combinations at an air temperature of 33 °C with 65% relative humidity. Rectal temperature, mean skin temperature, and heart rate became more stable through wearing the personal cooling garments. Notably, among the seven cooling conditions, those incorporating the cooling vest proved to be the most effective in mitigating heat strain (as indicated by conditions #5, #7, and #8 in Fig. 7.7). Furthermore, even when wearing a cooling scarf covering approximately 0.4% of the body surface area, a significant difference was observed when compared with the control.

As previously mentioned, ice-cooled garments pose challenges due to their weight and restrictions on body movement, particularly after melting when the cooling effect is diminished. The removal or exchange of cooling vests follow-

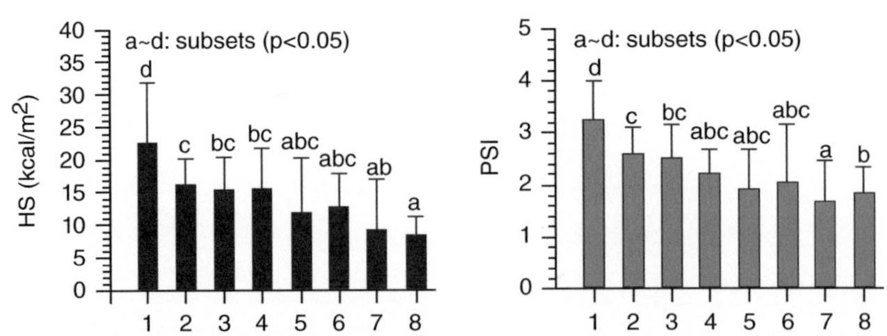

Fig. 7.7 Body heat storage (HS) and physiological strain index (PSI) during the various cooling applied [59]. #1: No cooling; #2: Neck cooling A; #3: Neck cooling B; #4: Cooling hat; #5: Cooling vest; #6: Hat + Neck B; #7: Hat + Vest; #8: Hat + Neck B + Vest. "a," "b," "c," "ac," "bc," or "abc" represent significant differences among the 8 conditions at $P < 0.05$

ing melting may not be practical for workers in personal protective equipment (PPE) because of their protective clothing. Moreover, progressive body fluid losses during exercise in heat increases body temperature [60]. Dehydration is associated with significant decrements in exercise tolerance time [61, 62]. Offering cold drinking water as a preventive measure against dehydration can be an effective intervention to counteract the impact of heat stress. In this context, the synergistic effects of skin cooling and cold fluid consumption can be anticipated for workers donning PPE in hot environments. We devised an innovative vest with dual functions, providing both skin cooling and the option for cold fluid intake for firefighters [42]. The cooling packs in the vest were filled with frozen sports drink, allowing firefighters to consume cold fluids as the packs melt (see Fig. 7.8a). During the recovery period, the rectal temperature declined for the drinking (DO), cooling (CO), and cooling and drinking (CD) conditions (see Fig. 7.8b), with a notable reduction observed in the CD condition (see Fig. 7.8c).

An alternative approach involves cooling specific body zones characterized by high tissue density. Koscheyev et al. [63] investigated zonal tissue heat transfer at varying cooling temperatures to enhance liquid cooling garments for spacesuits. The body zones found to be most effective for heat dissipation were the head, neck, hands, and feet, all of which possess higher tissue density compared to other areas of the body. Koscheyev et al. [64, 65] proposed physiological design principles for cooling garments as crucial factors in enhancing astronaut comfort and performance. The primary principle asserts that tissues with high density, exhibiting high conductivity and intricate vessel networks inclusive of magistral vessels, are preferable for facilitating rapid heat influx and efflux. Another significant principle highlighted the inefficiency of the body's cooling mechanism when subjected to extremely low temperatures on the body surface, because the reflex response of superficial vessels is to conserve heat. Accordingly, they advocated for a mild cooling regime, emphasizing its superior effectiveness over more intensive cooling approaches.

As elucidated by Koscheyev et al. [63], cooling the head proves more effective for heat dissipation than other body regions. Employing head cooling could be a practical strategy for workers, especially considering that many workers are required to wear protective helmets, even in hot weather. However, the presence of head hair may impede the removal of heat from the head due to the reduction in thermal conductivity caused by hair. Grahn et al. [66] highlighted that heat loss from glabrous skin (areas like the face and palms) during exercise in hot environments were more than five times greater than that from nonglabrous skin (areas like the upper back, lower back,

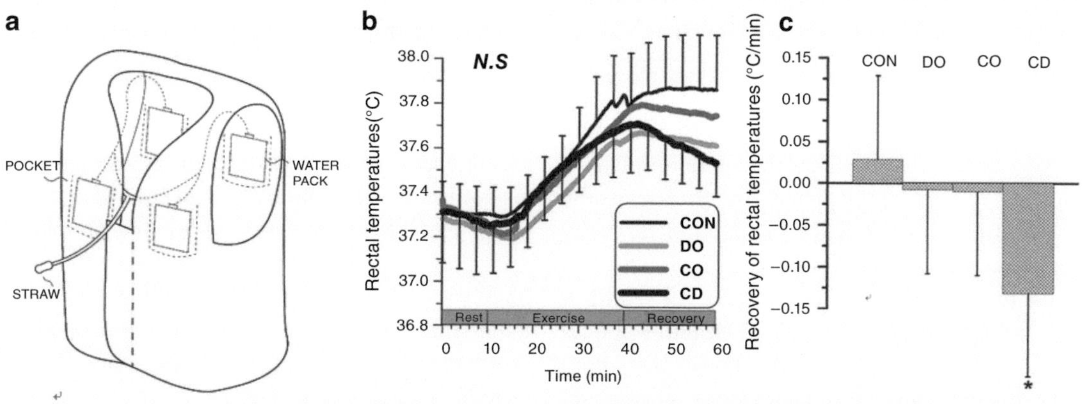

Fig. 7.8 A cooling vest with dual functions of skin cooling and drinking fluids (**a**), time courses of rectal temperature (**b**), and changes in rectal temperature for the four experimental conditions (**c**). CON, DO, CO, and CD represent "control," "drinking only," "cooling only," and "cooling with drinking" conditions [42]

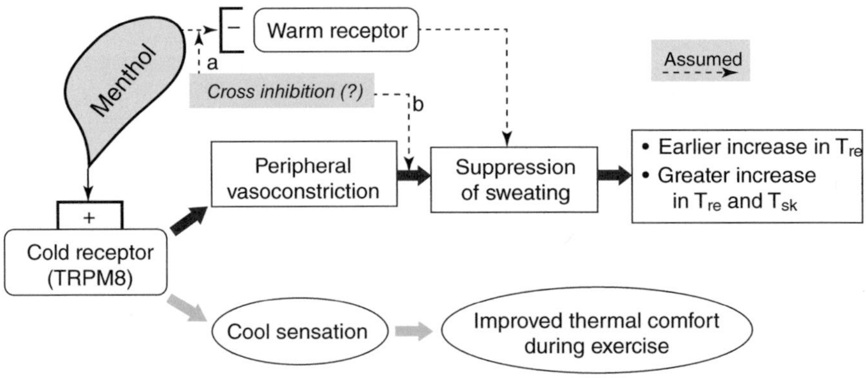

Fig. 7.9 Schematic diagram of the effects of 0.8% menthol application over the upper/whole body on physiological and psychological responses during exercise. This diagram is limited to the effects of menthol on cutaneous thermal receptors, when sprayed over the half of the entire body. (Source from Lee et al. [69])

abdomen, thigh, and upper arm) in humans. In our investigation examining the impact of head hair on heat removal at cooling temperatures of 10, 15, and 20 °C [67], interesting findings were observed. Heat removal in the normal hair condition did not differ from that in the cropped hair condition across all three cooling temperatures. However, subjects expressed a preference for the 15 °C cooling regimen over the 10 °C or 20 °C regimens. Thus, the critical factor in head cooling under heat stress is the selection of an effective cooling temperature rather than the length of male workers' hair.

In our investigation, we delved into the potential body cooling effects of menthol as a means to alleviate heat strain among firefighters [68, 69]. The practicality and feasibility of firefighters wearing cooling vests are questionable. As an alternative cooling method, we explored the application of menthol to the skin [69]. Our study examined the impacts of L-menthol application (0.8%) in relation to the surface area on physiological and psychological heat strains. The results revealed that the influence of menthol topically applied was dependent on the body region as well as the exposed body surface area. A very intriguing finding was that the 0.8% menthol applications over the upper/whole body had both advantageous and disadvantageous effects on thermoregulatory, cardiovascular, and psychological responses during exercise. Notably, the menthol application appeared to be beneficial to improve thermal sensation and comfort, whereas thermoregulatory functions during exercise were impeded, as evidenced by an earlier and greater increase in rectal temperature and a smaller total sweat rate at the end of exercise. These adverse reactions were associated with peripheral vasoconstriction in the initial phase and the delayed onset of sweating (Fig. 7.5). Consequently, a topical menthol application should be viewed as a perceptual cooling medium distinct from surface cooling countermeasures based on conductive/convective heat exchanges (Fig. 7.9).

7.5.4 Noninvasive Heat Strain Index for Workers

Over the past few decades, a multitude of studies [38, 48, 70–75] have been undertaken to assess physiological strain associated with the utilization of PPE in hot environments. A Physiological Strain Index (PSI) was devised to mitigate the risk of heat-related illnesses at an individual level [76]. The PSI is determined by incorporating measurements of rectal temperature (T_{re}) and heart rate (HR) to depict the amalgamated stress experienced by the cardiovascular and thermoregulatory systems: $5(T_{ret} - T_{re0})\cdot(39.5 - T_{re0})^{-1} + 5(HR_t—HR_0)\cdot(180—HR_0)^{-1}$.

Nevertheless, approaches such as PSI for predicting heat strain are impractical for workers donned in protective clothing and equipment, because the measurement of core body temperature remains constrained by practical limitations stemming from invasive data collection in occupational field settings. Emerging technologies, specifically ingestible pills, have gained prevalence in this domain [77, 78]. However, the disadvantage of the ingestible pills is that it takes several hours to migrate from the mouth to the small intestine, coupled with their elevated cost [79, 80]. Numerous factors, unrelated to heat stress, can induce an elevated heart rate (e.g., physical exertion, emotional stimuli) [11]. Consequently, to account for these considerations, the perceptual strain index (PeSI) has been formulated to gauge the physiological responses of individuals [81–84]. Nevertheless, the accurate and precise prediction of thermal strain remains challenging due to the substantial interindividual variation in responses.

Real-time biomonitoring technologies can be integrated into PPE. The incorporation of woven wires and sensors into the fabric not only furnishes firefighters with information regarding the ambient temperature they are exposed to but also enables continuous monitoring of their skin temperature. Furthermore, the inclusion of thermal sensors in both the interior and exterior of thermal protective clothing facilitates the regulation of microclimate temperature within the clothing itself.

A worker's core body temperature (e.g., rectal temperature) is intricately linked to metabolic heat production. In typical circumstances, the skin temperature at rest generally resides approximately 3–4 °C below the core body temperature of 37 °C. In light of this background, we have explored a simple and noninvasive parameter to estimate the heat strain experienced by workers wearing PPE in hot environments. Our findings indicate that auditory canal temperature emerges as one of the most promising candidates for estimating the rectal temperature of workers clad in protective clothing within nuclear facilities [85–87]. In the study conducted by Lee et al. [88], infrared tympanic temperature proved to be a valid thermal strain index when individuals wore impermeable protective clothing in hot environments. However, it is noteworthy that the application of auditory canal temperature is restricted to scenarios involving strenuous activities and the use of encapsulated personal protective clothing with a hood in hot conditions, as the ear canal is susceptible to the influence of air flow.

7.5.5 Cold Protective Clothing in Cold Environments

7.5.5.1 Cold Protective Clothing

Clothing designed for cold protection primarily aims to maximize air entrapment, utilizing a combination of three to four layers, because air serves as an excellent insulator. Cold protective clothing is typically constructed by integrating various material layers. This includes a base layer, interdependent for sweat absorption, facilitating enhanced moisture wicking to provide cooling through moisture dissipation. This layer also prevents the chilling effect. The mid-insulation layer offers thermal protection, and the outer protective layer is waterproof, windproof, and breathable. This outer layer allows for the evaporation of sweat and heat while concurrently impeding water entering into the clothing [89].

Siple [90] offers an early and comprehensive overview of cold weather clothing based on studies in the Antarctic, contributing to the development of the wind chill index. The calculation of thermal resistance or insulation required (IREQ) can be conducted through the body heat balance equation or using a thermal model. ISO 11079 [91] delineates a method for computing IREQ based on the temperature difference across the clothing (the variance between mean skin temperature and the mean temperature of clothing) divided by the metabolic rate, subtracting any heat loss from breathing and evaporation. Appropriate clothing selection involves choosing ensembles with basic clothing insulation values that, when adjusted for ventilation, align with the calculated IREQ. However, the compression of effective air layers within clothing by multiple layers can result in the total insulation of the clothing ensemble being less than the sum of the thermal insulation of each individual garment.

It is proposed that the optimal thickness of the air layer within clothing should range between 5 and 15 mm in terms of clothing insulation. A larger air layer may induce ventilation inside the clothing during body movement, thereby escalating convective heat loss from the skin.

Garments for protecting the peripheral body parts under cold stress are indeed imperative. The hands, particularly the fingers and thumbs, are susceptible to heat loss as they have a large surface area for heat exchange when compared with their mass. Covering the whole hand with mittens reduces available surface area and preserves heat but generally reduces manual dexterity. Determination of the insulation of gloves is provided in EN511 [92]. In extreme cold, gloves, unless actively heated, will not keep the hands warm. Wet gloves and wet hands should be avoided. Heated gloves provide advantage. In particular, wearing gloves in the cold is recommended for the elderly because they may not feel the cold, even though their hands continue to lose heat [93]. Although hand protection is usually traded with task requirements and manual dexterity, avoid tight gloves that restrict circulation. Bare feet will feel cold at skin temperature below around 15 °C and can be injured when cold and wet. The size of cold protective boots should be 2–3 sizes greater than usual in order to avoid compression of the legs and feet that may restrict circulation. Failure to protect the feet in the cold can lead to nonfreezing cold injury and frostbite. Head temperature remains high even in cold environments and has a high potential for heat loss, especially, in cold. Head hair provides good insulation and has a significant effect in protecting against the cold. Hats or head covers are essential if no hair is available. Ear muffs are essential even in moderately cold weather. The hat or cap should protect the back of the neck as well. The chin requires protection in windy cold. Beards can freeze. Wind chill can be considered to be the effect of wind on bare skin (e.g., the face, head, and hands) that creates a chilling sensation.

7.5.5.2 Winter Jacket Wear Trials in Cold Environments

We conducted wear trials to assess the efficacy of diverse cold protective clothing types in cold environments. Initially, we focused on the thermal insulation of conventional winter padded jackets under various conditions. In the study conducted by Baek et al. [94], we delved into the thermal physiological and psychological responses elicited by clothing ensembles, specifically long padded jackets with two distinct filling materials: goose down and lightweight polyester, in two cold environments characterized by different wind velocities, encompassing both windless and windy conditions. Our findings revealed that in cold and windless environments, polyester padded jackets exhibited similar thermal insulation values comparable to those of goose-down jackets. However, in cold and windy conditions, the thermal insulation of the goose-down padded jackets was greater than that of the polyester counterparts. This disparity may be attributed to the superior resilience of the goose-down filling to wind in comparison to the polyester filling.

An intriguing approach in winter padded jacket design involves incorporating air, resulting in what is termed an air-padded jacket. Theoretically, the thermal insulation of an air-padded jacket is expected to surpass that of a typical down padded jacket due to the higher thermal conductivity of still air compared to goose or duck down. To explore this, we conducted a comparison between the thermal insulation of air-padded and down-padded jackets, both possessing identical designs and outer materials, as detailed in studies by Kim et al. [95] and Kwon et al. [96]. Contrary to our initial expectations, the total thermal insulation for the air-padded jacket (1.16 clo) was found to be lower than that of both the heavy-down jacket (1.29 clo) and the light-down jacket (1.23 clo) [95]. This discrepancy may be attributed to the reduced air layer between the body's skin and the air-padded jacket, resulting from the inflexibility of the air-packs. Notably, it is recommended to wear an air-padded jacket one size larger than the normal size.

Fig. 7.10 Construction of air-filled jacket (upper illustrations) and sample jackets (lower photos). (Source from Kwon et al. [96])

In a subsequent study, Kwon et al. [96] assessed the thermal insulation of air-filled winter jackets based on the quantity of air filler (Fig. 7.10). The findings indicated that clothing insulation (Icl) for jackets with 0%, 50%, 70%, and 100% air, as well as 100% goose down, were 0.208, 0.243, 0.207, 0.176, and 0.315 clo, respectively. Notably, the 50% air-filled jacket consistently exhibited the highest clothing insulation per unit weight in all conditions. To optimize the thermal insulation of air-filled jackets, further exploration into the optimal amount of air filler is warranted.

7.5.6 Active Heating Garments in Cold Environments

Cold protective clothing typically entails an increased bulkiness of the clothing, which significantly limits human performance. In this sense, electrically heated garments could be a solution. They are capable of providing extra heat, which can reduce the bulkiness associated with conventional winter clothing materials, such as fur or down jacket. Recently, various active heating technologies have been applied to clothing, and most are based on electric heating systems. We have evaluated the impact of the distance of the heating unit from the body in a multilayered winter clothing system on thermal insulation and heating efficiency [97]. In addition, we investigated the effects of intermittent and continuous heating protocols using graphene-heated clothing and identify more effective body region for heating in a cold [98] (Fig. 7.11). The results showed that the electric power consumption of the intermittent heating protocol (2.49 W) was conserved by 71% compared to the continuous protocol (8.58 W). Concerning effective body region for heating, heating the upper back showed more beneficial effects on skin temperatures than heating the chest in cold environments.

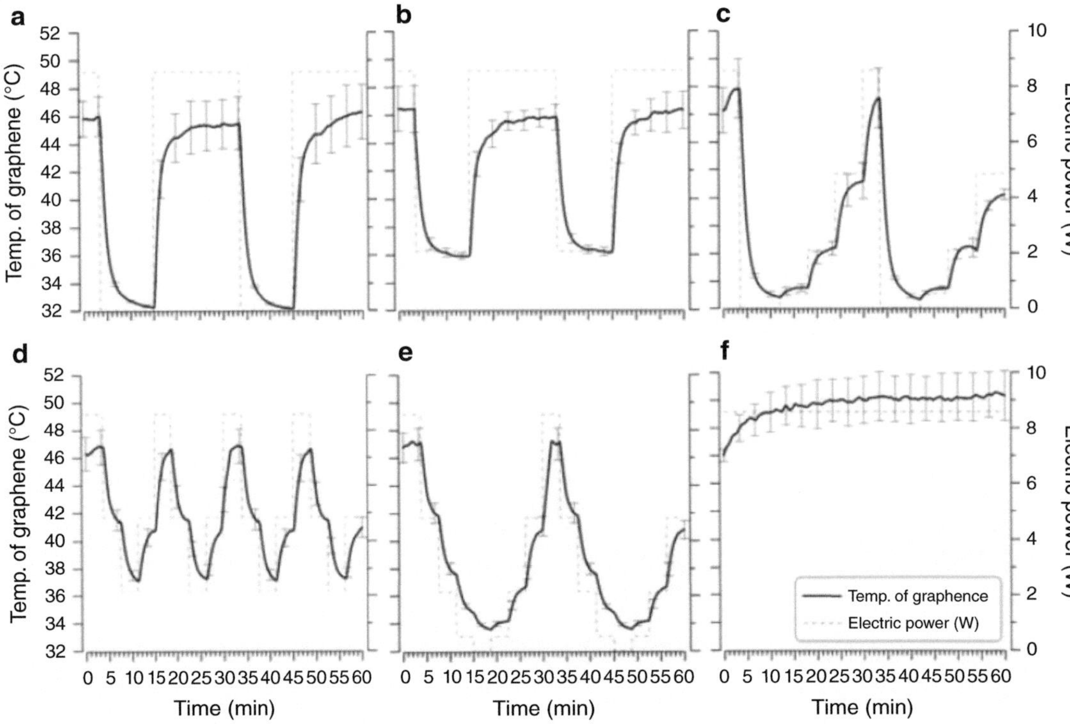

Fig. 7.11 Heating protocols using graphene heaters: Averaged surface temperature and sum of electric power of three pieces of graphene heaters according to continuous and intermittent heating protocols. (Source from Shin et al. [98])

In cold conditions, extremities such as hands, feet, and ears are vulnerable to frostbite, necessitating the maintenance of warmth. Surprisingly, compared to the active heating garments of the hands or feet, the topic of ear heating has received relatively less attention. The ears, being thin with low blood flow and a large surface area relative to mass, are prone to rapid heat loss. To address this, we conducted an assessment of the effects of newly developed earmuffs equipped with graphene heaters for police officers, particularly focusing on continuous and intermittent heating scenarios [99]. A noteworthy outcome of our study was the observed improvement in thermal sensation and overall body comfort through heating only the ears, constituting a mere 1% of the body surface, at an air temperature of −5 °C. Another intriguing discovery was the similar improvement observed in both intermittent heating and continuous heating conditions, as illustrated in Fig. 7.12 Notably, the intermittent heating protocol utilized only 70% of the electric power consumption compared to the continuous heating protocol, while maintaining similar subjective perceptions. This suggests the need for further development of various intermittent heating protocols, aimed at conserving battery power in cold environments.

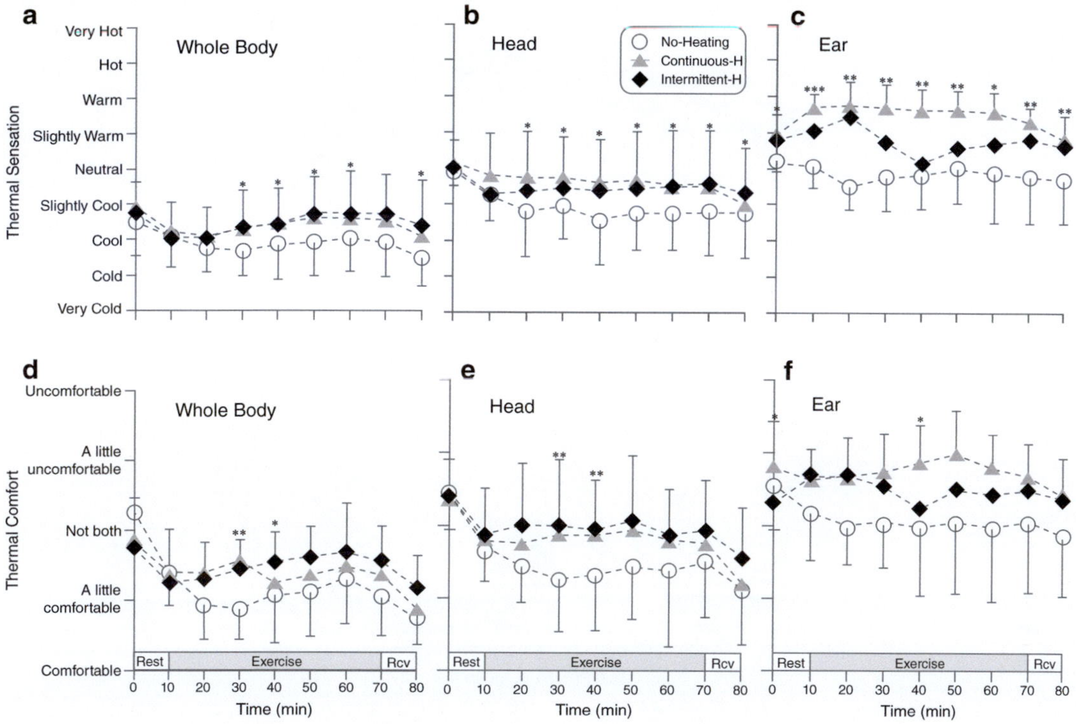

Fig. 7.12 Time course of thermal sensation (**a–c**) and thermal comfort (**d–f**) in cold environments. (No-Heating: wearing the earmuffs, but not heated; Continuous-H: surface of the earmuffs maintained at 45 °C; Intermittent-H: surface of the earmuffs maintained at 45 °C for 7 min, turned off for 3 min; $*P < 0.05$ and $**P < 0.01$). (Source from Lee et al. [99])

7.6 [Step 5] Mobility Tests and Protective Clothing in Hot and Cold Environments

The very bulk of the protective clothing renders the individual less efficient in movement due to the "friction drag" between the layers and actually increases the metabolic heat generated by the wearer on a given task [100]. This deterioration of mobility can be exacerbated in hot environments due to the high humidity inside the PPE. Wetted skin resulting from excessive sweating and the high water vapor resistance of PPE material can cause a sticky feeling and disturb joint movements, ultimately affecting task performance. While numerous international standards exist for requirements and test methods regarding protective properties, relatively few test methods for assessing the mobility of PPE have been reported. We have developed mobility protocols to assess firefighters' PPE, chemical PPE, healthcare workers' PPE, military soldiers' combat uniform, navy uniforms, or so. During the mobility tests, we continuously monitor heart rate, core body temperature, skin temperatures, clothing microclimate humidity, sweat rate, subjective perceptions, and/or completion time. The validity of the newly developed mobility test protocols is evaluated based on their distinguishing power among different PPE conditions [101].

We developed a mobility protocol to evaluate firefighters' PPE, which consisted of eight different tasks, including walking in front of a radiant heater (#7 in Fig. 7.13) [102]. Davis et al. [103] developed five representative tasks of firefighters, including ladder folding out and moving hoses, while Deakin et al. [104] proposed a firefighter physical fitness maintenance evaluation protocol with 10 different tasks. According to our protocol

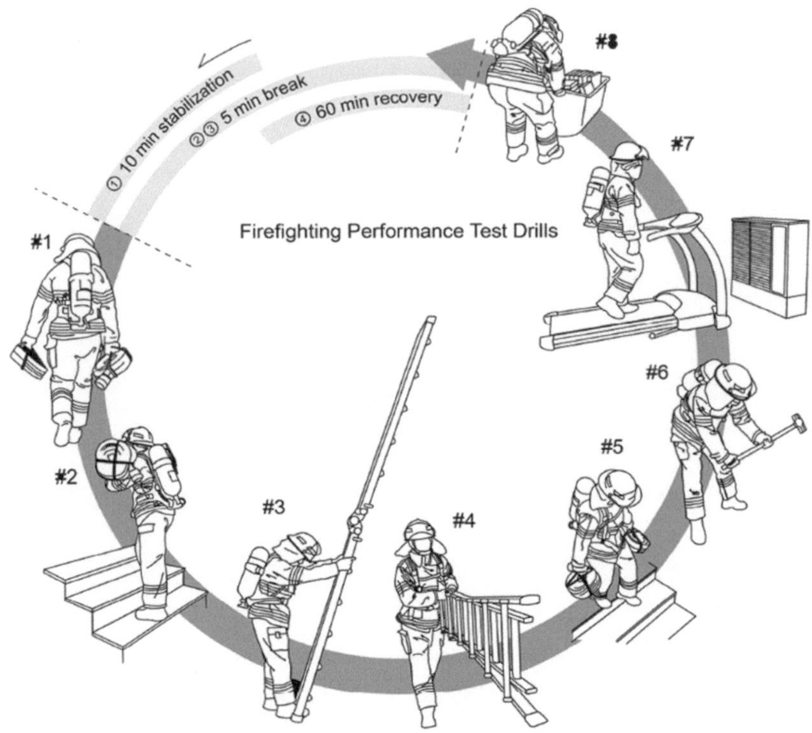

Fig. 7.13 A mobility test protocol for evaluating firefighters' protective clothing. Note: #1 Carrying hose: 40 m walking with two 65 mm rolled-up hoses (7 kg each) in both hands; #2 Going up the stairs: going up six flights of stairs (66 steps) with two 65 mm (7 kg each) in both hands; #3 Carrying and setting up a ladder: 20 m walking with a ladder (15 kg) and setting a ladder to the wall; #4 Withdrawing and carrying a ladder: taken down a ladder (15 kg) and 20 m walking with the ladder; #5 Going down the stairs: going down ten flights of stairs (110 steps) with two 65 mm (7 kg each) in both hands; #6 Simulated forcible entry: striking a rubber tire heavily using a 4.5 kg hammer three times; #7 Walking with facing radiant heat: walking on the treadmill in 5.5 km h^{-1} while facing on radiant heat; #8 Pulling a victim: pulling a 70 kg sledge 20 m. One circuit which consists of eight tasks was repeated three times with a 5-min break after each drill. Firefighters took off a helmet, a hood, and gloves with a zipper of bunker jacket lowered for the break. (Source from Kim and Lee [102])

below, rectal temperature and heart rate increased to 38.0 ± 0.3 °C and 185 ± 13 min, respectively, when wearing full PPE after completing three repetitions. This protocol can be used to evaluate the thermal comfort and mobility of firefighters' PPE [102].

Further, we developed a new mobility test protocol combined with posture analysis for chemical protective clothing, consisting of 16 postures or tasks, including a dexterity test (Fig. 7.14) [55]. Through this test, we discovered that the clothing microclimate humidity increased to over 85%RH, sometimes reaching 98%RH, which caused fogging inside the visor and impaired visibility. The mobility test can provide essential insights for improving PPE design, such as the hood, gloves, boots, clothing size, clothing mass, opening, or donning/doffing methods.

During the COVID-19 pandemic, Fig. 7.15 illustrates a mobility test protocol for evaluating healthcare workers' PPE. The mobility protocol consisted of 10 different tasks in addition to donning and doffing [105, 106]. Through the test, significant differences were observed among the various PPE conditions in terms of skin temperature, clothing microclimate, thermal sensation, thermal comfort, and humidity sensation of workers. Wearing a powered-air purifying respirator (PAPR) caused discomfort and increased noise levels as the tasks progressed while subjects felt cool during the initial minutes [105].

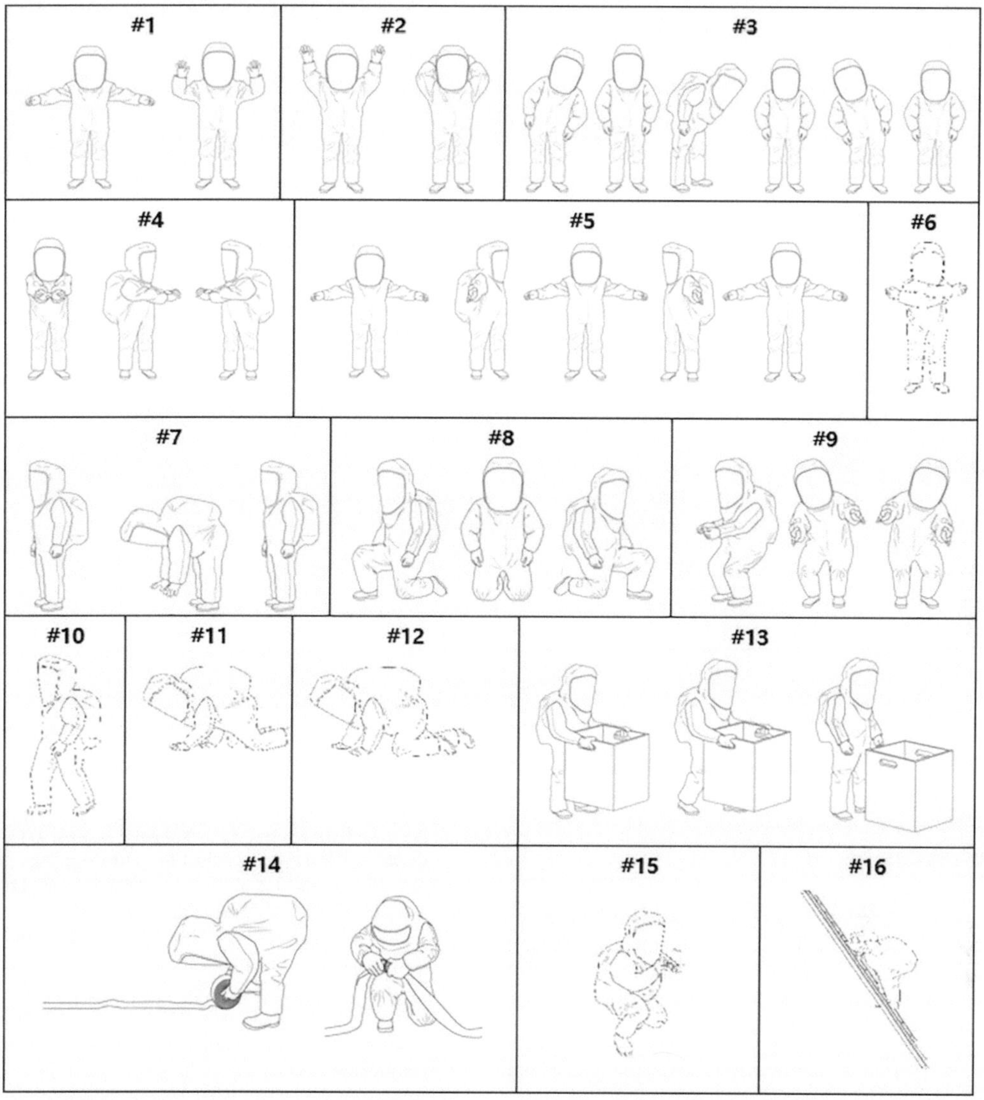

Fig. 7.14 Mobility test protocol of 16 simulated postures and tasks. #1. Upright posture: Straighten the arms to the sides and bend the elbows upward (5 s); #2. Upright posture: Raise both arms above the head and bend the elbows back (5 s); #3. Upright posture: Put the arms straight on both sides, bend the body to the left (5 s), and return to your normal posture. Bend the body forward (5 s) and go back to the normal position. Bend the body to the right (5 s) and return to the normal position; #4. Upright posture: Stretch the arms straight in front of the body, then pinch the thumbs together, twist the upper body to the left with your outstretched arms (5 s), and twist to the right again (5 s); #5. Upright posture: Straighten both arms horizontally on both sides of the body. In this position, turn the torso to the left (5 s) and return to the original position. Turn to the right (5 s) and return to the original position; #6. Upright posture: Make sure that both arms cross in front of the chest (right arm on the left arm) and straighten horizontally in opposite directions (5 s); # 7. Upright posture: Bend the upper body down to 90° and stretch the both arms down to the floor (5 s). Then, return to the original position; #8. Kneel with the left knee (right leg is a right angle) (5 s), then kneel with both knees (5 s), then kneel with the right knee (left leg is a right angle) (5 s), and then immediately stand up; #9. After being prone to the squat position (5 s), move the center of gravity to the right as much as possible (5 s), and then move the center of gravity back to the left (5 s). Immediately afterward, and stand up; #10. Walk on flat ground at normal speed for 1 min; #11. Crawl a distance of 6 m with both palms and knees; #12. Crawl back 6 m distance with both palms and knees; #13. Pick up one standard wooden carrying box with open handles (approx. 10 kg; 50 × 60 × 74 cm high), move five steps, place it on the floor, and put it in a straight position (5 s). Lift the box on the floor again, take five steps, and put it back in place. Back in a standing position; #14. Unscrew, the two fire hoses on the floor, connect and separate the two couplings, and roll up one fire hose again; #15. Remove one tightened screw (9 mm diameter) using a screwdriver (25 cm); #16. Go up to the fifth step of the ladder placed on the wall, pause for 5 s, and then descend again [55]

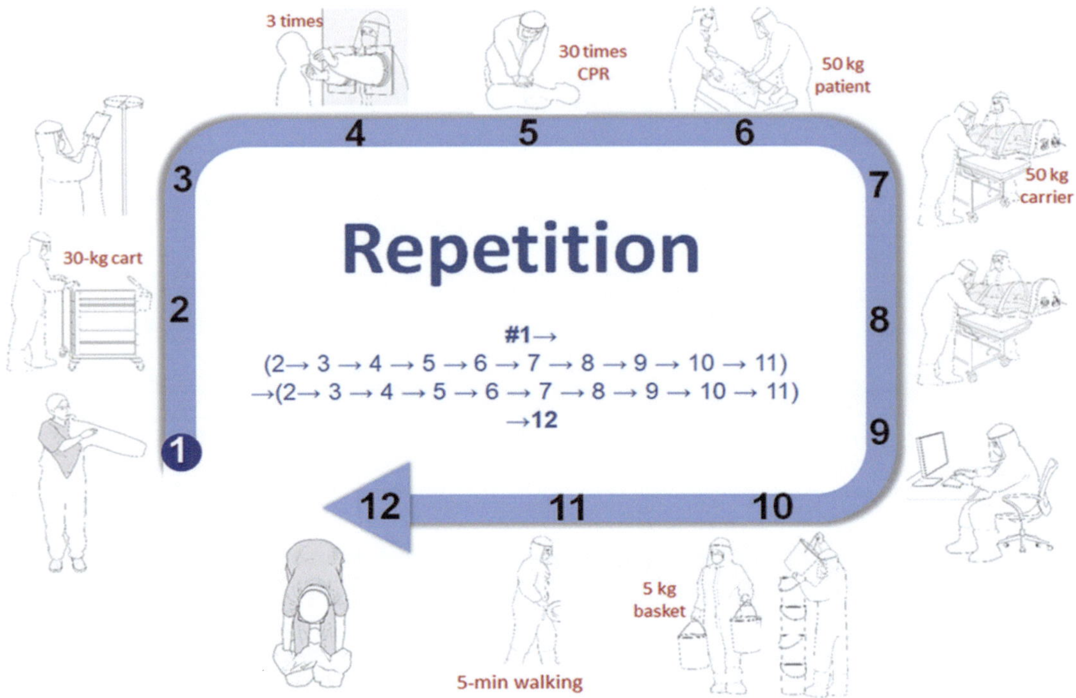

Fig. 7.15 Mobility test protocol of 10 simulated COVID-19 healthcare workers' tasks [105, 106]

References

1. Lee JY, Park JH, Park H, Coca A, Kim JH, Taylor NA, Son SY, Tochihara Y. What do firefighters desire from the next generation of personal protective equipment? Outcomes from an international survey. Ind Health. 2015;53(5):434–44.
2. Kim S, Kim DH, Lee HH, Lee JY. Frequency of firefighters' heat-related illness and its association with removing personal protective equipment and working hours. Ind Health. 2019;57(3):370–80.
3. Son SY, Lee JY, Tochihara Y. Occupational stress and strain in relation to personal protective equipment of Japanese firefighters assessed by a questionnaire. Ind Health. 2013;51(2):214–22.
4. Lee HH, Kim S, Kim DH, Kim DH, Lee JY. Requirements of improvement on personal protective equipment and experiences exposed to accidently high risk circumstance while firefighting: a questionnaire study. JKSLES. 2017;24(5):549–61. (in Korean)
5. Baek YJ, Kim DH, Bang SH, Lee JY. Korean firefighter questionnaire to improve the test methods of firefighting personal protective equipment. J.KSLES. 2019;26(3):322–44. (in Korean)
6. Kim DH, Kim S, Lee JY. An empirical investigation of firefighter personal protective equipment and burn injuries in Korea. Ind Health. 2022;60:2–15.
7. Kim DH, Nam K, Oh JW, Lee JY. Evaluation and usage behavior concerning firefighting gloves among firefighting officials in Seoul. Fash Text Res J. 2021;23(4):515–26. (in Korean)
8. Fourt LE, Hollies NRS. Clothing: comfort and function. New York: Marcel Dekker; 1970.
9. Hollies NRS, Goldman RF. Clothing comfort: interaction of thermal, ventilation, construction and assessment factors. Ann Arbor: Ann Arbor Science Publishers; 1977.
10. Renbourn ET, Ress WH. Materials and clothing in health and disease: history, physiology and hygiene: medical and psychological aspects. London: HK Lewis; 1972.
11. NIOSH. Criteria for a recommended standard. Occupational exposure to heat and hot environments. CDC NIOSH 2016.
12. ASHRAE 55. Thermal environmental conditions for human occupancy. Atlanta: American Society of Heating, Refrigerating and Air-conditioning Engineers, Inc.; 2023.
13. Fanger PO. Thermal comfort. Copenhagen: Danish Technical Press; 1970.
14. ISO 7730. Ergonomics of the thermal environment: analytical determination and interpretation of thermal comfort using calculation of the PMV and PPD indices and local thermal comfort criteria. International Organization for Standardization. 2005.

15. Wissler EH. Human temperature control: a quantitative approach. Springer; 2018.
16. Gagge AP, Burton AC, Bazett HC. A practical system of units for the description of heat exchange of man with his environment. Science. 1941;94:428–30.
17. ISO 9920. Ergonomics of the thermal environment: estimation of thermal insulation and water vapour resistance of a clothing ensemble. International Organization for Standardization. 2007.
18. Holmér I. Protective clothing in hot environments. Ind Health. 2004;44:404–13.
19. Kuklane K, Gao C, Holmer I. Calculation of clothing insulation by serial and parallel methods: effects on clothing choice by IREQ and thermal responses in the cold. Int J Ind Ergon. 2007;13(2):103–16.
20. Nilsson H. Analysis of two methods of calculating the total insulation. In: Nilsson H, Holmer I, editors. A European seminar on thermal manikin testing. Solna: National Institute for Working Life; 1997. p. 17–22.
21. Lee JY, Nakao K, Tochihara Y. Validity of perceived skin wettedness mapping to evaluate heat strain. Eur J Appl Physiol. 2011;111:2581–91.
22. Parsons KC. Human heat stress. Boca Raton: CRC Press; 2019.
23. Havenith G. Heat balance when wearing protective clothing. Ann Occup Hyg. 1999;43(5):289–96.
24. Baek YJ, Cho G, Hong YJ, Lee JY. Effects of air velocity on the thermal insulation of winter-padded clothing ensembles at 10°C air temperature -comparison of human wear trials with a thermal manikin-. J Korean Soc Cloth Text. 2021;45(4):703–13. (in Korean).
25. Havenith G, Heus R, Lotens W. Clothing ventilation, vapour resistance and permeability index: changes due to posture, movement and wind. Ergonomics. 1990;33:989–1005.
26. Havenith G. Individualized model of human thermoregulation for the simulation of heat stress. J Appl Physiol. 2001;90(5):1943–54.
27. Havenith G. Interaction between clothing insulation and thermoregulation. Exog Dermatol. 2002;1(5):221–30.
28. Nielsen R, Olesen BW, Holmer I. Effect of physical activity and air velocity on the thermal insulation of clothing. Ergonomics. 1985;28(12):1671–31.
29. IUPS. Glossary of terms for thermal physiology (3rd edition). Jpn J Physiol. 2001;51(2):245–80.
30. Bakri I, Lee JY, Nakao K, Wakabayashi H, Tochihara Y. Effect of fire fighters' self-contained breathing apparatus' weight and its harness design on the physiological and subjective responses. Ergonomics. 2012;55(7):782–91.
31. Bakri I, Tochihara Y, Lee JY, Wakabayashi H. Effect of personal protective equipment on the physiological responses of different body weight groups of firefighters. J Hum Environ. 2022;24(1):11–6.
32. Barr D, Gregson W, Reilly T. The thermal ergonomics of firefighting reviewed. Appl Ergon. 2010;41(1):161–72.
33. Lee JY, Bakri I, Kim JH, Son SY, Tochihara Y. The impact of fire fighter personal protective equipment and treadmill protocol on maximal oxygen uptake. J Occup Environ Hyg. 2013;10(7):397–407.
34. Lee JY, Yamamoto Y, Oe R, Son SY, Wakabayashi H, Tochihara Y. The European, Japanese, and US protective helmet, gloves, and boots for firefighters: thermoregulatory and psychological evaluations. Ergonomics. 2014;57(8):1213–21.
35. Lee JY, Kim S, Jang YJ, Baek YJ, Park JH. Component contribution of personal protective equipment to the alleviation of physiological strain in firefighters during work and recovery. Ergonomics. 2014;57(7):1068–77.
36. Tochihara Y, Lee JY, Son SY, Bakri I. Heat strain of Japanese firefighters wearing personal protective equipment: a review for developing a test method. Ergonomics. 2022;66(5):676–89.
37. Holmér I, Gavhed D. Classification of metabolic and respiratory demands in fire fighting activity with extreme workloads. Appl Ergon. 2007;38(1):45–52.
38. Holmér I, Gao C, Kuklane K. Test of firefighter's turnout gear in hot and humid air exposure. Int J Occup Saf Ergon. 2006;12(3):297–305.
39. Chou C, Tochihara Y, Saat M, Lee JY. Physiological strains of wearing aluminized and non-aluminized firefighters' protective clothing during exercise in radiant heat. Ind Health. 2011;49:185–94.
40. Baek YJ, Jung DH, Son SY, Lee JY. Comparisons between Shikoro-type helmet with no hood and typical fire protective helmets with hood in a hot and humid environment. Ergonomics. 2018;61(3):420–8.
41. Roh SH, Ko Y, Lee JY. Physiological and subjective burden when wearing fire protective boots between 3.2 to 5.3 kg. Fash Text. 2020;7:14.
42. Kim DH, Bae GT, Lee JY. A novel vest with dual functions for firefighters: combined effects of body cooling and cold fluid ingestion on the alleviation of heat strain. Ind Health. 2020;58:91–106.
43. Tochihara Y, Lee JY, Son SY. A review of test methods for evaluating mobility of firefighters wearing personal protective equipment. Ind Health. 2022;60:106–20.
44. ISO 16602. Protective clothing for protection against chemicals: classification, labelling and performance requirements. International Organization for Standardization. 2007.
45. OSHA. Personal protective equipment. Occupational Safety and Health Administration. 2020.
46. Dai X, Havenith G. Effect of clothing ventilation on thermal insulation and vapor resistance. In: Proceedings of the 13th international conference on environmental ergonomics; 2009. p. 20–5.
47. Ding D, Tang T, Song G, McDonald A. Characterizing the performance of a single-layer fabric system through a heat. Text Res J. 2011;81(9):945–58.
48. Havenith G, den Hartog E, Martini S. Heat stress in chemical protective clothing: porosity and vapour resistance. Ergonomics. 2011;54:497–507.

49. Xu X, Rioux TP, MacLeod T, Patel T, Rome MN, Potter AW. Measured body composition and geometrical data of four "virtual family" members for thermoregulatory modeling. Int J Biometeorol. 2017;61(3):477–86.
50. Xu X, Rioux TP, Pomerantz N, Tew S. Effects of fabric on thermal and evaporative resistances of chemical protective ensembles: measurement and quantification. Measurement. 2019;136:248–55.
51. Hanson MA. Development of a draft British standard: the assessment of heat strain for workers wearing personal protective equipment. Ann Occup Hyg. 1999;43(5):309–19.
52. Jung DG. Exploration and validity evaluation of physiological strain index for workers when wearing personal protective equipment in hot environments. Ph.D. Dissertation. Seoul National University. 2023.
53. Kim DH, Lee JY. Heat strain while wearing pesticide protective clothing in hot environments: effects of textile physical properties and ambient humidity. Int J Ind Ergon. 2023;93:103388.
54. Candas V, Libert JP, Vogt JJ. Human skin wettedness and evaporative efficiency of sweating. J Appl Physiol. 1979;46(3):522–8.
55. Salsabila S, Kim DH, Lee JY. Improving design and evaluating mobility of firefighting chemical and flame protective clothing for the national 119 rescue headquarters. Fash Text Res J. 2023;25(4):520–33.
56. Goldman RF, Joy RJT. Prevention of heat casualties in soldiers wearing CB protective clothing. Natick: Army Research Institute of Environmental Medicine; 1967.
57. Speckman KL, Allan AE, Sawka MN, Young AJ, Muza AR, Pandolf KB. A review: microclimate cooling of protective overgarments in the heat. Natick: US Army Research Institute of Environmental Medicine; 1988.
58. Choi JW, Kim MJ, Lee JY. Efficacy of cooling vests for alleviating heat strain of farm workers in summer. J Korean Soc Cloth Text. 2005;29(8):1176–87.
59. Choi JW, Kim MJ, Lee JY. Alleviation of heat strain by cooling different body areas during red pepper harvest work in WBGT 33°C. Ind Health. 2008;46:620–8.
60. Sawka MN, Montain SJ, Latzka WA. Hydration effects on thermoregulation and performance in the heat. Comp Biochem Physiol. 2001;128(4):679–90.
61. Noakes TB. Fluid replacement during exercise. Exerc Sport Sci Rev. 1993;21(1):297–330.
62. Sawka MN. Physiological consequences of hypohydration: exercise performance and thermoregulation. Med Sci Sports Exerc. 1992;24(6):657–70.
63. Koscheyev VS, Lee JY, Kim JH, Leon GR, Kwon S, Gernhardt ML. Cooling and thermal control strategies in the space suit for routine and emergency situations. SAE Technical Paper 2008-01-1993, 2008.
64. Koscheyev VS, Leon G, Coca A, Trevino R. Physiological design of a space suit cooling/warming garment and thermal control as keys to improve astronaut comfort, performance, and safety. Habitat. 2006;11(1):15–25.
65. Koscheyev VS, Coca A, Leon G. Overview of physiological principles to support thermal balance and comfort of astronauts in open space and on planetary surfaces. Acta Astronaut. 2007;60:479–87.
66. Grahn DA, Dillon JL, Heller HC. Heat loss through the glabrous skin surfaces of heavily insulated, heat-stressed individuals. J Biomech Eng. 2009;131(7):071005.
67. Shin S, Park JH, Lee JY. Does the hair influence heat extraction from the head during head cooling under heat stress? Ind Health. 2015;53(6):533–41.
68. Kim S, Lee JY. L-menthol attenuates the magnitude of cold-induced vasodilation on the extremities of young females. J Physiol Anthropol. 2018;37:14.
69. Lee JY, Nakao K, Bakri I, Tochihara Y. Body regional influences of L-menthol application on the alleviation of heat strain while wearing firefighter's' protective clothing. Eur J Appl Physiol. 2012;122(6):2171–83.
70. Epstein Y, Heled Y, Ketko I, Muginshtein J, Yanovich Y. The effect of air permeability characteristics of protective garments on the induced physiological strain under exercise-heat stress. Ann Occup Hyg. 2013;57(7):866–74.
71. Holmér I. Protective clothing and heat stress. Ergonomics. 1995;39:166–82.
72. Kim S, Lee JY. Skin sites to predict deep-body temperature while wearing firefighters' personal protective equipment during periodical changes in air temperature. Ergonomics. 2016;59(4):496–503.
73. McLellan TM. Heat strain while wearing the current Canadian or a new hot-weather French NBC protective clothing ensemble. Aviat Space Environ Med. 1996;67:1057–62.
74. Montain SJ, Sawka MN, Cadarette BS, Quigley MD, McKay JM. Physiological tolerance to uncompensable heat stress: effects of exercise intensity, protective clothing and climate. J Appl Physiol. 1994;77:216–22.
75. Taylor NAS, Lee JY, Kim S, Notley SR. Physiological interactions with personal-protective clothing, physically demanding work and global warming: an Asia-Pacific perspective. J Therm Biol. 2021;97:102858.
76. Moran DS, Shitzer A, Pandolf KB. A physiological strain index to evaluate heat stress. Am J Phys. 1997;275:129–34.
77. O' Brien C, Hoyt RW, Buller MJ, Castellani JW, Young AJ. Telemetry pill measurement of core temperature in humans during active heating and cooling. Med Sci Sports Exerc. 1998;30:468–72.
78. Ruddock A, Tew GA, Purvis A. Reliability of intestinal temperature using an ingestible telemetry pill system during exercise in a hot environment. J Strength Cond Res. 2014;28(3):861–9.
79. Gosselin J, Béliveau J, Hamel M, Casa D, Hosokawa Y, Morais JA, Goulet EDB. Wireless measurement of rectal temperature during exercise: comparing an

ingestible thermometric telemetric pill used as a suppository against a conventional rectal probe. J Therm Biol. 2019;83:112–8.
80. Notley SR, Meade RD, Kenny GP. Time following ingestion does not influence the validity of telemetry pill measurements of core temperature during exercise-heat stress. Temperature. 2021;8(1):12–20.
81. Tikuisis P, McLellan TM, Selkirk G. Perceptual versus physiological heat strain during exercise-heat stress. Med Sci Sports Exerc. 2002;34(9):1454–61.
82. Dehghan H, Sartang AG. Validation of perceptual strain index to evaluate the thermal strain in experimental hot conditions. Int J Prev Med. 2015;6:78.
83. Borg DN, Stewart IB, Costello JT. Can perceptual indices estimate physiological strain across a range of environments and metabolic workloads when wearing explosive ordnance disposal and chemical protective clothing? Physiol Behav. 2015;147:71–7.
84. Borg DN, Costello JT, Bach AJ, Stewart IB. Perceived exertion is as effective as the perceptual strain index in predicting physiological strain when wearing personal protective clothing. Physiol Behav. 2017;169:216–23.
85. Lee JY, Nakao K, Takahashi N, Bakri I, Son SY, Tochihara Y. Validity of infrared tympanic temperature as a thermal strain index while wearing impermeable protective clothing in hot environments. Ind Health. 2011;49:714–25.
86. Takahashi N, Lee JY, Wakabayashi H, Tochihara Y. Development and operational results of a real-time remote biological information monitoring device for the workers wearing protective clothes at a nuclear facility. Jpn J Health Phys. 2011;46:140–7. (in Japanese)
87. Takahashi N, Lee JY, Wakabayashi H, Tochihara Y. The rectal temperature estimation method based on tympanic temperature for workers wearing protective clothing in nuclear facilities (in Japanese). Health Phys. 2012;47(1):54–65. (in Japanese)
88. Lee JY, Ko ES, Lee HH, Kim JY, Choi JW. Validation of clothing insulation estimated by global and serial methods. Int J Cloth Sci Technol. 2011;23(2/3):184–98.
89. Wang F, Gao C. Protective clothing: managing thermal stress. Woodhead Publishing; 2014.
90. Siple PA. General principles governing selection of clothing for cold climates. Proc Am Philos Soc. 1945;89:200–34.
91. ISO 11079. Ergonomics of the thermal environment: determination and interpretation of cold stress when using required clothing insulation (IREQ) and local cooling effects. International Organization for Standardization. 2007.
92. EN511. Protective gloves against cold. European Standards. 2006.
93. Park J, Lee JY. Wearing effects of winter gloves in cold environment on physiological responses and subjective perception in elderly females. J Korean Soc Cloth Text. 2019;43(6):866–76.
94. Baek YJ, Cho K, Hong Y, Lee JY. Which jacket is thermo-physiologically and psychologically warmer between long-padded jackets filled with goose down and polyester in a windy and cold environment? J.KSLES. 2022;29(1):16–29. (in Korean)
95. Kim YB, Jang W, Kim K, Kim S, Baek YJ, Lee JY. Comparisons of thermal insulations between on air-cell packs embedded jacket and down jackets. J Korean Soc Cloth Text. 2015;39(1):39–46.
96. Kwon JY, Kim S, Baek YJ, Lee JY. Comparison and evaluation of clothing insulation of newly-developed air-filled baffle jackets and down padded jackets. Fash Text Res J. 2021;23(2):261–72. (in Korean)
97. Park H, Hwang SK, Lee JY, Fan J, Jeong Y. Impact of electrical heating on effective thermal insulation of a multi-layered winter clothing system for optimal heating efficiency. Int J Cloth Sci Technol. 2016;28(2):254–64.
98. Shin S, Choi HH, Kim YB, Hong BH, Lee JY. Evaluation of body heating protocols with graphene heated clothing in a cold environment. Int J Cloth Sci Technol. 2017;29(6):830–44.
99. Lee HR, Lim GY, Ghim Y, Hong BH, Lee JY. Heating the ears using graphene film earmuffs in a cold environment. Manuscript submitted for publication 2025.
100. Teitelbaum A, Goldman RF. Increased energy cost with multiple clothing layers. J Appl Physiol. 1972;32:743–4.
101. Quinn TD, Gutierrez B, Saez I, Santisteban A, Lee JY, Kim JH, Coca A. Comparison of three internationally certified firefighter protective ensembles: physiological responses, mobility, and comfort. Int J Ind Ergon. 2021;86:103232.
102. Kim S, Lee JY. Development of firefighting performance test drills while wearing personal protective equipment. J Korea Inst Fire Sci Eng. 2016;30(1):138–48. (in Korean)
103. Davis PO, Dotson CO, Maria DLS. Relationship between simulated fire fighting tasks and physical performance measures. Med Sci Sports Exerc. 1982;14(1):65–71.
104. Deakin JM, Smith JT, Stevenson JM, Wolfe LA, Lee SW. Development of a bona fide physical maintenance standard for CF and DND fire fighters. Final Report Submitted to Canadian Forces Personnel Support Agency. Queen's University, Kingston. 1996.
105. Kwon JY, Cho YS, Lee BH, Kim MS, Jun YM, Lee JY. Validity of a simulated practical performance test to evaluate the mobility and physiological burden of COVID-19 healthcare workers wearing personal protective equipment. Fash Text Res J. 2022;24(5):655–65. (in Korean)
106. Lim GY, Lee HL, Chun YM, Lee JY. Personal protective equipment for healthcare workers during COVID-19 pandemic: improvement of thermal comfort and development of a mobility test protocol. Korean J Community Living Sci. 2021;32(3):363–79.

8 Work in Artificial Severe Cold Environments

Yutaka Tochihara

8.1 Work in Cold Environments

Working in the cold decreases performances and productivity and increases the occurrence of accidents (e.g., Enander [1]). Moreover, it can have different adverse effects on human health, including respiratory, cardiovascular, musculoskeletal, and skin diseases (e.g., Mäkinen and Hassi [2]). Cold work is typically divided into outdoor and indoor cold work. Outdoor cold work is performed in winter forestry, fishing, agriculture, construction, etc. Indoor cold work includes work in the industry in winter and in cold stores.

There are a lot of field studies on the health problems of workers in indoor moderate cold environments [3–9] and moderate cold stores [10]. Moderate cold work mainly involves work in the food industry. Fresh foodstuffs are often processed in cold stores at temperatures of 0 to +10 °C, and many females work in these environments. Griefahn et al. [6] conducted a questionnaire survey on health problems and working conditions for more than 1000 male and female workers in moderately cold environments (−5 to 15 °C) in Germany. Workers' most frequently reported health problems were colds, low back pain, pain in the neck or shoulder, and rheumatism, in descending order. Inaba et al. [8] investigated subjective symptoms among female workers and their working conditions in a consumer cooperative in Japan. Approximately 70–80% of workers who classified cold storage goods complained of cold sensations in different regions, and shoulder and low back pain. The average temperature of their working environments was 4.8 °C. Oliveira et al. [9] conducted a questionnaire survey of 1576 workers in supermarkets in Portugal, working in cold climates, about their working conditions and health problems. Of the workers, 78.1% were women, and 37.3% reported several health problems. Repetitive work in the cold may cause musculoskeletal disorders, such as low back pain [11].

Indoor cold work is further divided into work in a natural environment and an artificial cold environment. Work in artificially cold environments performed in all seasons. Therefore, working in artificially cold environments can be more stressful for workers, except during winter. Cold stores are the most common artificial cold work environment. Cold stores operated at 10 °C or below are further classified into Classes C (10 to −20 °C) and F (below −20 °C). The total capacity of commercial cold stores in Japan was 28.2 million m^3, and 87% are kept at Class F. Most cold stores in Japan (82%) are operated between −20 and −30 °C. Previously, cold stores were common only in the United States, Europe, and Japan, but recently, owing to the development of frozen food chains, the capacity of cold stores has increased in other parts of Asia. Figure 8.1

Y. Tochihara (✉)
Department of Human Life Design and Science, Faculty of Design, Kyushu University, Fukuoka, Japan
e-mail: tochihara@kyudai.jp

Y. Tochihara (ed.), *The Thermal Environment*, https://doi.org/10.1007/978-981-96-3009-7_8

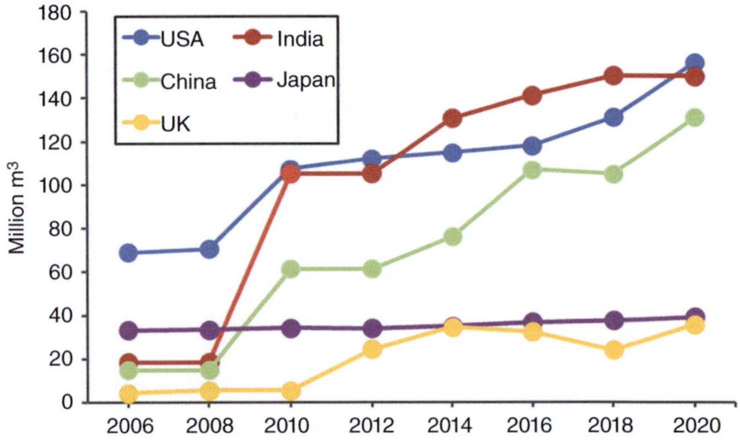

Fig. 8.1 Changes in total capacity of cold stores in the top five ranked countries. (Adapted from International Association of Refrigerated Warehouses [13] and Japan Association of Refrigerated Warehouses [12])

shows the changes in the total capacity of cold stores in the top five ranked countries [12, 13]. In the past 30 years, there has been no significant change in Japan, but capacity in China and India has increased considerably over the last 10 years. Therefore, even in tropical and subtropical countries, working in cold environments is not unusual.

However, the number of studies on working in severely cold stores (below −20 °C) is limited. We conducted several surveys to investigate the workload of workers in severely cold stores using questionnaires [14], time studies [15], and studies of on-site physiological responses and performance of workers [16]. We also conducted experimental studies using climatic chambers to simulate work in severely cold stores. We previously reported simulated experiments that focused on the effects of different temperatures, workloads [17], or exposures (exposure times and repetitions), on the following: thermal responses when the total cold exposure was kept the same under each condition [18]; thermal responses with and without cold protective clothing in a warm environment after severe cold exposures [19]; thermal responses from repeated exposures to severe cold with intermittent warmer temperatures [20]; and thermal responses at night [21]. Recently, Zlatar et al. [22] reviewed the scientific literature, including our four studies above, on the influence of the severe cold thermal environment (below −5 °C) on core and skin temperatures.

This chapter summarizes the characteristics of working in severely cold stores (below −20 °C), including our several surveys and experimental studies.

8.2 Surveys on Work in Artificial Severe Cold Environments

8.2.1 Questionnaire Survey and Health Examination

The questionnaire survey was administered to workers in 377 cold stores to investigate their working environments and conditions [14]. The most common temperature in cold stores was −20 to −30 °C. Between 30,000 and 40,000 workers worked in cold stores in Japan. The most frequent physical complaints of cold store workers were "lumbago (low back pain)," followed by "It's easy to catch a cold," "neuralgia," and "rheumatism" (Fig. 8.2). Workers in cold stores faced several health problems. The extreme coldness and large temperature difference between the inside and outside of cold stores [6, 23] should also be considered a cause of health problems, in addition to working conditions. Tanaka et al. [24] also conducted a questionnaire survey by mail with labor managers of 500 cold stores in Japan. They reported that the loading methods in cold stores included forklifts, manual handling, and automatic machines, and the use of forklifts appeared to be widespread. They reported that as

Fig. 8.2 Physical complaints in 377 cold stores. (Adapted from Tanaka et al. [14])

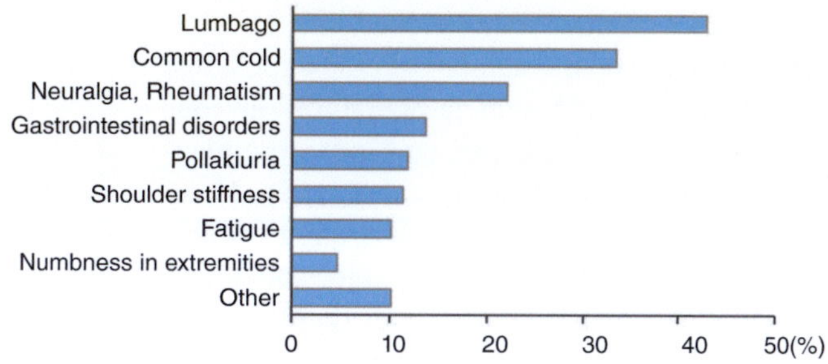

the set temperature of the cold storage rooms reduced, the working time in the cold decreased, and the cycle became more frequent.

Chen et al. [25] investigated the health problems of workers in cold stores (−10 to −25 °C), moderately cold stores (+5 to −5 °C), and ordinary stores (20 to 30 °C) in China using a self-administrated questionnaire. They reported that the number of complaints of low back pain and knee pain in cold stores were significantly higher than those in ordinary stores. They concluded that insufficient protective clothing and too long a time staying inside the cold stores might be the key factors causing musculoskeletal symposium and health problems among severely cold store workers. Dovrat and Katz-Leurer [26] conducted a cross-sectional survey on low back pain in severely cold (−20 °C) and ordinary (20 to 25 °C) store workers who had been employed for at least 1 year in food stores in Israel. Workers in severely cold stores had increased odds of low back pain symptoms than those in ordinary stores. Penzkofer et al. [27] conducted a questionnaire study in Germany on age-related stress and strain associated with working in the cold between 3 and −24 °C. The participants were 63 young and 65 middle-aged order-pickers in 24 cold stores. Regardless of age, order-picking work in the cold leads to frequent complaints especially in the upper part of the body. Younger workers reported greater cold discomfort than middle-aged workers.

Health problems related to working in extremely cold stores (−42 to −62 °C) in a freeze-drying coffee company in Colombia were investigated using questionnaires and physical examinations [28]. Workers in extremely cold stores were at risk of cold-related health problems (respiratory symptoms, musculoskeletal complaints in the lower back and neck, etc.) and injuries owing to the lowered concentration level. They wore heavy protective clothing (2.4 clo, 7.5 kg); however, it may not protect the respiratory system. Exposure to extreme cold conditions may trigger respiratory symptoms [29].

In summary, from questionnaire studies, workers in severely cold stores are at a greater risk of health problems, especially musculoskeletal disorders.

8.2.2 Onsite Field Survey

Tymanski [30] investigated the actual working conditions of severely cold store workers in the summer in Poland using a time study. Temperatures outside and inside stores were 20 to 30 °C and 4 to −38.5 °C, respectively. The workers entered the cold store up to 7 times per day, and stayed there for 1–47 min each time. Williamson et al. [31] measured the hand and foot skin temperatures and thermal discomfort of eight workers in a severely cold store (−27 °C) in the UK. The lowest skin temperatures of the hand and foot were 6.8 °C and 11.1 °C, respectively. The authors concluded that, when proper protective clothing was worn, 40% of the total cycle time (24 min/h) would not cause an acceptable risk of discomfort.

We conducted a field study on work in severely cold stores in Japan [15, 16]. The participants were 10 forklift-truck workers (Group R) in two cold stores (−20.3 to −23.2 °C) and eight forklift-truck workers (Group C) working in an ordinary store (12 to 15.2 °C). The investigators followed the workers from the start to the end of the workday. They checked time spent in the cold store and the number of cold exposures, etc. Heart rate and skin temperatures at four points (cheek, chest, finger, and toe) were measured every minute during work [15]. Hand tremor, handgrip strength, pinch strength, counting task performance, flicker value, and blood pressure were measured five times (before work, at 10 a.m., before lunch, at 3 p.m., and after work) per day. Blood samples were collected before lunch [16]. The working hours in Group R were longer than those in Group C. Group R' work was irregular, so they could not rest regularly. Figure 8.3 shows the cold exposure frequencies per day for Group R. The mean cold exposure time for Group R was 125 min/day. The mean frequency of entering cold stores was 73 times per day, much greater than those previously reported [31]. Although the duration of cold exposure per stay was very short (<5 min), the workers entered cold stores very frequently,

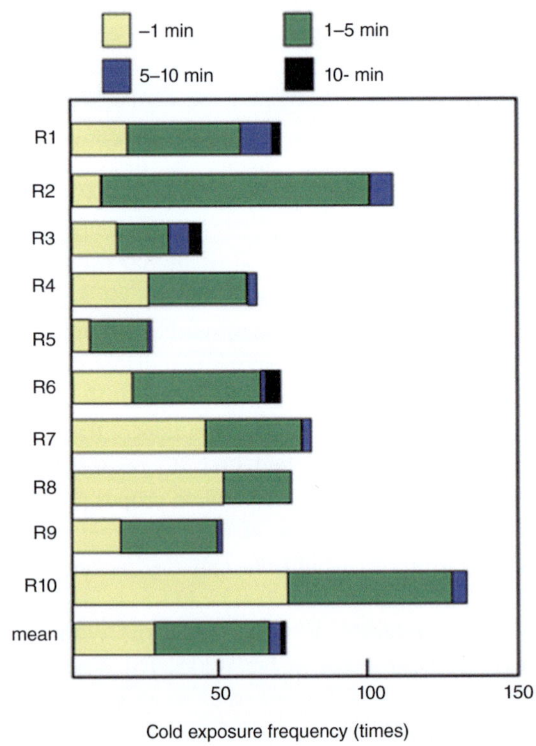

Fig. 8.3 Cold exposure frequencies in a day and distribution of cold exposure time per each stay. (Adapted from Tochihara et al. [15])

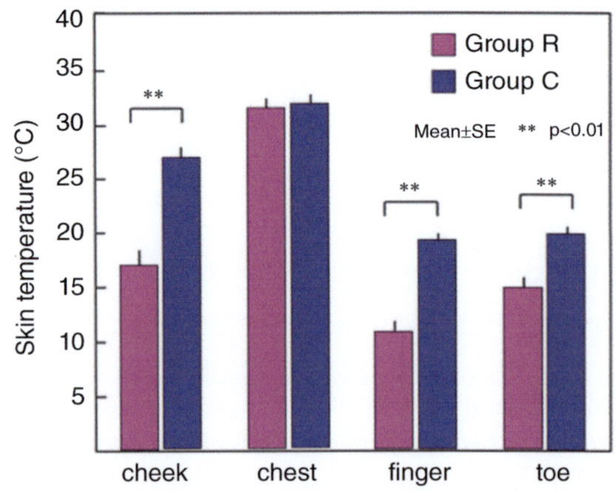

Fig. 8.4 The lowest skin temperatures during work in both groups. (Adapted from Tochihara et al. [15])

and the skin temperatures at the peripheral parts decreased markedly. The averages of the lowest skin temperatures at four points on the skin surface were measured for both groups. The lowest skin temperature at the chest showed no significant difference between groups. In contrast, the lowest skin temperatures at the cheek, finger, and toe were significantly lower in Group R than in Group C. The mean values of the lowest skin temperatures at the cheek, finger, and toe in Group R were 16.9, 11.0, and 15.1 °C, respectively (Fig. 8.4). Therefore, when we set the threshold limit values for working at cold temperatures, we must regulate not only the consecutive cold exposure time but also the total cold exposure time in an hour.

The free fatty acid (FFA) was significantly higher in Group R than in Group C. There were no significant differences in handgrip strength, pinch strength, counting task performance, or flicker value between Groups R and C. However, the changes in hand tremors and diastolic blood pressure were significantly greater in Group R than in Group C (Fig. 8.5). There was a significant relationship between FFA and hand tremor values measured for the second time only in Group R. The work loads of Group R would be increased by not only the extreme coldness but also by the large temperature difference between the inside and outside of the cold stores. The actual forklift work in these cold stores did not cause a distinct reduction in manual performance but caused an

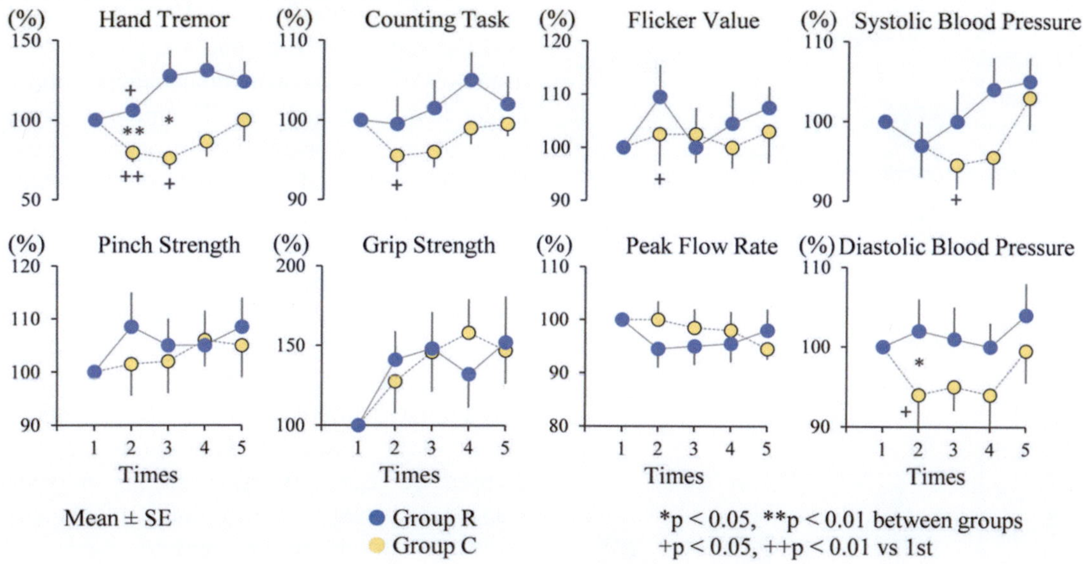

Fig. 8.5 Changes in physiological responses and manual performance during work in both groups. 1: before work, 2: at 10 a.m., 3: before lunch, 4: at 3 p.m., 5; after work. The value before the work is expressed as 100%. (Adapted from Tochihara et al. [16])

increase in stress which would be expressed as an increase in catecholamine excretion.

In Portugal, Oliveria et al. [32] investigated physical parameters (air temperature, globe temperature, relative humidity, and air velocity), estimated metabolic rate, and thermal insulation of clothing (I_{clr}) in 21 cold stores (−15 to −25 °C). The average air temperature, metabolic rate, and clo values of them were −17.7 °C, 164.7 w•m^{-2}, and 1.5 clo, respectively. They calculated the neutral required clothing insulation ($IREQ_{neutral}$) and minimal required clothing insulation ($IREQ_{min}$) using physical factors in ISO11079 2007 [33]. Only approximately 10% of the cold stores had an estimated Iclr value in the area between $IREQ_{neutral}$ and $IREQ_{min}$, where the selected clothing ensemble provides sufficient insulation. In Brazil, Zlatar et al. [23] measured continually core temperature and eight points of skin temperatures during work in a severely cold store (−25 °C) for four male workers. The lowest hand, forehead, core, and mean body temperatures were 14.1, 18.6, 37.4, and 34.9 °C, respectively. These values are in good agreement with those reported previously [15].

8.3 Experiments on Work in Artificial Severe Cold Environments

8.3.1 Manual Loading Work

Physiological responses and manual performance were measured during loading work in a severely cold environment [17]. Eight male students wearing standard cold-protective clothing (3.4 clo) occupied a 20 °C room for 20 min and were then transferred to a severely cold room (−25 °C) for 30 min. This exposure pattern was repeated three times, for a total time of exposure to extreme cold of 90 min. In the severely cold room, the subjects transported 9 kg (Condition R1) and 18 kg (Condition R2) of goods for 10 min, or performed no work (Condition C). At the end of the cold exposure, the rectal, mean skin and mean body temperatures in Conditions R2 and R1 were higher than those in Condition C. Cold stress declined due to an increase in heat production during work as observed when the subjects noted a “less cold sensation” with loading work. Figure 8.6 shows the plasma concentration of noradrenaline before and after cold exposure and

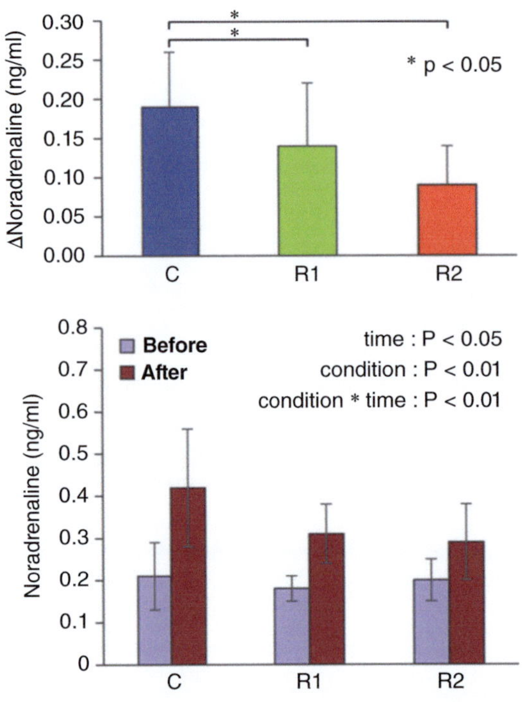

Fig. 8.6 Plasma concentration of noradrenaline before and after cold exposure and increase in plasma concentration of noradrenaline under three conditions. Values are means±SE. C: Control, R1: 9 kg, R2: 18 kg. (Adapted from Kim et al. [17])

the increase in noradrenaline under the three conditions. The increase in plasma noradrenaline levels was depressed according to an increase in work. In contrast, the skin temperature of the finger did not change. Moreover, the precision of manual work decreased as the load weight increased [17].

German ergonomists [34–36] reported three experimental studies using the same experimental procedure for two groups in actual cold stores. Each group of 15 workers participated in a moderately cold store (+3 °C) or a severely cold store (−24 °C) manual handling work—lifting and carrying up to 15 kg of frozen goods. They worked in the cold for 80, 100, and 120 min, separated by 20-min warm up breaks (+20 °C room). Tympanic temperature, skin temperatures, heart rate, blood pressure, and metabolic rate were measured during the seven working hours. First, Kluth et al. [34] focused on age differences in physiological responses of core and skin temperatures working in the cold at +3 and −24 °C. One group consisted of young workers (20–35 years old), and the other consisted of middle-aged workers (40–65 years old). In severely cold stores, a complete warming of core temperature during the breaks was often not possible for both groups, and the core temperature of the young group decreased up to 1.5 °C, while the decrease in the middle-aged group was up to 2.2 °C. There were no age-related differences in skin temperatures at +3 and −24 °C. Second, Baldus et al. [35] also focused on age differences in physiological responses between young and middle-aged females. Working in the severely cold store caused a decline of core temperature up to 1.1 °C for younger females and 1.3 °C for older females. In both sexes, the physiological strain indicated by a decrease in the core temperature of middle-aged workers was significantly greater than that of young workers in severely cold stores. Finally, Kluth et al. [36] reported age-related differences among male workers in the physiological responses of heart rate and blood pressure as well as core and skin temperatures. Although there were no age differences in blood pressure, the heart rate indicated a substantially higher strain in older workers.

8.3.2 Repeated Cold Exposures

Numerous studies have demonstrated the effects of long-term exposure to cold on health (e.g., Holmér [37]). However, investigations of the effects of repeated cold exposures on health, physiological responses and performance are limited in literature [38]. Moreover, these types of cold exposure have rapidly become more common in the industry. For example, jobs that involve working in cold stores have been changing from manual jobs, such as those performed by porters, to sedentary jobs, such as those performed by forklift-truck workers [24]. Therefore, although workers are not exposed to severe cold for a long time, they frequently go in and out of cold storage areas [15, 31].

A study was conducted to investigate the effects of different exposure rates on thermal responses, with the total cold exposure time being the same under each condition [18]. After resting in a warm room (25 °C) for 10 min, six male students wearing standard cold protective clothing (3.3 kg, 2.4 clo) entered an adjoining cold room (−25 °C). Cold exposure for 5-, 10-, and 20-min was repeated 12, 6, and 3 times, respectively. A similar duration of rest at 25 °C followed each cold exposure (Fig. 8.7). The total cold exposure time (60 min) was the same for all three conditions. The rectal temperature, skin temperatures, blood pressure, plasma 17-hydroxycorticosteroid (17-OHCS), counting task performance, and subjective responses were measured. At the end of the cold exposure, skin temperatures were higher for those who underwent shorter exposures than those exposed to the other conditions, except for foot skin temperature. However, there were no differences among the three experimental conditions in the decrease in rectal temperature (Fig. 8.7) and urinary excretion of 17-OHCS, which are good indices of cold stress.

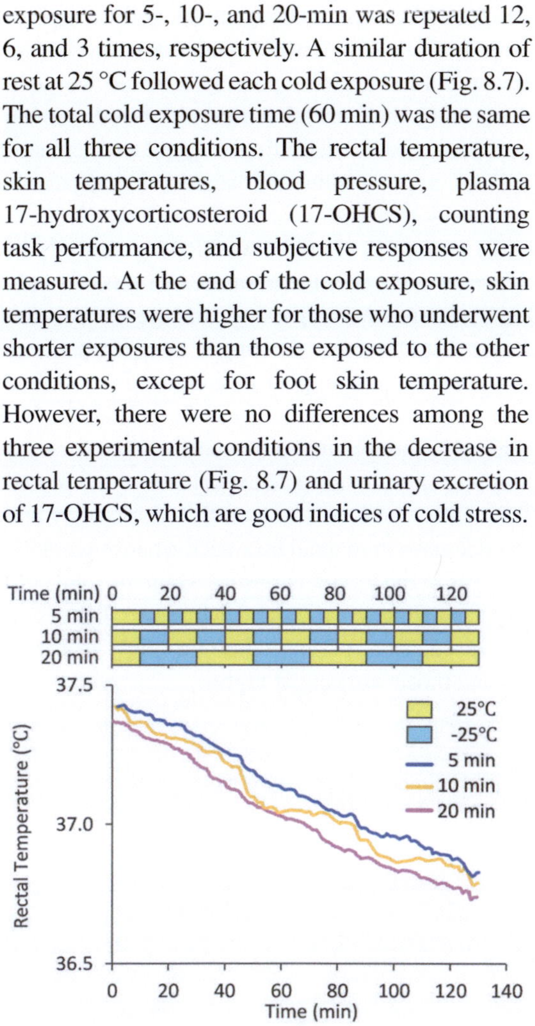

Fig. 8.7 Changes in rectal temperature under three conditions. Values are means of 6 subjects. (Adapted from Tochihara et al. [18])

Moreover, the increase in blood pressure and decreased performance of the counting task due to cold did not differ among the three conditions. Although the cold exposure time for each stay was short, when cold exposures were repeated frequently, the cold stress on the whole body and the decrease in manual task performance were the same as in the longer exposure when the total cold exposure time was the same.

8.3.3 Work at Night

With the recent development of chain stores that sell frozen food, the potential for working in cold stores has increased to 365 days a year and 24 h a day, usually in shifts. Since much of this work involves frozen fish/meat, many workers work night shifts to deliver them to the markets in the early morning [10]. We evaluated human physiological responses and the performance of manual tasks during exposure to severe cold at night and in the afternoon [21]. Thirteen male students, wearing standard cold protective clothing (3.3 kg, 2.4 clo), were exposed to severe cold in the afternoon (3–5 p.m.) and at night (3–5 a.m.). The schedule used in this study was chosen for two reasons: first, it tested responses at the most common time for the delivery of frozen foods, and second, it tested responses when the core temperature reached its maximum and minimum levels.

The study participants were kept in a severely cold room (−25 °C) for 20 min; thereafter, they were placed in another room (10 °C) for 20 min. This pattern was repeated three times, with a total cold exposure time of 60 min. Their rectal temperature, skin temperatures, manual dexterity, blood pressure, and thermal comfort were measured during the experiments. At the beginning of the experiment, rectal temperatures in the afternoon were significantly higher than those at night because of the participant's circadian rhythm. The decrease in rectal temperature at night was significantly greater during cold exposure than during cold exposure in the afternoon (Fig. 8.8). Although there were no significant differences in the mean skin temperature between the afternoon and night, the finger (Fig. 8.9) and toe skin temperatures at night were significantly higher. These higher skin temperatures on the peripheral parts of the body would imply increase heat loss, which could cause a decrease in rectal temperature. This phenomenon represents a change in the set-point. Since the peripheral skin temperatures of participants are higher at night than in the afternoon, they are less likely to feel cold or pain sensations at night.

While the peripheral skin temperatures at night were higher than those in the afternoon, manual performance was significantly lower at night than in the afternoon, from the beginning of the experiment as shown in Fig. 8.10. Consequently, we suggest that manual performance decreases during night work. Since the coefficients of correlation between mean body temperature and manual dexterity were significantly greater than those between finger skin temperature and manual dexterity, we suggest that the decrease in manual performance at night was more affected by the decrease in whole-body temperature, expressed by mean body temperature. In conclusion, we found that exposure to severe cold at night negatively affected workers, decreasing their rectal temperature and manual performance. These findings suggest an increased risk of hypothermia and accidents individuals who work at night.

In summary, when the workload increased, the decrease in the core temperature in a severely cold environment was suppressed, and the workers felt less cold. However, the finger skin temperature did not change owing to an increase in workload and caused a greater decrease in manual performance. Although the cold exposure time for each stay was short, when cold exposures were repeated frequently, the cold stress on the whole body and the decrease in manual task performance were the same as in the longer exposure when the total cold exposure time was the same. Moreover, there is an increased risk of hypothermia and accidents among those working at night.

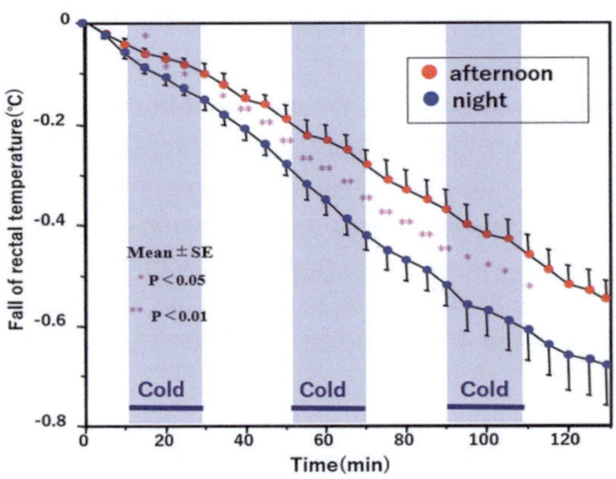

Fig. 8.8 Changes in rectal temperature in the afternoon and at night under −25 °C and 10 °C conditions. (Adapted from Ozaki et al. [21])

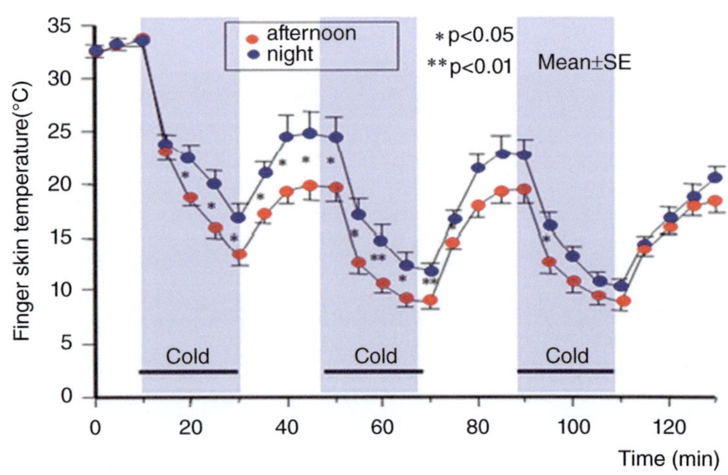

Fig. 8.9 Changes in finger skin temperature in the afternoon and at night. (Adapted from Ozaki et al. [21])

Fig. 8.10 Changes in manual performance in the afternoon and at night. (Adapted from Ozaki et al. [21])

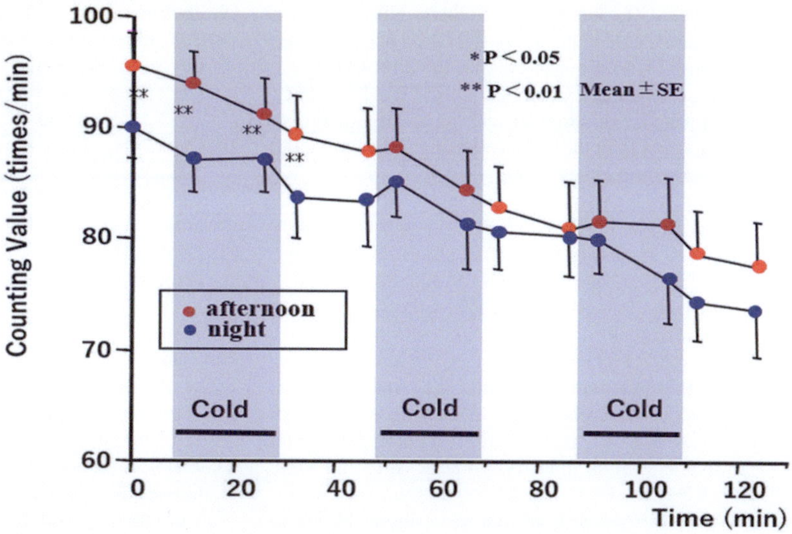

8.4 Future Research

As the population ages, the number of middle-aged and older workers working in severely cold stores is expected to increase. Many women have worked in moderately cold stores, but rarely in severely cold stores. However, the number of women working in severely cold conditions is expected to increase since 1999 due to the revision of the law in Japan under Ministry of Health, Labour and Welfare [39], which makes it possible for women to work in extreme cold conditions (below −10 °C). Several research groups [27, 34–36] investigated physiological responses such as core and skin temperatures of young and middle-aged males and females working in the cold at +3 and −24 °C. However, to date, there are no field or experimental studies targeting people aged 65 years and older who work in severely cold conditions. It will be necessary to conduct research targeting women and middle-aged and older people.

Many studies have investigated the effects of severe cold on the human body. Most of their results, however, were obtained from experimental participants, such as students and soldiers, who were not expose to severe cold conditions in their ordinary lives. It may be possible, therefore, that there are differences between those who are or are not habitually exposed to cold [40]. Therefore, an experimental study comparing cold store workers' physiological and subjective responses and the general public under severely cold conditions is necessary.

In addition, ultralow-temperature cold stores (below −40 °C) are emerging to preserve the color of tuna for *sashimi* [24] and to freeze coffee [28]. Further research on working under extremely cold conditions is needed.

From these field and experimental studies, we could measure the physiological strain placed on severely cold store workers and evaluate their work environments. However, the burden of labor on women, middle-aged, and older people under severely cold conditions, the effects of cold adaptation, and the burden of labor under ultralow temperatures will be important subjects for future research.

References

1. Enander A. Performance and sensory aspects of work in cold environments: a review. Ergonomics. 1984;27(4):365–78.
2. Mäkinen TM, Hassi J. Health problems in cold work. Ind Health. 2009;47(3):207–20.

3. Lundqvist GR, Jensen PL, Solberg HE, Davidsen E. Moderate cold exposure in the Faroe fishing industry. Scand J Work Environ Health. 1990;16(4):278–83.
4. Chiang HC, Ko LC, Chen SS, Yu HS, Wu TN, Chang PY. Prevalence of shoulder and upper-limb disorders among workers in the fish-processing industry. Scand J Work Environ Health. 1993;19(2):126–31.
5. Ohlsson K, Hansson GA, Balogh I, Stromberg U, Palsson B, Nordander C, et al. Disorders of the neck and upper in women in the fish processing industry. Occup Environ Med. 1994;51(12):826–32.
6. Griefahn B, Mehnert P, Brode P, Forsthoff A. Working in moderate cold: a possible risk to health. J Occup Health. 1997;39(1):36–44.
7. Bang BE, Aasmoe L, Aardal L, Andorsen GS, Bjørnbakk AK, Egeness C, et al. Feeling cold at work increases risk of symptoms from muscles, skin, and airways in seafood industry workers. Am J Ind Med. 2005;47(1):65–71.
8. Inaba R, Mirbod SM, Kurokawa J, Inoue M, Iwata H. Subjective symptoms among female workers and winter working conditions in a consumer cooperative. J Occup Health. 2005;47(5):454–65.
9. Oliveira AVM, Gaspar AR, André JS, Quintela DA. Subjective analysis of cold thermal environments. Appl Ergon. 2014;45(3):534–43.
10. Enander A, Ljungberg A, Holmer I. Effects of work in cold stores on man. Scand J Work Environ Health. 1979;5(3):195–204.
11. Oksa J, Ducharmeand MB, Rintamäki H. Combined effect of repetitive work and cold on muscle function and fatigue. J Appl Physiol. 2002;92(1):354–61.
12. Japan Association of Refrigerated Warehouses. Statistical charts of refrigerated warehouses [Internet]. 2021. Available from: www.jarw.or.jp/koow/statistics.
13. International Association of Refrigerated Warehouses. URL: https://en.wikipedia.org/wiki/International_Association_of_Refrigerated_Warehouses.
14. Tanaka M, Tochihara Y, Matsui J, Ohnaka T, Yamazaki S, Yoshida K, et al. Status and problems of the worker in cold stores from a viewpoint of occupational hygiene (in Japanese). J Sci Labour. 1979;55:101–6.
15. Tochihara Y, Ohkubo C, Uchiyama I, Komine H, Honda Y, Aoki N, et al. A survey on work loads of forklift-truck workers in cold storages. Bull Inst Public Heal. 1990;39:29–36.
16. Tochihara Y, Ohkubo C, Uchiyama I, Komine H. Physiological reaction and manual performance during work in cold storages. Appl Hum Sci. 1995;14(2):73–7.
17. Kim TG, Tochihara Y, Fujita M, Hashiguchi N. Physiological responses and performance of loading work in a severely cold environment. Int J Ind Ergon. 2007;37(9–10):725–32.
18. Tochihara Y, Ohnaka T, Tuzuki K, Nagai Y. Effects of repeated exposures to severely cold environments on thermal responses of humans. Ergonomics. 1995;38(5):987–95.
19. Tochihara Y, Ohnaka T, Tsuzuki K, Nagai Y. Comparison of thermal responses with and without cold protective clothing in a warm environment after severe cold exposures. J Therm Biol. 1993;18(5–6):639–42.
20. Ozaki H, Enomoto-Koshimizu H, Tochihara Y, Nakamura K. Thermal responses from repeated exposures to severe cold with intermittent warmer temperatures. Appl Hum Sci. 1998;17(5):195–205.
21. Ozaki H, Nagai Y, Tochihara Y. Physiological responses and manual performance in humans following repeated exposure to severe cold at night. Eur J Appl Physiol. 2001;84(4):343–9.
22. Zlatar T, Costa JT, Vaz M, Baptista JS. Influence of severe cold thermal environment on core and skin temperatures: a systematic review. Work. 2019;62:337–52.
23. Zlatar T, Junior BB, Bezerra LM, Araujo Brito MM, Costa JT, Vaz M, et al. Safety and health risks for workers exposed to cold thermal environments: a frozen food processing industry perspective. Work. 2021;70(2):645–55.
24. Tanaka M, Nakamura K, Shimai S, Takahashi H. Work at cold storage. J Therm Biol. 1993;18(5–6):647–50.
25. Chen F, Li T, Huang H, Holmér I. A field study of cold effects among cold store workers in China. Arctic Med Res. 1991;50(Suppl 6):99–103.
26. Dovrat E, Katz-Leurer M. Cold exposure and low back pain in store workers in Israel. Am J Ind Med. 2007;50(8):626–31.
27. Penzkofer M, Kluth K, Strasser H. Subjectively assessed age-related stress and strain associated with working in the cold. Theor Issues Ergon Sci. 2013;14(3):290–310.
28. Piedrahita H, Oksa J, Malm C, Rintamäki H. Health problems related to working in extreme cold conditions indoors. Int J Circumpolar Health. 2008;67(2–3):279–87.
29. Koskela HO. Cold air-provoked respiratory symptoms: the mechanisms and management. Int J Circumpolar Health. 2007;66(2):91–100.
30. Tymanski S. A research into the degree of exposure of fish cold-stores dockers to temperatures inherent in their work. Bull Inst Marit Trop Med Gdynia. 1982;33(3–4):181–8.
31. Williamson DK, Chrenko FA, Hamley EJ. A study of exposure to cold in cold stores. Appl Ergon. 1984;15(1):25–30.
32. Oliveira AVM, Gaspar AR, Quintela DA. Occupational exposure to cold thermal environments: a field study in Portugal. Eur J Appl Physiol. 2008;104(2):207–14.
33. Ergonomics of the thermal environment : determination and interpretation of cold stress when using required clothing insulation (IREQ) and local cooling effects. In: International Organization for Standardization (ISO). ISO11079. Geneva; 2007.
34. Kluth K, Penzkofer M, Strasser H. Physiological responses of core and skin temperature of two

age groups to working in the cold at −3°C and −24°C. Occup Ergon. 2009;8(4):147–57.
35. Baldus S, Kluth K, Strasser H. Order-picking in deep cold—physiological responses of younger and older females. Part 2: body core temperature and skin surface temperature. Work. 2012;41 Suppl 1:3010–7.
36. Kluth K, Penzkofer M, Strasser H. Age-related physiological responses to working in deep cold. Hum Factors Ergon Manuf. 2013;23(3):163–72.
37. Holmér I. Evaluation of cold workplaces: an overview of standards for assessment of cold stress. Ind Health. 2009;47(3):228–34.
38. Tochihara Y, Ohnaka T, Yamazaki SYK. Effects of repeated cold exposures on thermal responses of man. Ann Physiol Anthropol. 1989;8:169–71.
39. Ministry of Health Labour and Welfare. Ordinance on industrial safety and health 2021 [Internet]. 2021. Available from: https://jsite.mhlw.go.jp/ishikawa-roudoukyoku/library/ishikawa-roudoukyoku/seido/kinto/file/aramashi_jyosei.pdf.
40. Nagai Y, Nakayama ETY. Thermal responses of cold storage workers during repeated exposures to severe cold. In: Proceedings of the second international conference on human-environment system; 1998. p. 227–30.

9 Residential Thermal Environment for the Elderly

Nobuko Hashiguchi

9.1 Elderly and Room Temperature

9.1.1 Cold Environment

The room temperature, amount of clothing, and thermal sensations of older people in their homes were measured during the winter of 1993 in Japan [1]. The elderly living room was 11 °C at 7 am and 15.6 °C at 7 pm, lower than those of young adults. Nevertheless, many of the elderly reported that they were warmer. It has been suggested that this is related to a higher clo value (heavier clothing) in the elderly than in young adults. The housing and health guidelines of the World Health Organization strongly recommend that the room temperature in winter should not fall below 18 °C. However, the results of the winter thermal environment measurements of 331 houses in 11 regions of Japan conducted by Ohnaka et al. [2] showed that most houses tended to have room temperatures below 18 °C, including living rooms, bathrooms, and corridors, except in one cold region (Sapporo), where the houses were fully equipped with heating systems (Fig. 9.1). Tochihara et al. [1] examined age differences in physiological and psychological responses to exposure to 10 or 35 °C for 49 min with standard clothing. Older women complained significantly less of the cold immediately after exposure than younger women, suggesting that older adults have delayed sensitivity to the cold (Fig. 9.2). Some studies have examined older adults' warm and cold points to evaluate their temperature discrimination ability. Murata et al. [3] examined the density of cold and pain points in elderly people and young adults. They found that cold points were 32 of 100 skin points for the elderly and 66 for the young, indicating a significant age difference. The age-related decrease in the number of cold points may be one reason for the delayed response to cold stimulation.

In addition, to stay comfortable in a residential environment, it is necessary to maintain appropriate temperatures according to the season and minimise differences in temperature between verticals and between rooms (heat shock). In winter, people use various heaters, such as stoves and air conditioners, to maintain thermal comfort in the room. Warmed air usually stagnates at the top, near the ceiling, and temperatures at lower heights are lower; therefore, vertical temperature gradients tend to increase, particularly during winter. When air-conditioning heating was performed in an experimental room at a room temperature of 25 °C, the air temperatures at 10 and 110 cm above the floor were 21 and 25 °C, respectively, resulting in a vertical temperature difference of approximately 4 °C [4] (Fig. 9.3). Although this may depend on the heating system model, room size, and other factors, we usually

N. Hashiguchi (✉)
Department of Health Science, Faculty of Medicine, Kyushu University, Fukuoka, Japan
e-mail: hashiguchi.nobuko.053@m.kyushu-u.ac.jp

Y. Tochihara (ed.), *The Thermal Environment*, https://doi.org/10.1007/978-981-96-3009-7_9

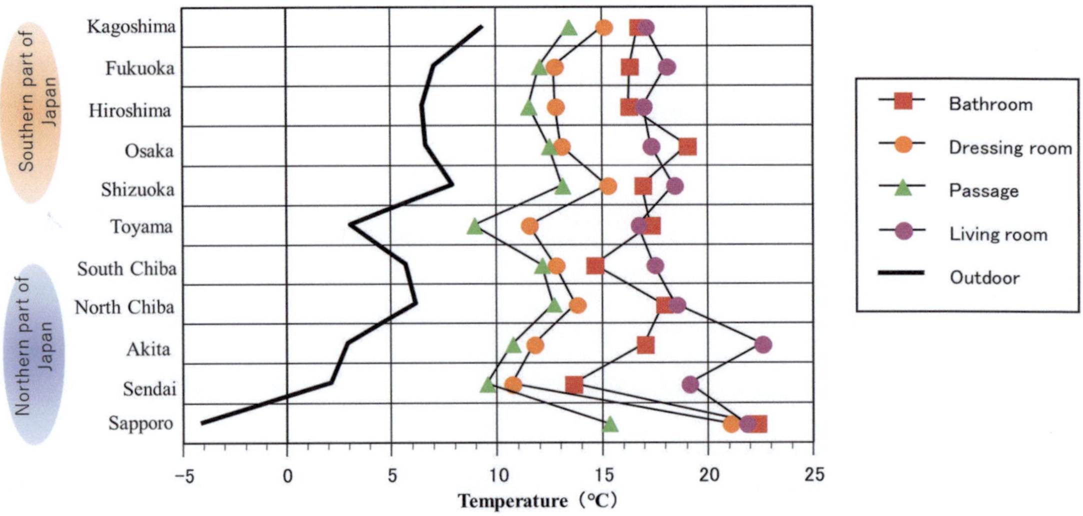

Fig. 9.1 Mean air temperatures in the bathroom, dressing room, corridor, living room, and outdoor during bathing in 11 regions in Japan. (Modified from Ohnaka et al. [2])

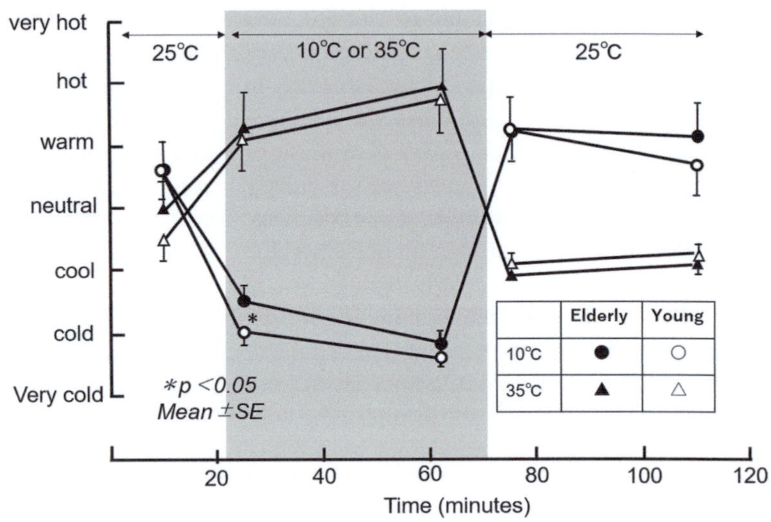

Fig. 9.2 Changes in thermal sensation of the elderly females and college-aged females. (Adapted from Tochihara et al. [1])

live in an environment with such vertical temperature differences. Hashiguchi et al. [5] examined the physiological and psychological responses of elderly to differences in vertical temperature. Eight healthy elderly male subjects and eight healthy young male subjects were exposed to an air box in an artificial climate chamber for 90 min with the upper and lower temperatures controlled at the boundary below the umbilical region. The upper temperature was 25 °C, and the lower temperatures were 16, 19, 22, and 25 °C. Systolic blood pressure was significantly higher in the elderly than in the young at lower temperatures below 22 °C (Fig. 9.4). Regarding subjective responses, both the elderly and young participants perceived cold with decreasing lower temperatures. However, the elderly reported less thermal discomfort even as the lower temperature decreased compared to the younger subjects (Fig. 9.5).

It has been suggested that a vertical temperature gradient of 3 °C or more, even if the room temperature is 25 °C, can increase blood pressure in the elderly. This is also a more serious

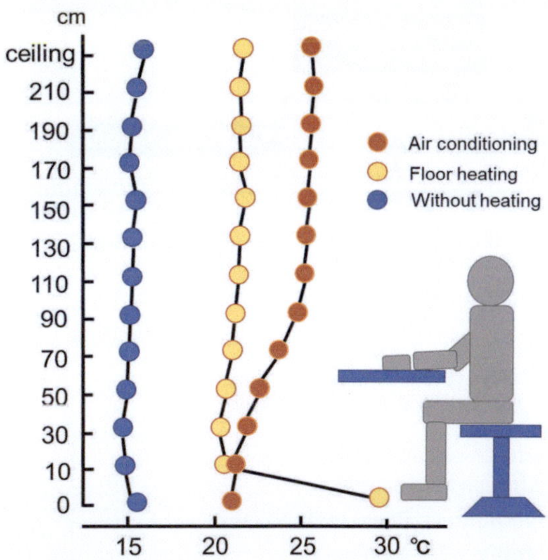

Fig. 9.3 The vertical distributions of the air temperature under three respective heating system conditions. (Adapted from Hashiguchi et al. [4])

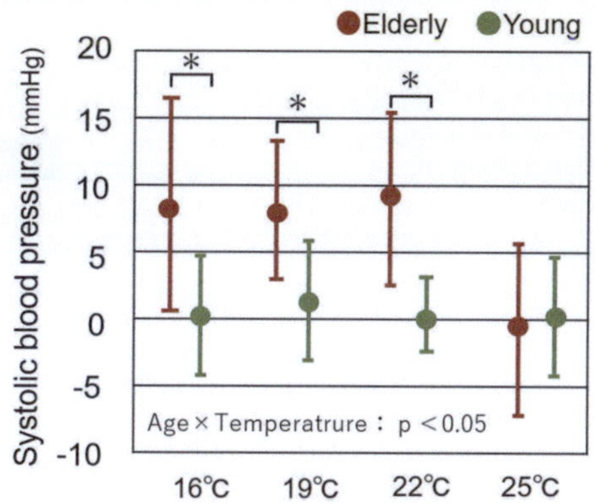

Fig. 9.4 The differences in systolic blood pressure before and during exposure to the vertical distributions of the air temperature. (Adapted from Hashiguchi et al. [5])

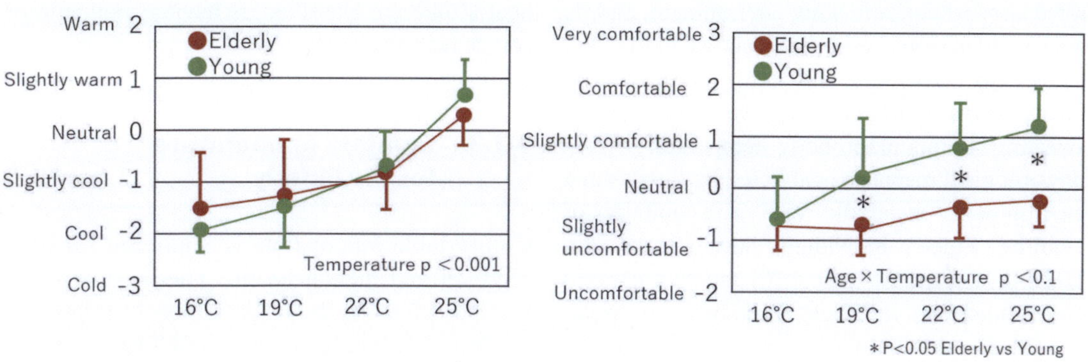

Fig. 9.5 Thermal sensation (left) and thermal comfort (right) at the lower body in the early phases of the exposure to the vertical distributions of the air temperature. (Adapted from Hashiguchi et al. [5])

problem for the elderly because they do not subjectively perceive it as uncomfortable. Health problems in cold environments include cerebrovascular and cardiovascular diseases. Cold stimuli induce the peripheral constriction of blood vessels, resulting in increased blood pressure. This increase in blood pressure can cause cerebral haemorrhage, angina pectoris, and myocardial infarction. The daily mean ambient temperature and relative humidity showed a statistically significant negative effect on the incidence of emergency transport events due to stroke (Ohshige 2006). Elderly people generally have reduced vascular flexibility due to arteriosclerosis, and their blood pressure increases significantly under the same cold stimulus. However, as mentioned above, aging causes a decrease in sensitivity to cold environments and stimuli in the lower part of the body. Thus, it is difficult for them to make appropriate behavioural adjustments, making their health problems more severe in cold environments.

9.1.2 Hot Environment

Heat stroke is one of the most serious health problems associated with exposure to hot envi-

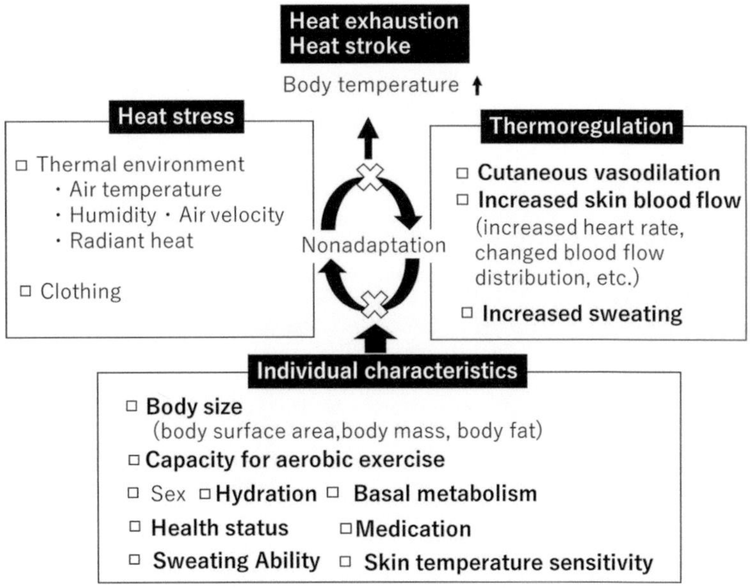

Fig. 9.6 The thermoregulatory response in a hot environment and its associated personal attributes

ronments. The relationship between the thermoregulatory response in a hot environment, and its associated personal attributes is shown in Fig. 9.6. Under hot environments, skin blood vessels dilate, and skin blood flow increases, resulting in sweating to maintain body temperature. This physiological response is affected by the thermal environment, metabolic rate, and amount of clothing. Elderly individuals have a poor skin blood flow response to heat [6] and an increased core temperature threshold at the onset of sweating [7]. In addition, they cannot cope with heat stress because of a lack of proper perception of room temperature. Skin temperature sensitivity also decreases with age [8–12]. This also affects behavioral thermoregulatory responses. It has been reported that those over 70 years old spend less time using air conditioners than those under 59 years old [13] and cannot adapt to a hot climate. Consequently, the elderly are more likely to experience heat exhaustion and heat stroke due to increased body temperature. The elderly account for the largest number of emergency transports due to heat stroke during summer [14]. Individual characteristics such as body size, aerobic capacity, health status, and medications also affect the body's ability to regulate temperature in a hot environment. Therefore, heat exhaustion responses vary widely among individuals; however, the burden of heat stress is generally greater in the elderly. Most elderly people transported for heat stroke are classified as having moderate or severe heat stroke [15].

9.1.3 Comfort Temperature for the Elderly

Comfortable indoor time is important for both physical and mental health. The thermal comfort perceived by humans is closely related to six factors: air temperature, humidity, air velocity, radiant heat, clothing amount, and metabolic rate. In addition to these thermal and environmental conditions, such as season and weather, as well as external factors, such as the duration of exposure, and internal human factors, such as health status, basal metabolism, age, sex, and body size, also affect thermoregulatory and psychological responses (Fig. 9.6). American Society of Heating, Refrigerating and Air-Conditioning Engineers (ASHRAE) defined the comfort range as SET* (Standard New Effective Temperature), where satisfaction is 80% or higher, and SET* = 22.2–25.6 °C. However, these are not the exact comfort zones for the elderly. Rohels and Johnson [16] investigated the comfort temperature ranges of human subjects. In that study, the elderly and students were exposed to 15 conditions of 24.3–26.5 °C ET for

3 hs in 0.6 clo clothing. The results showed no difference in comfort temperature between the elderly and younger adults. Fanger [17] also stated that although the elderly have a lower metabolic rate than younger people, they also have a lower transpiration rate; thus, a balance between heat production and heat release is achieved, and the comfort temperature of the elderly is not higher than that of the young. In contrast, a study conducted by Iwashige et al. [18] on Japanese elderly participants showed slightly different results. They measured the ambient temperatures of 101 subjects aged 61–85 years in six prefectures of Japan during their daily lives. The results showed that the comfortable room temperature range was 25–29 °C in summer and 20–25 °C in winter, slightly higher than the ASHRAE recommendations. Kawashima et al. [19] proposed standard room temperature values that consider the elderly based on residential thermal environments and lifestyle behaviour surveys conducted throughout Japan. For the elderly, the proposed standard values are 1–4 °C higher in all rooms in winter and 1–2 °C lower in living rooms and bedrooms in summer than for younger adults. Although it is difficult to define a range of comfortable room temperatures suitable for all people, it is beneficial for elderly to use these recommended values as a reference for comfortable living. In addition, it is insufficient to determine the thermal sensation of the elderly in a comfortable temperature range. The thermal sensations of the elderly in cold and hot conditions also need to be focused. Collins et al. [20] conducted an experiment in which subjects themselves controlled the temperature of an artificial climate chamber and recorded the changes in room temperature. The mean comfortable temperature for both groups was 23 °C, with no significant age difference. However, the elderly controlled the temperature fewer times, indicating that the temperature variation was significantly larger than that in younger subjects. Ohnaka et al. [21] also conducted a similar study, in which control was given to a room temperature perceived as comfortable after exposure to 35, 25, or 10 °C. The initial room temperature selection was nearly constant for young people, regardless of the previous exposure temperature. In contrast, the elderly tended to set excessively high temperatures after exposure to the cold.

In summary, under hot and cold conditions beyond the comfortable temperature range, the elderly have difficulty adjusting their room temperature to a comfortable level because of their low thermal sensitivity. Regarding the vertical temperature difference, ISO 7730 recommends that the temperature difference between 10 and 110 cm above the floor should not be more than 3 °C. This recommendation is primarily based on evaluating thermoregulatory responses and subjective comfort. Even if elderly people do not have diseases such as hypertension, the results of Hashiguchi's study [5] indicate that they may also suffer from cardiovascular stress.

9.2 Humidity Environment

Humidity is an important environmental parameter that significantly affects thermal comfort [22] and human health [23–28]. The 'Japanese Law for Ensuring Sanitation in Buildings' [29] states that relative humidity (RH) should be maintained between 40% and 70% in office spaces. However, indoor humidity often deviates from these ranges. In summer, the indoor RH easily exceeds the recommended upper limit [30, 31]. Conversely, RH is often below the recommended lower limit, particularly when heating is required in winter [32, 33].

9.2.1 High Humidity

Many people suffer from allergies in modern society. Mite activity, which causes allergies, increases in proportion to the relative humidity. Similarly, fungi that cause allergies, such as asthma and inflammation of the nose, are also activated when humidity is high. The use of humidifiers in low-humidity environments can be beneficial for the treatment of respiratory infections. However, it can spread to allergenic protozoa and fungi. Therefore, it is important to use humidifiers carefully when controlling the humidity. Thus, high humidity causes thermal discomfort and allergic diseases, such as asthma, rhinitis, hypersensitivity pneumonitis, atopic dermatitis, and respiratory disease [34, 35]. Bornehag et al. [36] reviewed 61 health effects of indoor humidity and reported odds ratios between 1.4 and 2.2 for humidity levels and respiratory symptoms (cough, wheezing, and asthma). High-humidity environments also increase the risk of developing sick-building syndrome. Saijo et al. [35] found that the occupants of houses with high condensation and mould were more likely to develop various symptoms of sick-building syndrome (skin, eyes, nose, throat/respiratory, mental, and nervous systems). In a high-humidity environment, the perspiration evaporation rate decreases. This increases skin discomfort and the risk of heat stroke because heat dissipation by vapourisation is prevented [34, 37]. The combination of high humidity and high air temperature harms human health. This combination reduces the body's evaporative cooling, causing discomfort, heat stroke, heat exhaustion, and, in some cases, death [34].

9.2.2 Low Humidity

Studies on the effects of low humidity on the occupants during the heating season have been conducted in offices [31, 38, 39], homes [40, 41], and factories [42]. In addition, studies using intervention (humidification) surveys have been conducted in offices [43–45], hospitals [32], and factories [33]. The effects of low humidity on subjective responses [46] and human physiology have also been studied in artificial climate chambers on the skin [47–50], eyes [51], and several body parts [52–55]. There have also been studies on physiological changes during a 6-hour exposure to low humidity and low air pressure to simulate the in-flight environment [56]. An overview of previous studies shows that low-humidity indoor spaces in winter cause many problems, such as eye irritation, skin dryness, and respiratory infection, and dryness, and humidification has been reported to be effective in reducing these symptoms. Humidity increases the survival rate of fungi, mites, bacteria, and viruses, which are triggers of infection and allergy. In addition, relative humidity affects the rate of chemical emanation from building materials and is involved in the reaction of airborne chemicals with water vapour, which, in turn, changes the concentration of airborne toxic chemicals [34].

The incidence of infectious disease is related to the number of infected persons, the number of susceptible persons, exposure time, ventilation rate, aerosol settling rate, and the survival rate of pathogens attached to the aerosol. Relative humidity is related to aerosol settling and survival rates of airborne pathogens. In other words, relative humidity greatly contributes significantly to the occurrence of infectious diseases. Generally, humidity and ventilation rates are low in winter, and people tend to stay inside for a long time. When the relative humidity is low, the suspended aerosols settle slower and remain in the air longer, resulting in a higher contact rate with the contaminating aerosols. Incidentally, in summer, the humidity is high, but the size of the aerosol increases as it absorbs water, which accelerates the settling rate.

For example, adenovirus infections are more likely to occur during summer because the virus prefers high relative humidity, whereas influenza viruses prefer low relative humidity and are more prevalent during winter. This difference in the survival rates of the viruses at different humidity levels was dependent on the molecular structure of the virus. About influenza viruses, one of the most common infectious diseases, the relationship between relative humidity, survival, and infectivity has been studied in detail [57]. The inactivation rate of the virus increased rapidly when the relative humidity exceeded 40%. Arundel et al. [34] reported that, based on the results of several studies examining relative humidity and influenza virus transmission, the influenza infection rate was the highest at a relative humidity below 40% and decreased rapidly when the relative humidity reached 40–50%. In addition, by reviewing previous studies on the survival of other bacteria and viruses, he reported that the range of relative humidity that minimises the overall survival infectivity of organisms is approximately 40–70% RH. Low humidity affects not only the survival of viruses but also nasal mucosal function in humans, contributing to respiratory tract infections.

Sunwoo et al. [53] experimented with 16 healthy adults to determine the effects of different humidity conditions on their bodies. They showed that a 90-min exposure to 10% RH prolonged the saccharin clearance time (SCT). This indicated a decrease in nasal mucosal function. Thus, when humidity is low, the ability of the nasal mucosa to eliminate viruses once they enter the nasal cavity is more likely to be impaired, which may put people at risk of developing respiratory infections. Epidemiological studies have also examined the relationship between respiratory infections and humidity. Low humidity has been reported to increase the incidence of infections in schoolchildren [58] and military soldiers [59]. As these were epidemiological studies, the involvement of confounding variables could not be ruled out. However, given that low-humidity environments can increase the survival rate of viruses and bacteria and increase SCT, controlling the humidity within an appropriate range, is an effective strategy to reduce respiratory infections.

The effects of low humidity are not only related to allergies and infections. Low humidity environments in offices and hospitals have been reported to increase complaints about irritation of the eyes and upper respiratory tract along with air quality (Office: [38, 39, 43], Hospital: [32, 60]). Numerous studies have examined the relationship between low humidity and mucous membranes of the eye. Low humidity interferes with the tear film stability and decreases tear production. The instability of the eye tear film also leads to the formation of dry spots and increased sensitivity to irritation from air containing formaldehyde and other pollutants. Sunwoo et al. [53] also found that 90-minute exposures to 10 and 30% RH significantly reduced the number of blinks, while 50% RH did not change this, and it is clear that low humidity promotes dryness of the eye.

Finally, low humidity also affects the skin. Egawa et al. [47] found that exposure to 10% RH for 6 hours decreased the skin's stratum corneum's moisture content. Sunwoo et al. [53] investigated the effects of exposure to 10, 30, and 50% RH on skin physiology. They reported that 10 and 30% RH increased transepidermal water loss (TEWL) and decreased skin moisture content. Skin dryness can cause itching of the skin and sometimes even sleep disturbances. The skin is the largest human organ. The skin is divided into two layers: epidermis and dermis. The barrier function of the epidermis controls passive water loss from the body, suppresses the absorption of external chemicals, and prevents microbial infection. Katagiri et al. [61] showed that free amino acids, one of the natural moisturising factors, decreased under low RH conditions. Denda et al. [62] and Yoshikawa et al. [63] reported that ceramides, which comprise approximately half of the extracellular lipids, are altered by low humidity during winter. These reports show skin dryness under low RH conditions is attributed to changes in coenocytes and extracellular lipids in stratum corneum.

9.2.3 Humidity Environment and Health in the Elderly

Comparing the percentage of the world's total population aged 65 years and older in 2018, Japan (28.1%) had the highest percentage, followed by Italy (23.3%), Portugal (21.9%), and Germany (21.7%) [64]. The proportion of elderly people will continue to increase, especially in developed countries. Consequently, the elderly are increasingly exposed to high- and low-humidity environments at home and elsewhere (e.g. offices, factories, and hospitals). However, few studies have examined the effects of high and low humidity on the elderly. Many studies have reported that the elderly have decreased skin physiology and neurological, circulatory, immune, and thermoregulatory functions associated with aging [1, 65–68]. Therefore, physiological and psychological responses to different humidity conditions are also expected to depend on age [54].

In addition, little research has been conducted on the effects of summer dehumidification and winter humidification on thermal comfort and health regarding differences between age groups. Considering that older adults tend to spend more time indoors, it is important to investigate the effects of indoor humidity on the elderly. Hashiguchi et al. [32] surveyed humidity in hospitals during winter and examined the effectiveness of humidification installations. The temperature and humidity in the hospital rooms and nurses' stations before and after the installation of the humidifiers are shown in Fig. 9.7. Before the humidifier was installed, the humidity in the hospital room was 32–33% RH, which is close to the lower limit of the recommended humidity range; however, after the humidifier was installed, the humidity increased to approximately 44% RH. At the nurses' station, the increase in humidity after the installation of the humidifier was smaller than that in the patient rooms because of the room's structure (open space). The subjective ratings before and after the humidifier installation are shown in Fig. 9.8. Patients (average age: 71.0) showed no significant change in all subjective reports of thermal sensation, thermal discomfort, and air dryness before and after humidifier installation. In contrast, the staff (average age: 38.7) showed a significant decrease in dryness after installing the humidifier. These results suggest that aging affects the perception of humidity differences.

Humidity levels generally tend to be higher in summer and lower in winter, and people dehumidify and humidify to remain comfortable. Hashiguchi et al. [69] compared the effects of indoor dehumidification (−20% RH) and humidification (+20% RH) on the physiological and psychological responses of elderly and younger people. The experimental conditions are shown in Fig. 9.9. The following results were obtained:

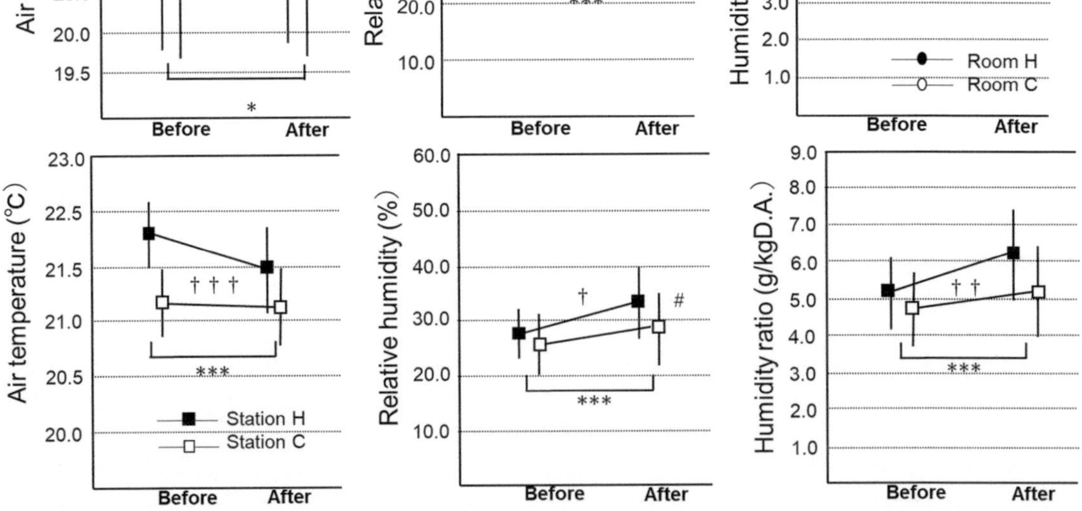

Fig. 9.7 Air temperature and humidity in the sick room and nurse stations before and after setting up humidifiers. (Adapted from Hashiguchi et al. [32])

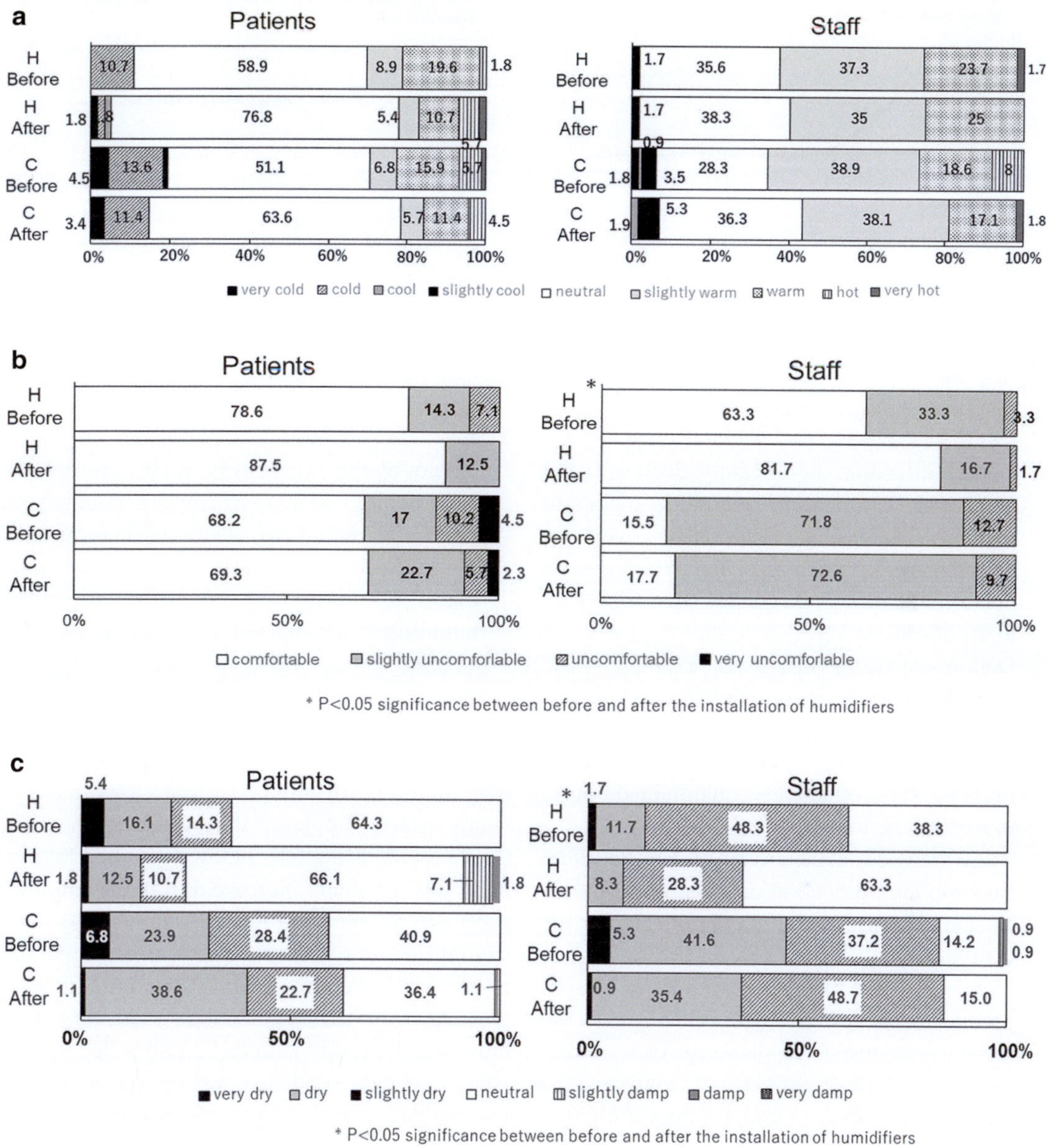

Fig. 9.8 (**a**) Percentage of thermal sensation of patients (left) and staff (right). (**b**) Percentage of thermal comfort of patients (left) and staff (right). * $P < 0.05$, significance before and after humidifier installation. (**c**) Percentage of humidity sensation of patients (left) and staff (right). * $P < 0.05$, significance before and after humidifier installation. (Adapted from Hashiguchi et al. [32])

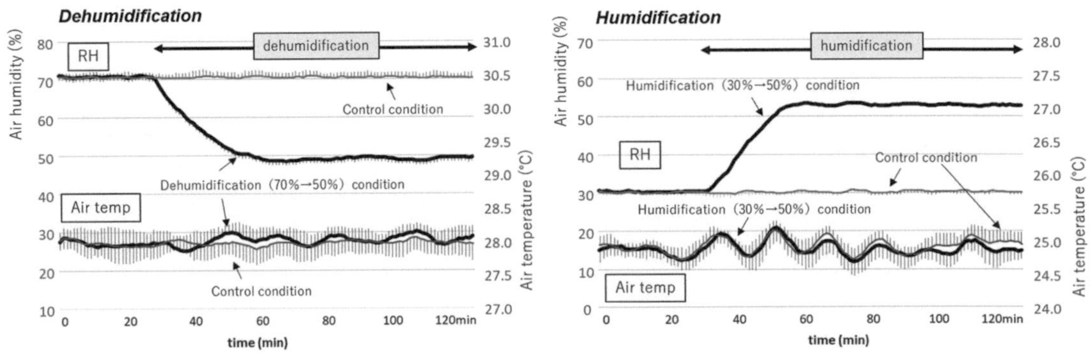

Fig. 9.9 The actual measurement results of the room temperature and relative humidity during the experiments. (Adapted from Hashiguchi et al. [69])

1. Dehumidification significantly decreases the mean skin temperature over time owing to increased insensible evaporative sweating. The change in mean skin temperature with dehumidification was greater than that with humidification.
2. Dehumidification increased TEWL, while humidification decreased TEWL.
3. Humidification significantly shortened the SCT in both age groups. Humidification (+20% RH) effectively maintained nasal mucosal function.
4. The differences in the effects of dehumidification and humidification on the above physiological responses were not significant for the different age groups.
5. However, there were age differences in subjective responses. Both the elderly and the young recognised the decrease in humidity caused by dehumidification. However, the change in humidity due to humidification could be recognised by the younger group but not by the elderly (Fig. 9.10). The elderly may be less able to recognise changes in humidity owing to dehumidification and humidification subjectively. This suggests that the elderly may be unable to adjust humidity to an appropriate range using their senses.

In summary, a review of the health effects of relative humidity shows that maintaining a relative humidity of 40–60% minimises adverse health effects. However, during winter, the relative humidity often falls below 40%. In addition, humidity can easily increase during the summer, especially in enclosed rooms. Intentionally controlling humidity can be beneficial for reducing respiratory infections in winter, the onset of allergies in summer, and living a healthy life. However, humidity cannot be controlled by the elderly on their subjective perception. An objective evaluation using a hygrometer or other means is recommended.

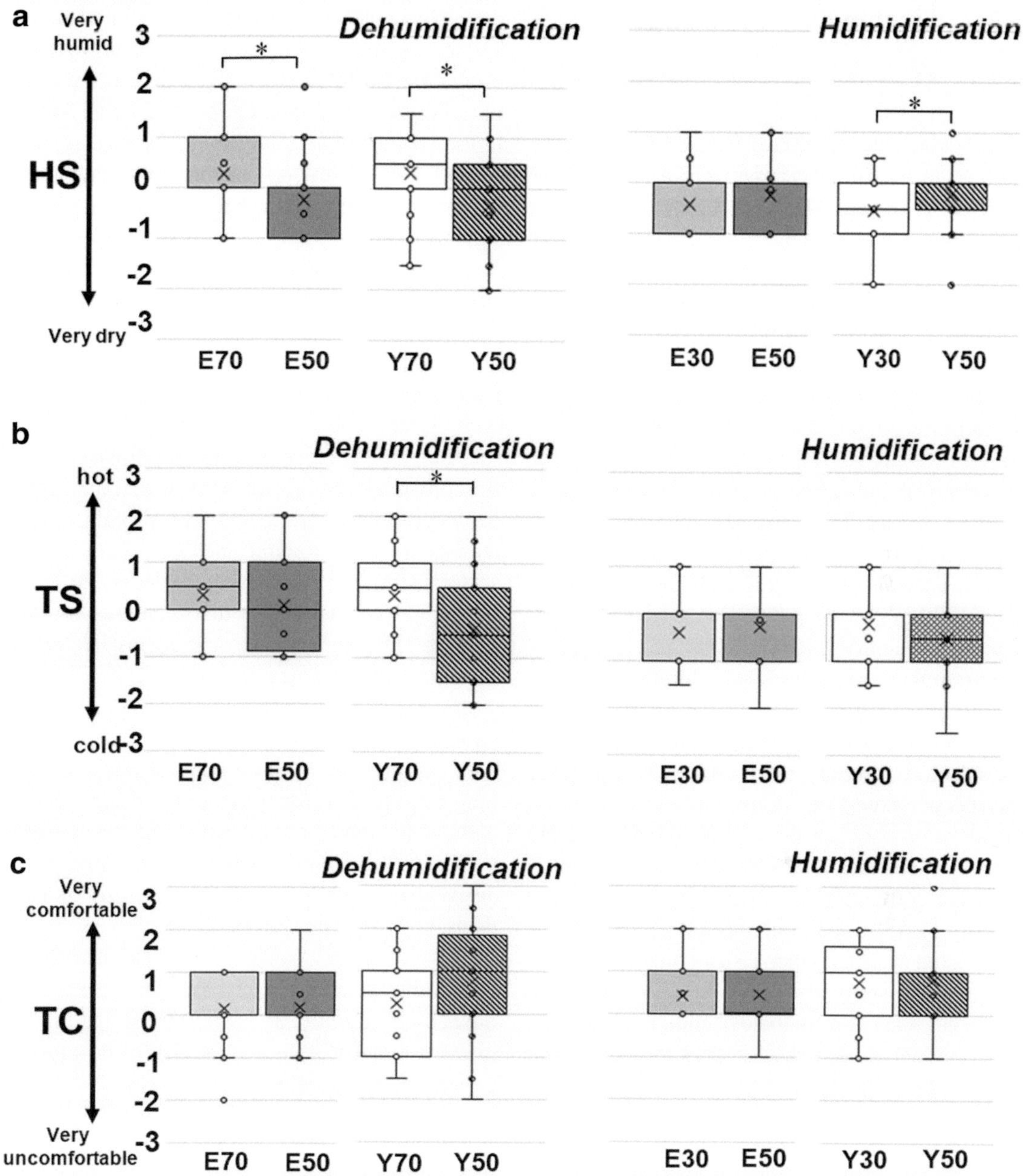

Fig. 9.10 Subjective responses based on the data at two time points (100 and 120 min after exposure). (**a**): Humidity sensation. (**b**): Thermal sensation. (**c**): Thermal comfort. Values are the median and interquartile range. (Adapted from Hashiguchi et al. [69])

9.3 Sleeping Environment

9.3.1 Sleep Disturbance and Health

Insufficient sleep at night decreases daytime task performance and delays recovery after physical activity. Disrupted sleep increases fatigue and decreased physical vitality [70]. The OECD Gender data portal 2021 survey found that sleep duration for those aged 15–64 years was 8 h and 29 min for all participating countries, while Japan had the shortest sleep duration at 7 h and 22 min. Health Status Survey of Japan in 2021 (https://www.mhlw.go.jp/content/11131500/000904748.pdf) showed that 67.7% of the population slept for less than 7 h, indicating an overall lack of sleep.

Itani et al. [71] studied the relationship between short sleep duration and mortality risk in a meta-analysis of 36 studies. The relative risk of death with short sleep duration was 1.12 (95% confidence interval [CI]: 1.08–1.16), a significantly higher risk. Short sleep duration is a risk factor for the onset of other health outcomes such as diabetes, hypertension, cardiovascular disease, stroke, and coronary artery disease. A 13-year prospective study of young adults also found a relationship between short sleep (6 h or less) and obesity (odds ratio 7.4, 95% CI: 1.3–43.1) and a negative correlation between sleep duration and obesity [72]. In addition, short sleep duration is associated with a higher risk of metabolic syndrome [73]. Short sleep may be a risk factor for diseases such as diabetes and hypertension because it is more likely to lead to obesity and hyperlipidaemia. However, there are also some findings that long sleep duration increases the risk of mortality, cardiovascular disease, stroke, coronary heart disease, and obesity [74]. In other words, obtaining an appropriate amount of sleep (neither too long nor too short) is important for maintaining good health.

In modern society, many occupations require 24-hour workdays, and many people work shifts. Shift work contributes to the high prevalence of sleep disorders. There have been many reports on health problems faced by shift workers. Vogel et al. [75] suggested that the disruption of circadian rhythms owing to shift work can have various adverse effects on physical and mental health. These conditions include breast cancer, diabetes, premature birth, miscarriage, low birth weight, intrauterine growth retardation, menstrual irregularities, infertility, ischaemic heart disease, and stroke. The following mechanisms have been hypothesised to explain health problems due to shift work [76]. Sleep restriction causes neuroendocrine stress, immune responses, and oxidative stress. Circadian rhythm disturbances can induce immune responses and oxidative stress. Neuroendocrine stress resulting from sleep restriction causes hypertension due to elevated cortisol and catecholamines and is a risk factor for the development of cardiovascular disease. Immune responses elevate leukocytes, inflammatory cytokines, and C-reactive protein levels, leading to inflammation and promoting atherosclerosis. Immune reactions also decrease natural killer (NK) cell activity, which decreases immune defense and increases the risk of possible carcinogenic events.

Insomnia is also well-known to be associated with depression and other psychological disorders. Symptoms of insomnia include taking a long time to fall asleep (trouble falling asleep), waking up many times during the night (mid-wake), and waking up early in the morning (early awakening). The frequency of insomnia in depression is reported to be as high as 84.7% [77]. A systematic review and meta-analysis of longitudinal studies on insomnia and depression onset by Baglioni et al. [78] reported that insomnia significantly increases the risk of developing depression in the future.

9.3.2 Health and Sleep of the Elderly

Elderly are often suffering from sleep disorders [79]. Specifically, they were more likely to have difficulty falling asleep, feeling deep sleep, and waking up early. They also tend to be easily awakened by small stimuli, such as noise, and have difficulty getting continuous sleep due to itching, pain, and frequent urination at night. As the elderly have a low basic metabolism and are less physically active during the day, deep sleep is reduced. Thus, sleep disorders are common in the elderly, but many cases go undiagnosed and untreated. It is important to focus on sleep in the elderly because impaired sleep quality may lead to a functional decline in physical and mental health. Previous studies reported various associations between sleep and health in older adults. Nakazaki et al. [80] examined the association between sleep duration and arterial stiffness in 86 older participants (mean age: 73.6). They found that sleep shorter than 5 h was associated with significantly thicker carotid intima-media thickness than sleep longer than 7 h, indicating that sleep insufficiency is associated with atherosclerosis, which contributes to cardiovascular disease.

In recent years, a relationship between sleep and frailty has been reported. The Japanese Geriatrics Society defines frailty as ‘a state of being at risk of outcomes such as functional disability, need for nursing care, and possible death’ and a concept that includes not only physical problems such as loss of flexibility in movement due to muscle decline and increased tendency to fall, but also mental and psychological problems such as cognitive dysfunction and depression, and social problems such as living alone and economic deprivation. This is considered a preliminary stage, leading to a need for long-term care. The prevalence of frailty increases with age, but it is influenced by multiple factors other than aging. To prevent frailty in community-dwelling older people, it is important to intervene from multiple perspectives, including appropriate exercise habits, management of nutrition and chronic diseases, and prevention of infections. One such intervention is sleep. It has been suggested that frailty has three domains: physical (physical frailty), psycho-psychological and cognitive (psycho-psychological frailty), and social (social frailty). Each interacts with the other, leading to negative health outcomes.

Studies have investigated the relationship between sleep and physical frailty in the elderly. Ensrud et al. [81] investigated sleep disturbances and physical frailty in community-dwelling elderly. They reported that lower subjective sleep quality, lower sleep efficiency, and longer sleep latency were associated with frailty. There are also reported to be associated with decreased walking speed and grip strength [82]. It has also been reported that subjective sleep quality decline, sleep inefficiency, nocturnal midnight awakenings, and nocturnal hypoxaemia are associated with future flail development and death [83]. Sleep duration, not only short but also long, has been reported to be associated with physical frailty [84] and linked to many indicators of physical frailty, including decreased walking function, weight loss, and muscle decline. Furthermore, it has been observed that those who sleep longer than 7–8 h and have daytime sleepiness have a hazard ratio of approximately 2.3 times greater for the occurrence of a need for long-term care compared to those who sleep for 7–8 h and do not have excessive daytime sleepiness [84]. Sarcopenia is a decline in the amount of skeletal muscle mass and ligament strength that occurs with aging and is an inclusive component of frailty. Short or long sleep durations have also been found to be associated with sarcopenia in systematic reviews [85, 86].

In the elderly, there are many studies on the relationship between sleep and dementia. Sleep characteristics in patients with dementia, especially Alzheimer’s disease, include decreased sleep duration, decreased sleep efficiency, an increased proportion of shallow sleep, increased nocturnal arousal, and decreased rapid eye movement (REM) sleep. Patients with dementia are more likely to have impaired sleep–wake rhythms because they are often less active during the day and less likely to go out and participate in social activities, leading to insufficient sunlight exposure, which in turn reduces the stimulation of the biological clock. Alzheimer’s patients have also been reported to have decreased melatonin secretion and abnormal secretion timing. Disorders of the biological clock, including abnormal melatonin secretion, lead to abnormalities in sleep–wake phases circadian rhythm disorders. Longitudinal studies have reported that daytime sleepiness and the presence of sleep disturbances are associated with a twofold higher odds ratio for the onset of later cognitive decline than those without [87]. It has also been reported that having daytime sleepiness increases the risk of dementia onset, according to a longitudinal study conducted over 3 years [88]. Maintaining good sleep in the elderly is important for preventing dementia.

Studies have also been conducted on the relationship between sleep disturbance and mortality in the elderly. A prospective 2-year study by Manabe et al. [89] of older patients residing in geriatric care homes also showed that mortality was higher with impaired sleep at night, sleepiness during the day, and delayed sleep latency. In addition, a study was conducted to determine whether sleep disturbance can predict future mortality [90]. This study was based on EEG parameters rather than on a subjective evaluation of sleep with subsequent follow-up. After controlling for age, sex, and baseline medical burden, those with a sleep latency of 30 min or greater at baseline were shown to have a 2.14 times higher risk of mortality, while those with a sleep efficiency of less than 80% had a 1.93 times higher risk. This clearly shows that sleep disturbance in older adults can predict future mortality.

Sleep and health are closely related, not only in older adults. However, in older people, sleep disturbances can also be a risk factor for the onset of age-related diseases, such as dementia and frailty, suggesting that interventions to optimise sleep quality in the elderly are more necessary than in other age groups.

9.3.3 Thermal Environment and Sleep

Sleep is prompted by the accumulation of sleep substances, such as prostaglandins, which increase drowsiness. After sleep, these substances decrease and wakefulness increases. Endocrine activity also has a biological rhythm of approximately 24 h, called the circadian rhythm. Circadian rhythms facilitate sleep. Core temperature is also linked to sleep–wake rhythms, falling during the night and rising upon awakening. Approximately 30 min before falling asleep, heat dissipation is promoted by an increase in the peripheral skin temperature, leading to a decrease in the core body temperature. Skin temperature is closely related to sleepiness while falling asleep; the earlier the skin temperature rises, the shorter the sleep latency [91]. Van den Heuvel et al. [92] showed that a significant increase in peripheral skin temperature and decrease in core body temperature are precursors of sleep onset. In other words, changes in skin temperature are considered input signals for sleep [93]. Environmental factors that influence sleep include sound and light. Given that thermoregulation significantly influences sleep, the thermal environment of the bedroom is an important factor for good sleep [94]. Various studies have shown that sleep is disturbed in environments with low temperatures and humidity, even in healthy people with no sleep disturbances.

9.3.3.1 Bedroom Thermal Environment and Sleep in Young Adults

Table 9.1 summarises the experimental studies that examined the effects of temperature, humidity, and airflow on sleep in young adults. In most studies, polysomnography (PSG) was used to measure sleep quality objectively. Many of these studies were conducted in experimental laboratories and controlled for the thermal environment and other factors affecting sleep quality. In addition, many studies have been conducted on healthy individuals with no sleep disturbances.

High temperatures have generally been shown to decrease total sleep time, REM sleep, and slow wave sleep (SWS) duration and increase sleep onset latency (SOL) and arousal [95, 96]. A slight increase in air temperature above the preferred temperature (2–3 K) decreases subjective sleep quality [97]. Okamoto-Mizuno et al. [98] showed that nude, high temperature, and high humidity environments suppressed the reduction in skin temperature which induced sleep onset and stage 3, SWS, and REM.

Tsuzuki et al. [99] reported that heart rate and sweating were higher and the secretion of melatonin metabolites was lower during sleep in high-temperature and high-humidity environments than at moderate temperatures. They also suggested that a rapid decrease in melatonin secre-

Table 9.1 Studies on the effect of thermal environment on sleep

Author (year)	Subject *N*	Age (SD)	Thermal environment conditions Air temperature/RH, etc.	Sleep assessment	Main results
Haskell (1981)	6M	18–30	① 21 °C ② 24 °C ③ 29 °C ④ 34 °C ⑤ 37 °C	EEGs, EMG, EOGs	21 °C was the most disruptive condition, and that cold Tair was generally more disruptive to sleep than warm
Candas (1982)	5M	Students	① From 25 to 41 °C ② From 25 to 17 °C	EEGs, EMG, EOGs	The decreases in air temperature affected these sleep stages more frequently than did the air temperature increases
Okamoto-Mizuno (1999)	7M	22.7 (1.63)	① 29 °C/50% ② 29 °C/75% ③ 35 °C/50% ④ 35 °C/75%	EEG, EMG, EOG, Trec, SK, SEI	Wakefulness is significantly higher at 35 °C/75%. Lower sleep efficiency index and stage 3, 4 and REM at 35 °C/75% than at 29 °C/50% and 35 °C/50%
Tsuzuki (2004)	9M	25 (3.8)	① 26 °C/50% ② 32 °C/80%	EEG, EMG, EOG, Trec SK, urinary melatonin	Wakefulness at sleep stage 1 is significantly higher at 32 °C/50%; at stages 2 and 4, sleep efficiency is significantly lower at 32 °C/50%
Okamoto-Mizuno (2004)	10M	69.2 (1.35)	① 26 °C/50% ② 32 °C/50%	EEG, EOG, EMG, SK Trec	Wakefulness was significantly increased at 32 °C than at 26 °C. The total amount of wakefulness increased and rapid eye movement decreased at 32 °C compared to 26 °C
Okamoto-Mizuno (2005)	8M	26.2 (3.31)	Experiment 1 ① 26 °C/50% ② 26 °C/50% → 32 °C/80%	EEG, EOG, EMG, SK Trec	Wakefulness increased significantly over the last 4 h under 26–32 compared to 26
			Experiment 2 ① 26 °C/50% ② 32 °C/80% → 26 °C/50%		Wakefulness increased both in first and during the last 4 h, and slow wave sleep decreased in the first 4 h under 32–26 compared to 26
Tsuzuki (2008)	A: 9M	25 (3.8)	① 26/50/0.2 m/s ③ 32/80/0.2 m/s	EEG, EOG, EMG, SK Trec	Total duration of wakefulness decreased significantly under the 32/80 with airflow compared to 32/80 with still air
	B: 8M	26.2 (4.5)	① 26/50/0.2 m/s ② 32/80/1.7 m/s		
Okamoto-Mizuno (2009)	9M	23.1 (2.1)	① 17 °C/50% ② 10 °C/50% ③ 3 °C/50–80%	EEG, EOG, EMG, ECG	The normalized LF(LF/(LF + HF)) significantly decreased in 3 °C and 10 °C than in 17 °C during SWS.
Li Pan (2012)	4M, 4F	21.0 (2.0)	① 17 °C ② 20 °C ③ 23 °C	EEG, SK subjective questionnaire	23 °C condition was the most satisfactory temperature for sleeping. The relative power of the sleep band was the highest, the duration of sleep onset latency was the shortest, and the slow-wave sleep was the longest at 23 °C
Li Lan (2013)	18M, 18F	6–72 35.08 (24.75)	① With personalized ventilation (22.9 °C/43%) ② Without personalized ventilation(22.9 °C/52%)	ECG, SK actigraphy	Compared with no PV, the VLF power and the LF/HF ratio throughout the sleep significantly decreased when the PV was turned on

(continued)

Table 9.1 (continued)

Author (year)	Subject N	Age (SD)	Thermal environment conditions Air temperature/RH, etc.	Sleep assessment	Main results
Li Lan (2014)	9M, 9F	23.3 (1.8)	① 23 °C ② 26 °C ③ 30 °C	EEG SK subjective questionnaire	The duration of sleep onset (SOL) was longer, the duration of slow wave sleep (SWS) was lower, and the subjective sleep quality was lower at 23 °C or 30 °C compared with 26 °C
Tsuzuki (2015)	8M	64 (1)	① Spring ② Summer ③ Autumn ④ Winter	Actigraphy sleep diary	Sleep efficiency was worst in the summer than in the other seasons.
Li Lan (2016)	6M, 6F	23.1 (2.1)	26 °C ② 26–27–28–27–26 °C ③ 28–27–26–27–28 °C	EOG, EEG, EMG, subjective questionnaire	The subjects had better next-day performance in the fall–rise condition than in the other two conditions. Simulation results show that compared with the reference condition, the total cooling load of a simulated bedroom is reduced by 27.8% in the rise–fall and 34.3% in the fall–rise condition
Li Lan (2016)	9M, 9F	23 (1.6)	① 26 °C(constant) ② 25–26–27–28–27–26 °C (rise–fall) ③ 28 °C–27 °C–26 °C–27 °C–28 °C(fall–rise)	EOG, EEG, EMG, subjective questionnaire	The subjects took longer time to fall sleep in the rise–fall condition, subjective reported poor sleep quality in the rise–fall condition than in either of the other conditions.
Li Lan (2017)	16M	23.5 (1.9)	① Back and head cooling ② back cooling ③ neck cooling ④ no cooling (31–32 °C/51–58%)	EEG, EOC, EMG, SK, subjective response	Thermal comfort and sleep quality were improved, with the largest improvement in both back and head cooling. The mean sleep efficiency was improved from 84.6% in the no cooling to 95.3% and 92.8%, which is characteristic of very good sleep, in both back and head cooling and the back cooling
Li Lan (2019)	10M, 8F	71 (5)	① 30 °C + max0.8 m/s, mean 0.7 m/s (ceiling fan) ② 30 °C + max 0.8 m/s, mean (task fan) ③ 27 °C + 0.2 m/s	EEG, EOC, EMG, urine cortisol SK, subjective questionnaires	Elderly could maintain sleep quality at an air temperature that was 3 K higher than the neutral temperature. (TST, REM sleep ①③>②)
Yan Yan (2022)	8M, 8F	72 (6)	① 27 °C with mechanical ventilation ② 27 °C without mechanical ventilation ③ 30 °C with mechanical ventilation ④ 30 °C without mechanical ventilation	EEG, EOC, EMG, HR, BP, SpO2 Tsk, saliva cortisol lysozyme, subjective questionnaires	TST, SE, duration of RME sleep decreased at 30 °C, in comparison with 27 °C The duration of REM sleep increased by 3.7 min, and deep sleep increased by 10.3 min in comparison with the absence of ventilation

tion was associated with decreased sleep efficiency. Sleep quality has also been evaluated using airflow at high temperatures [100].

These studies suggest high-temperature environments increase arousal and decrease slow-wave and REM sleep. REM and SWS are thought to be reduced, and arousal is increased to maintain thermoregulatory functions. In addition, when high humidity is added to a high-temperature environment, the conditions worsen in a sleep environment. High temperatures and humidity increase invalid perspiration, resulting in moist skin and discomfort. Even in a hot environment, if the humidity is low, the evaporation of sweat promotes heat dissipation, and the quality of sleep can be maintained.

However, studies on the effects of low temperatures are fewer than those of high temperatures. Even at low temperatures, because the cold can be protected by bedding and sleeping clothes used at bedtime, low-temperature environments can prevent worse sleep quality than high-temperature environments [101, 102]. However, the effects of low-temperature environments on cardiac autonomic activity and blood pressure during sleep are associated with the onset of cardiovascular disease. Therefore, it is necessary to be cautious about room temperature during winter. Electric blankets and other heating devices can be used when the room temperature is low in winter. Electric blankets may prevent a decrease in core body temperature even in low- and high-temperature environments. In such cases, arousal or humidity in the bed may increase due to sweating, which may harm sleep. Therefore, using heating devices appropriately regarding their adaptive range and timing is important. Reviewing these studies found that both high and low room temperatures (with a reference temperature of 26 °C) caused a decrease in sleep quality, either objectively or subjectively.

9.3.3.2 Bedroom Thermal Environment and Sleep in the Elderly

Lan et al. [103] investigated the effects of elevated airflow on sleep quality in 18 elderly. They reported that even a small thermal load in the elderly leads to decreased sleep quality and sympathetic hyperactivity, and that an airflow of 0.8 m/s can maintain sleep quality and thermal comfort. Yan et al. [104] investigated the effects of ventilation on sleep quality under 27 and 30 °C temperatures in 16 subjects over 65 years of age. 30 °C conditions showed that sleep efficiency and REM sleep decreased and arousal time increased, respectively. The results showed that deep sleep and REM sleep durations increased when ventilation was present and that high pollutant concentrations decreased sleep quality. They stated that good ventilation and avoidance of high temperatures in the bedroom were important for improving sleep quality.

Ohnaka et al. [105] studied the sleep status and bedroom thermal conditions of 312 elderly individuals and 266 young adults during summer. The mean (SD) bedroom temperature and relative humidity for the younger subjects were 27.6 (2.6) °C and 70.4 (6.6)%, whereas those for the elderly were 26.3 (2.2) °C and 72.6 (8.2)%, respectively. These values are slightly higher than the recommended summer bedroom temperature and humidity. This suggests that many do not sleep in environments that ensure quality sleep. This study also revealed a relationship between room temperature and body movement during sleep, indicating a positive correlation between higher room temperatures and higher body movement rates. Comparing younger and elderly persons, the rate of body movement during sleep in the elderly was significantly higher under 28 °C conditions [105] (Fig. 9.11). These results suggest that although the effects of high and low bedroom temperatures are not as significant in the elderly as in younger people, controlling room temperature may reduce body movements (reduce arousals) during night-time sleep. As noted above, the elderly have several factors related to sleep disturbances, many uncontrollable due to aging. Maintaining an appropriate thermal environment is an effective method of improving sleep quality.

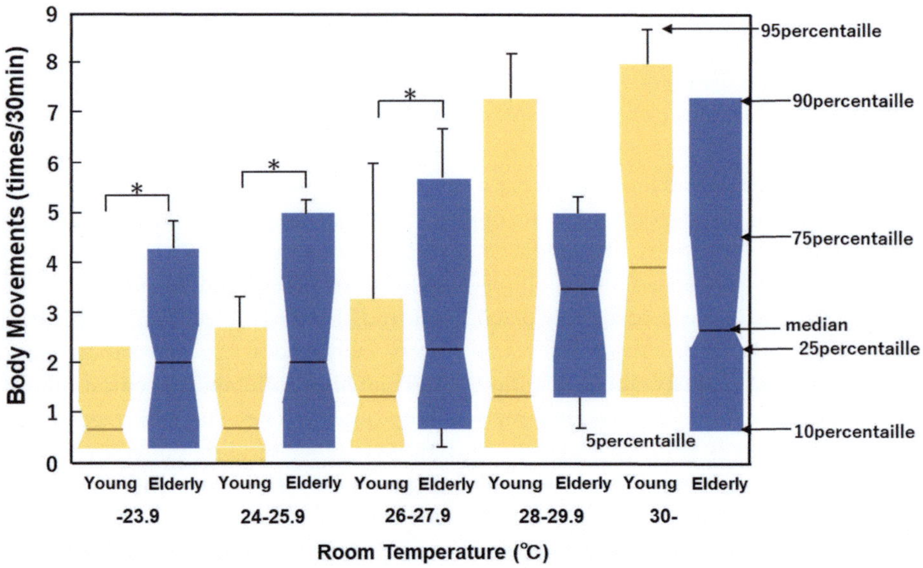

Fig. 9.11 Distribution of body movements in the elderly and young under the conditions of room temperature of less than 24 °C, 24 to 25.9 °C, 26 to 27.9 °C, 28 to 29.9 °C, above 30 °C. (Adapted from Ohnaka et al. [105]). *$P < 0.05$ significant differences between the elderly and young individuals

9.4 Bathing Environment

9.4.1 Bathing and Health

Bathing habits vary widely depending on country and culture. There are various bathing methods such as showering, soaking in a bathtub, and taking a sauna bath in a hot or humid environment. In particular, Japanese prefer bathing. In addition, the Japanese bathing not only cleanses their bodies but also relaxes and warms them. Japanese people bathe uniquely, soaking their shoulders in relatively hot water. In our survey of elderly Japanese men and women, 87.5% bathed daily in summer and 63.5% in winter (Table 9.2). Several studies have shown that bathing is beneficial for health. Regarding psychological effects, Yamasaki et al. (2022) [106] reported that habitual daily hot spring bathing were independently significantly associated with a history of depression. Hayashi et al. [107] evaluated the physical and psychological effects of bathing on patients with terminal cancer. Fatigue improved on bathing days compared with non-bathing days, and anxiety and depression improved after bathing. Numann et al. [108] examined the association between warm baths and depressive symptoms in patients with depression in a randomised clinical pilot trial. In a pilot study, a warm bath at 40 °C

Table 9.2 Bathing habits in the elderly in Japan

	Summer		Winter	
	N	%	*N*	%
Bathing frequency				
Everyday	175	87.5	127	63.5
Other	25	12.5	73	36.5
Frequency of soaking in bathtub				
Almost everyday	70	35.0	162	81.0
Sometime	64	32.0	23	11.5
Shower only	66	33.0	15	7.5
Preferred water temperature	(N = 134)		(N = 185)	
Hotter	9	6.7	32	17.3
Slightly hot	37	27.6	115	62.2
Slightly warm	88	65.7	39	21.1
Warm	19	14.2	4	2.2
Bathtub water temperature	(N = 130)		(N = 183)	
<= 39 °C	72	55.4	10	5.5
>39 °C, <=42 °C	56	43.1	144	78.7
>42 °C	2	1.5	29	15.8
Duration of soaking in the bathtub	(N = 132)		(N = 184)	
<= 5 min	59	44.7	51	27.7
>5 min, <=10 min	43	32.6	69	37.5
>10 min	30	22.7	64	34.8

twice a week for 4 weeks suggested the possibility of improving depressive symptoms. Thus, various studies have shown that bathing provides psychological benefits not only for healthy people but also for terminal patients and patients with mental diseases.

Regarding the effects on cardiovascular disease, bathtub bathing prevents cardiovascular disease because it improves hemodynamic function. Ukai et al. [109] conducted a follow-up study on the relationship between bathing habits, coronary heart disease (CVD), and stroke in 30,076 subjects aged 40–59 years from 1990 to 2009. During the follow-up period, 2097 CVD events were recorded, including 328 cases of coronary heart disease and 1769 cases of stroke. Using the Cox proportional hazards model, the multivariate hazard ratio (95% CI) for almost daily or daily vs. 0–2 times/week was 0.72 (0.62–0.84, trend $p < 0.001$) for all CVD and 0.74 (0.62–0.87, trend $p = 0.005$) for overall stroke. In summary, the frequency of bathtub bathing is inversely associated with CVD risk in middle-aged and older Japanese individuals. Katsuyama et al. [110] conducted a cross-sectional study on the relationship between bathing habits and cardiovascular risk factors in 1297 patients with diabetes. Multiple regression analysis revealed that the frequency of warm bathing was a significant determinant of haemoglobin A1c, body mass index, and diastolic blood pressure. This finding suggests the benefits of lifestyle interventions in patients with type 2 diabetes.

Regarding the relationship between bathing and physical function, Yagi et al. [111] conducted a 3-year prospective cohort study of 13,786 subjects to determine the association between bathing frequency and impaired motor function in the elderly. Functional impairment was observed in 1203 patients (8.7%). Compared to the low-frequency group, hazard ratios for the moderate- and high-frequency groups were 0.91 (95% confidence interval [CI]: 0.75–1.10) and 0.72 (95% CI: 0.60–0.85) in the summer and 0.90 (95% CI: 0.76–1.07) and 0.71 (95% CI: 0.60–0.84) in winter. In other words, more frequent bathtub bathing was associated with lower incidence of functional disabilities. Therefore, bathtub bathing is beneficial to the health of older adults.

Liao [112] reviewed the relationship between passive heating of the body by bathing and sleep. Body passive heating, which involves immersion in warm water at 40–41 °C to mid-chest level for 30 min in the evening, increases rectal temperature, delays the onset of hypothermia, and increases wave-out sleep. As the elderly subjectively perceived better sleep and earlier sleep onset, they concluded that bathing facilitated night-time sleep in older with insomnia. Haghayegh et al. [113] conducted a systematic review and meta-analysis of 13 studies. A warm bath at 40–42.5 °C was reported to be associated with improved subjective sleep quality and sleep efficiency, and a 10-minute bath 1–2 h before bedtime significantly reduced SOL. Sung and Tochihara [114] also examined the timing of bathing (40 °C, 20 min) to promote sleep in eight adult male subjects. They reported bathing 1.5 h before sleep promoted better sleep quality than bathing immediately.

9.4.2 Sudden Death During Bathing

Bathing offers various health benefits. However, several bathing-related accidental deaths have been reported in Japan. According to the Japanese Ministry of Health, Labor, and Welfare, the total number of accidental deaths at home in 2021 was 13,352, of which 5398 were due to drowning. This number is much higher than that for suffocation and falls, which are also household accidents, and much higher than the number of deaths due to traffic accidents (Table 9.3). Various factors, including cardiac and cerebrovascular dis-

Table 9.3 Number of accidental deaths at home in Japan in 2021

	Total	Over 65 years	(%)
Deaths in houses	13,352	11,723	87.8
Asphyxiation	3317	2907	87.6
Drowning	**5398**	**5054**	**93.6**
Falls	2486	2184	87.9
Flames and smoke	728	551	75.7
Toxic substance	293	94	32.1

eases and drowning, can cause accidental deaths associated with bathing. Therefore, it was difficult to determine the number of sudden deaths that occurred during bathing.

According to Otsuka et al. [115], the causes of death due to bathing-related accidents listed on death certificates mostly include cardiopulmonary arrest and cardiac and cerebrovascular diseases, with drowning being the least common cause of death. Estimates based on emergency medical cases suggest that 17,000 [116], and 19,000 [117], people die annually in Japan due to bathing. It has been warned that 27,000 people will die if no measures are taken [118]. Suzuki et al. [119] stated that verifying not only fatal accidents but also non-fatal accidents is necessary to determine the cause of accidents and examined transport cases related to bathing in three prefectures. The results showed that of the 4592 transport cases, 1528 involved cardiovascular arrest, 935 involved survivors in need of help, 1553 involved acute illnesses, and 577 involved injuries. Furthermore, there is a relationship between the level of consciousness and body temperature at the time of transport, and an increase in body temperature due to immersion in hot water is an important factor in bathing-related accidents.

It is well known that bathing accidents are more common among the elderly over 65 [117]. This is largely due to the increase in the elderly population but is also related to the decline in various physiological functions, such as circulation and thermoregulation, with aging. Chiba et al. [120] reported that ECG changes such as supraventricular extrasystole and ventricular tachycardia may occur during bathing, regardless of the season, even in elderly patients with no clinical problems by biochemical analyses. Hashiguchi et al. [121] examined the differences in physiological burden between shower baths, half-body baths with a water level up to the décolletage, and full-body baths with a water level up to the shoulders. According to this study, the heart rate increased twice as much during bathing, even in healthy adults (Fig. 9.12), and it is obvious that a large cardiac strain occurred during immersion because of the thermal effect

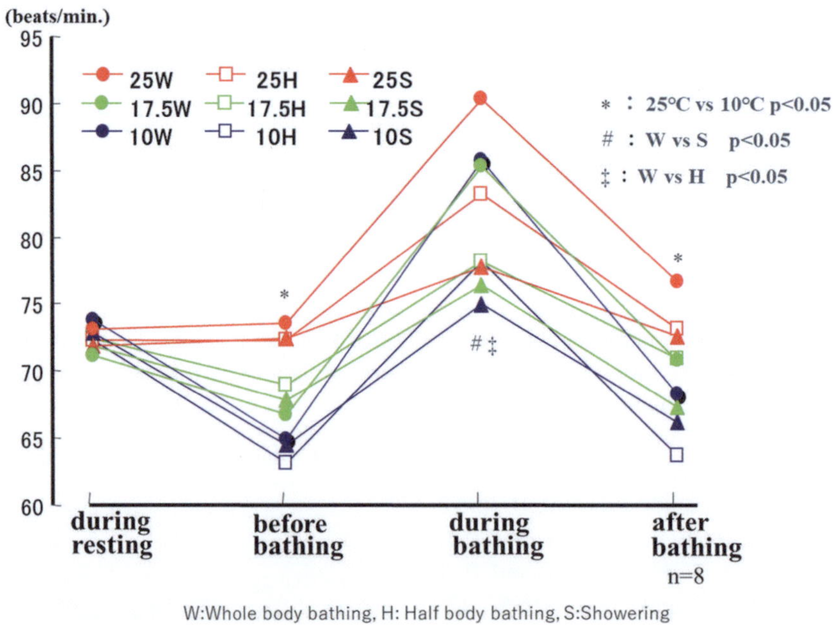

Fig. 9.12 Changes in heart rate during bathing (8 min, 41 °C). *W* whole body bathing, *H* half body bathing, *S* showering. (Adapted from Hashiguchi et al. [121])

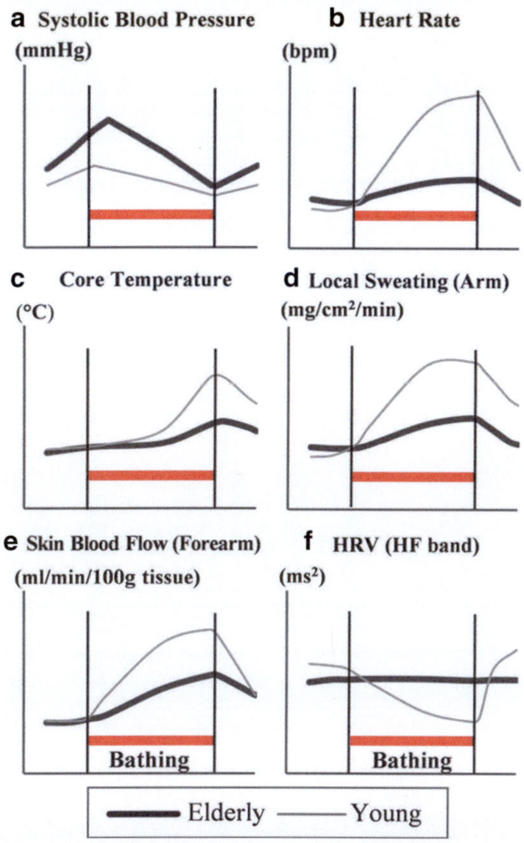

Fig. 9.13 Age-related differences in physiological reactions during Japanese style bating. (Adapted from Tochihara et al. [123])

of hot water and hydrostatic pressure. Miwa et al. [122] reported that the physiological response of older adults to bathing is slower and greater than that of younger people. Based on the findings of previous studies, Tochihara [123] presented a schematic representation of the physiological response to bathing in the Japanese style (Fig. 9.13 ABF). Systolic blood pressure rises sharply immediately after bathing, called the startle response, and then decreases as the subject is immersed in bath water. The change in blood pressure was greater in the elderly. Conversely, the change in heart rate was lower in older people group. The increase in the heart rate of older people was small, whereas that of the young increased significantly in the latter half of the bathing period. The HF band of the young adults decreased significantly during bathing, reflecting the heart rate results. .The compensatory response of the baroreceptor reflex to a decrease in blood pressure was lower in older people than in young individuals. They suggested that this drop in blood pressure causes syncope during bathing, which leads to drowning. Suzuki et al. [119] also found that functional impairment without organic diseases, such as stroke or cardiac disease, represented more than 80% of all emergency cases. Transient loss of consciousness, synocpe, and dehydration accounted for 30.6%, 9.4%, and 20.4% of cases, respectively. The problem is that even a transient loss of consciousness or fainting, which results in face immersion in hot water, can lead to an increase in the number of deaths and serious illnesses. The report indicates that 84% of emergency cases occurred in the bathroom, 60% of which were in the bathtub, and 29% of which were immersed in bath water. Regarding thermoregulatory responses to bathing, the elderly showed a smaller increase in core body temperature, and the amount of sweating and skin blood flow was also lower in older people than in the young (Fig. 9.13cde). Tochihara [123] explained that this is because older people have more subcutaneous fat, which slows down the transmission of the warming effect of hot water, making it difficult for them to warm up. In addition, it has been suggested that the decrease in sweating rate is related to a slower onset of sweating and a lower amount of sweating at a single sweating point due to aging. Decreased cutaneous blood flow is thought to be caused by decreased cutaneous vasodilatation.

In summary, many surveys and experimental studies have shown that bathing, although a routine activity, causes significant physiological strain, especially in elderly. Experimental studies on the physiological and psychological effects of bathing are summarised in Table 9.4. Six studies focussed on the effects of aging, an important factor in death associated with bathing. There are four studies on effects of different water temperatures. One study examined the effects of hot-water

Table 9.4 Experimental studies on thermal environment and bathing

	Subjects		Bathing					Research Objects			
Author (year)	*N* / Male or Female	Age Mean (SD) Range	Tw (°C)	Duration(min)	Body part to be immersed	Bath room temperature (°C)	Outcome	T room	T water	Age	Other
Kubota K (1992)	7M	26–36	47	3	1 m deep	–	Tsub, plasma-β-endorphin		●		Hot spring (unique bath method)
			42	10							
Ohnaka (1995)	4M	20.8 ± 1.5	40	10	Neck		Trec, Tskin, HR, BP, TS				Vs showering (standing or sitting)
	4M	48.0 ± 5.9									
Kanda (1995)	22M, 20F	65–88	40.8 ± 0.32	–	–	13.5	HR, BP, Tskin	○		●	Season (winter vs summer)
			40.1 ± 0.22			29.5					
Kanda (1996)	6F	20.5 (20–21)	40	5	–	10 17.5 25	Trec, Tskin, HR, BP, TS	●			
Allison TG (1998)	6M		40 vs 41.5	21			BP, sweat rate, Tesop		●		
Sung EJ (2000)	9F	29.6 (21–40)	40	20	Shoulder	10	EEG, EOG, EMG				Comparison with foot bath, sleep
Nagasawa Y (2001)	9M	27.4 (9.6)	40	10	Shoulder	24	BP, PR, ECG (HRV)			●	
	9M	74.6 (107)									
Hashiguchi (2002)	8M	23.5 (0.8)	40	10	Neck	10, 17.5, 25	Trec, Tskin, HR, BP,	●			
Kataoka Y (2005)	14M	28–42	38 vs 41	15	Chest	22–27	ECG(HRV)		●		

Chiba T(2005)	G1 30	72 (60–80)	40.88 (0.77)	5	–	20.5 (2.32)	ECG, TPR, CO, COMP	○		●	Season (winter vs summer)
	G2 20	74 (61–89)	40.06 (0.93)			25.1 (1.79)					
	G3 21	22 (19–30)	40.62 (0.71)			19.71 (1.45)					
Tochihara Y (2012)	8M	69.4 (1.1)	40	8	Neck	10,15,20,25	Trec, Tskin, BP, TS	●		●	
	8M	22.8 (0.6)									
Ishikawa J (2016)	28 (13M, 15F)	65.5 (9.7)	40.6 (1.5)	33.6 (13.6)	–	–	BP (central BP, 24-h BP)				BP change after bathing vs non-bathing day
Ono J (2017)	11M	69.5 (2.9)	39 vs 42	8	Armpits	20	Trec, Tskin, HR, BP		●	●	
	10M	24.1 (1.8)					TS, TC				
Masuda Y (2019)	7M, 5F	23.6 (1.4)	41	20	Subclavian	25 vs 15	Tty, HR, tsk, MAP, LDF	●			
Miwa (2020)	10M	20.4 (0.5)	41	15	Nipple	23	Tty, SR, SkBF			●	Duration time
	10M	69.7 (0.9)									

temperature and aging. Furthermore, because most bathing accidents occur during winter, six studies examined the effects of differences in bathroom temperature and season. Two studies evaluated the effects of both room temperature and aging on the elderly at the same time.

9.4.3 Bathtub Water Temperature

One of the most important factors influencing the physiological effects of bathing is the water temperature. In our survey of elderly Japanese, 34.7% of them in summer and 79.5% in winter preferred hot or slightly hot water temperatures in summer and winter, respectively. Kanda et al. [124] measured bathwater temperatures in the homes of 42 elderly people. According to that study, the average ± SE temperature of the hot water during bathing in summer and winter was 40.1 ± 0.22 °C and 40.8 ± 0.32 °C, respectively, with a reported tendency toward higher temperatures in winter. Higher water temperatures increase the thermal effects and cause greater haemodynamic loading during bathing. Kataoka and Yoshida [125] reported the effects of bathing on autonomic nervous system function in males aged 28–42 years using an analysis of heart rate variability. They compared two conditions of hot water temperature: 39 °C and 41 °C. The heart rate did not change significantly at 39 °C, while at 41 °C, the heart rate was significantly higher from 7 min after the start of bathing. HF, an indicator of parasympathetic activity, showed no significant change in both water conditions LF, a sympathetic indicator, decreased significantly at 41 °C, and the decrease was greater than at 39 °C. Thermal stimulation at 41 °C promoted peripheral vasodilation and increased the heart rate to maintain blood circulation volumes. It is suggested that suppression of vagal activity occurs more quickly at 41 °C than at 39 °C. Based on these findings, it was recommended that the bathing duration be no longer than 10 min at 39 °C and 5 min at 41 °C. Allison et al. [126] examined the effects of bath temperatures of 40 and 41.5 °C on heart rate, blood pressure, and oesophageal temperature and found that at 40 °C and above, there was no significant difference in the cardiac.

Ono et al. [127] examined the differences in hot-water temperature and age. This study compared the physiological and psychological responses to bathing among 11 older and 10 younger men at 39 and 42 °C bath water temperatures. The bathroom temperature was 20 °C, the bathing time was 8 min, and the water level was up to the armpits. The rectal temperature, skin temperature, heart rate, blood pressure, and subjective votes were measured in this study. Blood pressure rose significantly higher than baseline immediately after immersion in tub water, and the increase was greater in the elderly than in the younger subjects. At the 8th minute of bathing, blood pressure decreased owing to peripheral vasodilation caused by the thermal effect of the bath water. After bathing and dressing, the participants' blood pressure was higher than that at baseline. Although there was no significant difference in blood pressure variability for the elderly between water temperatures of 39 and 42 °C, the variability was greater for the elderly than for the younger subjects. The heart rate of the younger subjects was higher during bathing at 42 °C, but the increase in heart rate was lower in the elderly. Younger subjects adapted to the thermal effect of the higher temperature of the bath water by increasing their heart rate to maintain the volume of circulating blood, thus keeping the change in blood pressure small (Fig. 9.14). Thermal sensation (9-point scale: very cold to very hot) and thermal comfort (4-point scale: comfortable to very uncomfortable) were also evaluated before and after bathing. Younger males were able to perceive the difference in the temperature of the hot water after 8 min of bathing, while the elderly were not able to perceive the temperature difference. In addition, the younger participants reported significantly higher thermal discomfort after 8 min of immersion in water at 42 °C compared to 39 °C, while most of the elderly

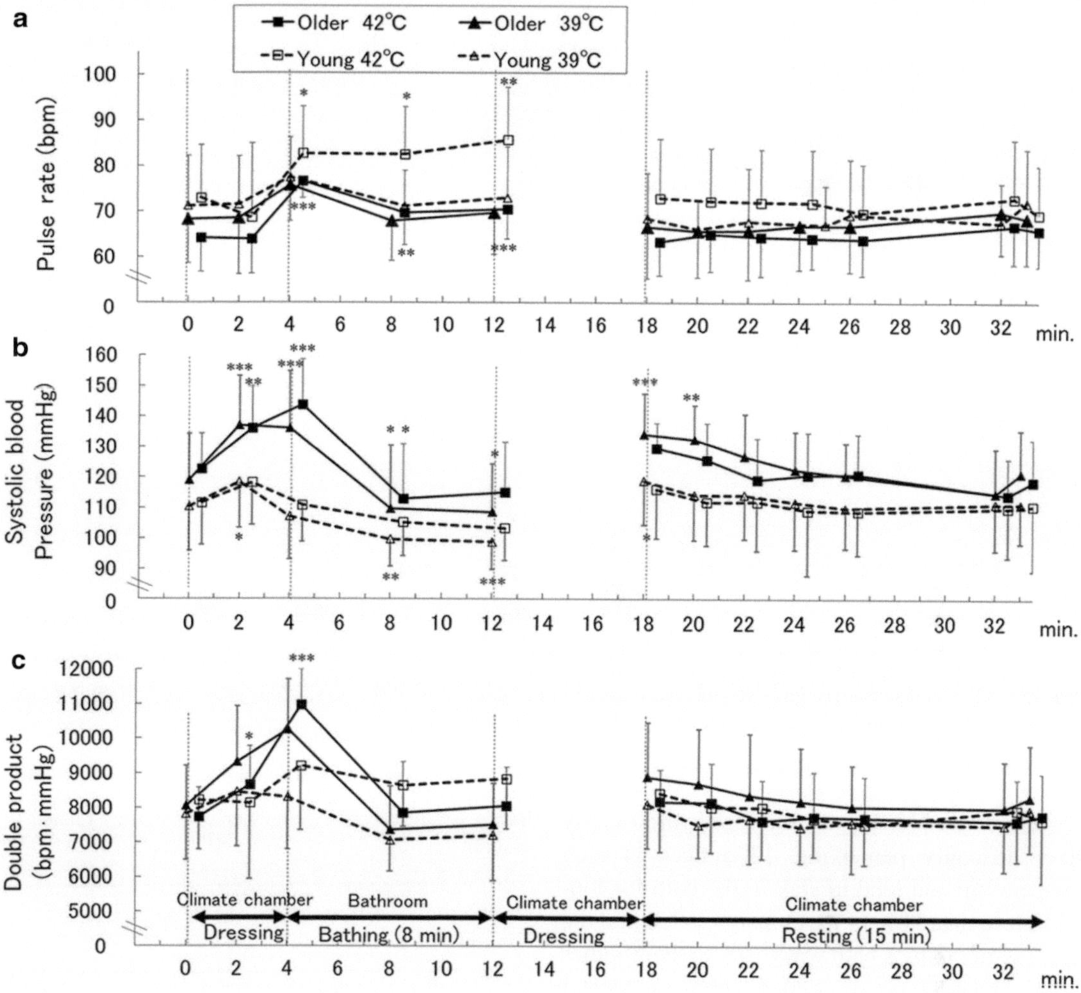

Fig. 9.14 Alterations in cardiovascular parameters. (**a**) Change in pulse. (**b**) Change in systolic blood pressure. (**c**) Change in double product (systolic blood pressure × pulse rate). (Adapted from Ono et al. [127]). Values are mean ± SD. *$P < 0.05$, **$P < 0.01$, ***$P < 0.001$ versus 0 min

reported being comfortable between both temperature conditions (Fig. 9.15). This suggests that even though the elderly were exposed to a strong circulatory dynamic strain due to the higher water temperature, they were subjectively unable to perceive it.

In summarising the previous studies described above, first, about the temperature of bath water, there seems to be a threshold between 40 and 39 °C that separates the effects on the cardiovascular system, in particular. Second, it is recommended that the duration of immersion in hot water be approximately 5 min when the water temperature is 40 °C or higher. This is because Nagasawa et al. [128] found a decrease in LF/HF and a decrease in sympathetic tone in elderly persons after 4 min of immersion at 40 °C for 10 min, and Kataoka and Yoshida [125] reported that this was to avoid hypotensive syncope during immersion. However, a study by Tai et al. [129] on 1094 elderly subjects showed that the average measured temperature of the bath water was in the range of 40.7–40.9 °C and that they generally bathed for 11–13 (median) minutes in bath water of about 41 °C.

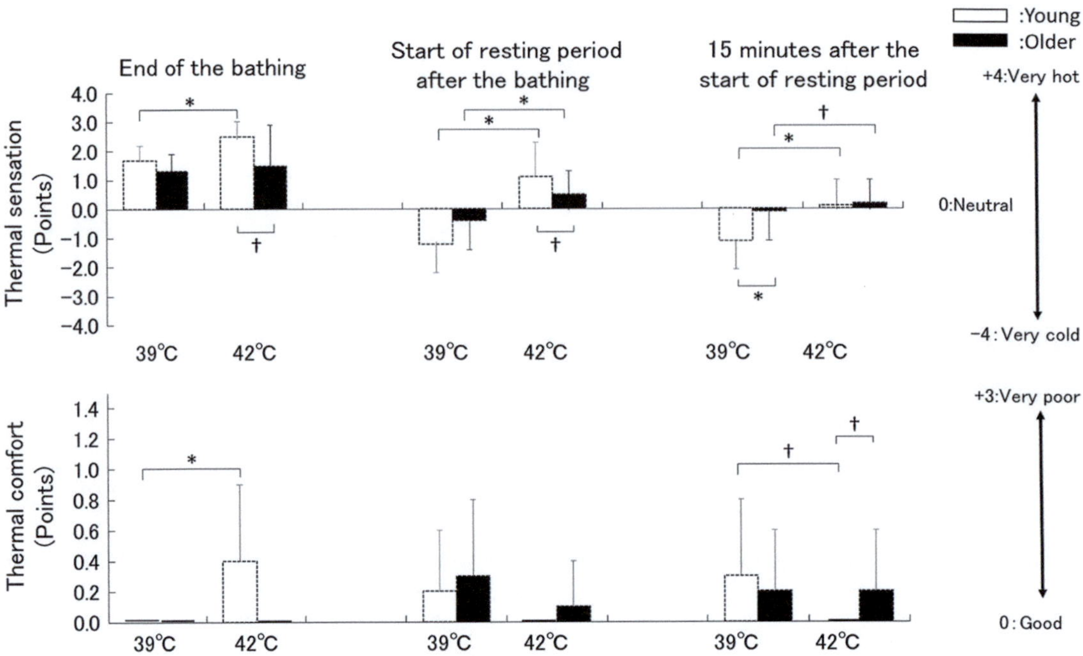

Fig. 9.15 Thermal sensation comfort scores. (Adapted from Ono et al. [127] $n = 11$ older people, $n = 10$young men. $*P < 0.05$ and $\dagger P < 0.1$)

Ono et al. [127] also found that the elderly have difficulty perceiving differences in bath temperature and tend to stay in the bath for longer. More than 80% of participants stated that the purpose of taking a bath was to warm their bodies [124], In winter, people tend to bathe in hot water for long periods to warm themselves. In our survey of elderly persons, only 37.6% answered that they set the water temperature below 41 °C to prevent bathing accidents (Table 9.5). It is difficult to change habits that have been cultivated over the years. The survey also suggested that contributing to the practice of preventive actions for bathing accidents is the prevention of the strong benefits of taking action and confidence to continue the action. It is important not only to make people aware of accidents associated with bathing but also to educate them about the benefits of these preventive behaviours and the preventive behaviours that are easy to continue to reduce the number of fatalities associated with bathing accidents.

Table 9.5 Preventive behaviour of accidents related to bathing

	N	%
Understanding of accidents related to bathing		
Yes	194	97.0
No	6	3.0
Preventive behaviours perceived		
Heating of the room and bathroom	164	84.5
Hot-water temperature below 41 °C	83	42.8
Notify family members before taking a bath	102	52.6
Half-body bathing	71	36.6
No knowledge	5	2.6
Preventive behaviour being practiced		
Heating of the room and bathroom	96	49.5
Water temperature below 41 °C	73	37.6
Notify family members before taking a bath	85	43.8
Half-body bathing	48	24.7
Not implemented	41	21.1
Benefits of taking preventive action		
Not at all	2	1.0
Not so much	4	2.1

(continued)

Table 9.5 (continued)

	N	%
Neither	7	3.6
Somewhat agree	72	37.1
Very much	109	56.2
Confidence in continuing the preventive behaviour		
Not at all	1	0.5
Not so much	23	11.9
Neither	31	16.0
Somewhat agree	99	51.0
Very much	40	20.6

9.4.4 Bathroom and Dressing Room Temperatures

Bathing accidents occur more frequently in winter. In a study by Suzuki et al. [119], cardiovascular arrest and acute illness cases in emergencies related to bathing were transported to hospitals, especially in the coldest months. In our study, the percentages of elderly people who soaked in bathtub water were 35.0% and 81.0% in the summer and winter, respectively. The bath water temperature was 42 °C or higher for 1.5% in summer and 15.8% in winter. Bathtub bathing was 10 min or longer in 22.7% of the summer and 34.8% of the winter months (Table 9.2). In short, the soaking frequency in the bathtub increased, the water temperature increased, and the soaking duration tended to be longer, particularly in winter. Ohnaka et al. [2] surveyed the thermal environments of houses in 11 prefectures throughout Japan. According to a survey, 10 of the 11 prefectures had room temperatures of 15 °C or lower in dressings and bathrooms in various regions of Japan during winter. The remaining one region (Sapporo) is located at the northernmost latitude and is cold; therefore, many of the homes are equipped with heating systems. This is related to the fact that the room temperature in the dressing rooms and bathrooms was also kept above 20 °C (Fig. 9.1; [123]). The number of fatal drownings in these areas was lower than in other areas. Takasaki et al. [130] found that significantly more people in cold areas did not feel cold while bathing than those in warm areas. These studies suggested that the thermal environment of the bathroom, rather than the ambient outdoor temperature, is associated with bathing accidents.

Several field and laboratory studies have reported reasons for the increase in bathing accidents during winter [121, 131–133]. In addition, Kanda et al. [124] and Chiba et al. [120] examined the effects of seasonal differences. Hashiguchi et al. [121] examined the physiological and psychological effects of different bathrooms and changing room temperatures on eight adult male subjects, including full-body baths with immersion up to the shoulders, half-body baths with immersion in hot water up to the epicardial region, and showers. The increase in heart rate during an 8-min half-body bath at 40 °C was significantly lower than that during a full-body bath. The double product value, an index of the cardiac load, was also significantly lower in the half-body bath than in the full-body bath, regardless of the bathroom temperature (Fig. 9.16). Half-body bathing is recommended because high water levels increase the cardiac burden owing to hydrostatic pressure. However, in the half-body bath, which places less burden on the circulation, there was a significant difference in the rectal temperature after the bath between the bath temperatures of 25 and 10 °C. During half-body bathing, a low bathroom temperature does not allow the body to warm sufficiently. In contrast, in the case of full-body bathing: immersion up to the shoulders, the effect of different bath temperature on the change in rectal temperature was insignificant (Fig. 9.17).

From the survey studies by Ohnaka et al. [2] and Takasaki et al., [130] and the experimental study by Hashiguchi et al. [121], it is clear that the thermal environment of the bathroom produces bathing habits that easily lead to a physiological burden. This is why deaths due to bathing accidents are more common during the winter season; conversely, there are fewer deaths in areas with warm bathroom temperatures.

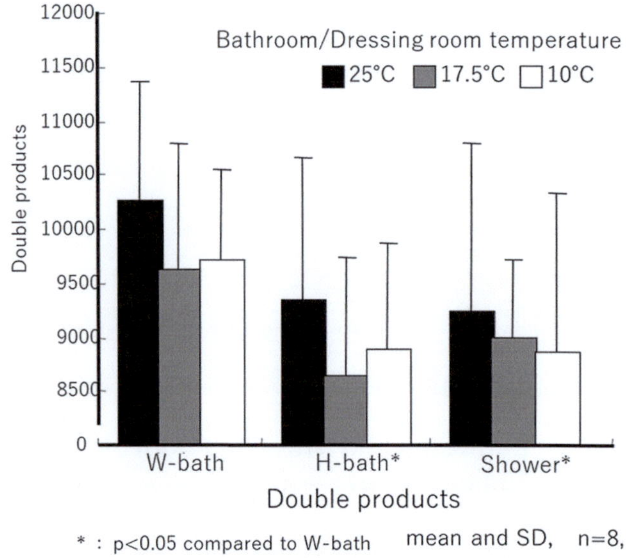

Fig. 9.16 Double products under nine conditions (3 bathing methods×3 room temperature conditions)

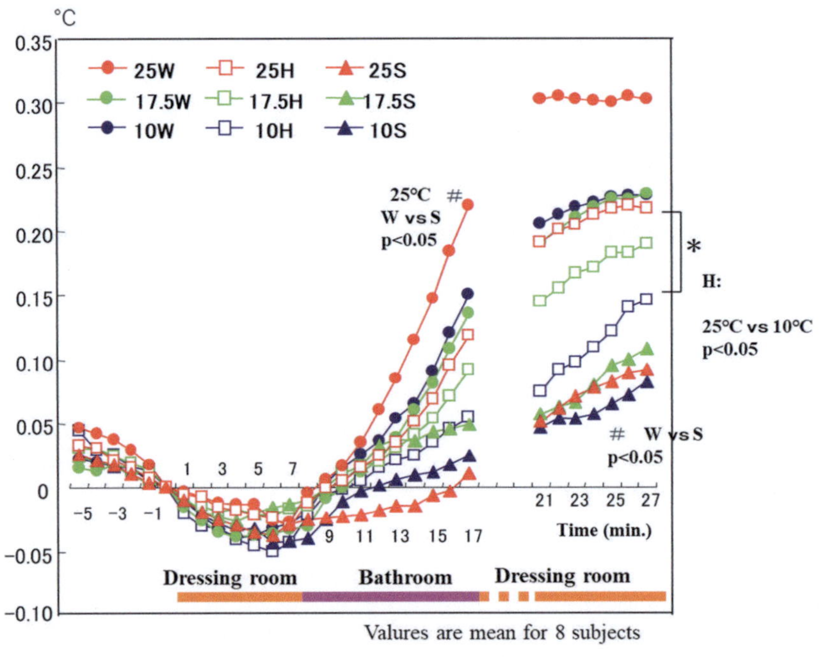

Fig. 9.17 Rectal temperature under nine conditions (bathing methods: W-bath (*W*), H-bath (*H*), showering (*S*), room temperature: 25, 17.5, 10 °C). (Adapted from Hashiguchi et al. [121]). $*P = 0.05$ indicates a significant difference between W and S at 25 °C. $*P = 0.05$ indicates the significant difference between 25 and 10 °C at H-bath

Previous studies have examined the recommended dressing rooms and bathroom temperatures. Kanda et al. [131] observed changes in blood pressure, body temperature, and thermal comfort during bathing in adult female subjects at three room temperatures: 10, 17.5, and 25 °C. It has been reported that there is a negative correlation between room temperature and the range of increase in blood pressure starting before bathing, and a negative correlation between skin tem-

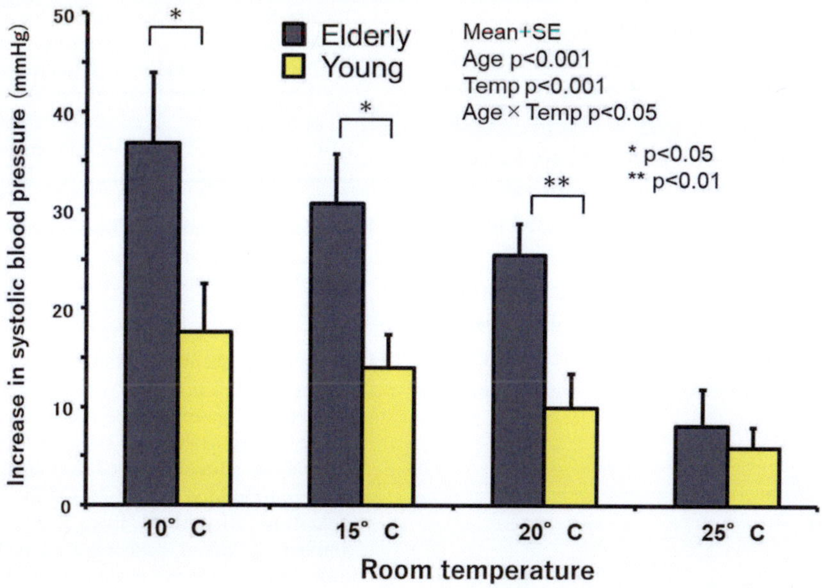

Fig. 9.18 Age-related differences in systolic blood pressure increases in the dressing room under four room temperature conditions. (Adapted from Tochihara et al. [132])

perature and increase in blood pressure. From those results, they conclude that the recommended bathroom temperature is 25 and 17.5 °C is acceptable. However, this was a finding from a study of healthy young adults. Therefore, Tochihara et al. [132] examined the thermoregulatory responses and blood pressure changes during bathing in healthy elderly subjects at four different dressing room/bathroom temperatures of 10, 15, 20, and 25 °C. The increase in systolic blood pressure in the dressing room was significantly higher in elderly subjects than in younger subjects at room temperatures of 10, 15, and 20 °C. In contrast, at 25 °C, there was no significant age difference at 25 °C (Fig. 9.18). These results indicate that bathroom and dressing room temperatures of 20 °C are insufficient for the elderly, and that a few degrees higher is recommended for safer bathing.

Furthermore, field studies have been conducted on elderly individuals, including those with hypertension and heart disease [120, 124]. Kanda et al. [124] examined the changes in blood pressure during bathing in winter and summer among 42 older men and women. The participants of this study were patients with cardiac and hypertensive diseases. The temperatures in the dressing rooms during winter and summer were 13.5 and 29.5 °C, respectively. The increase in systolic blood pressure when undressing in the dressing room was greater in winter than in summer. Blood pressure variation remained 8–19 mmHg higher in those with cardiovascular disease than in those without and was more than 160 mmHg at the time of undressing, indicating that they were at higher risk. Similarly, Chiba et al. [120] compared physiological and biomedical parameters during bathing in winter and summer in elderly subjects with those in younger people. The elderly individuals included those with hypertension or arrhythmias. Bathroom temperatures (mean ± SD) during winter and summer were 20.5 ± 2.3 and 25.11.7 ± 9 °C, respectively. It has been reported that bathing the elderly in winter increases cardiac output and more oxygen consumption of cardiac muscles compared with summer or younger persons. They also warned that persons with a prior history of arrhythmias were at risk of dangerous arrhythmias, such as ventricular tachycardia during bathing. Tochihara et al. [1] compared elderly and young subjects for warm and cold sensations during exposure to 10 and 35 °C. The results showed that older adults complained less of cold immediately after the start of exposure (Fig. 9.2). This suggests that cold perception is delayed with age. In addition, the elderly are less likely to complain about heat and cold in their perception and are less able to control indoor temperatures appropriately [21, 134]. The lack of recognition of the need to heat bathrooms and dressing rooms, even in cold environments, is one reason the risk of bathing accidents has not been reduced. Environmental adjustments should be made to encourage the installation of thermometers and enable objective evaluations.

References

1. Tochihara Y, Ohnaka T, Nagai Y, Tokuda T, Kawashima Y. Physiological responses and thermal sensations of the elderly in cold and hot environments. J Therm Biol. 1993;18:355–61.
2. Ohnaka T, Takasaki Y, Tochihara Y, Nagai Y, Ito H, Yoshitake S. A survey on thermal conditions of bathroom during winter in Japan (in Japanese). J Hum Living Environ. 2007;14(1):11–6.
3. Murata S, Iriki M. Body temperature of the aged—the effect of aging on the cutaneous sensory points. Jpn J Geriatr. 1974;11(3):157–63.
4. Hashiguchi N, Tochihara Y, Ohnaka T, Tsuchida C, Otsuki T. Physiological and subjective responses in the elderly when using floor heating and air conditioning systems. J Physiol Anthropol Appl Hum Sci. 2004;23(6):205–13.
5. Hashiguchi N, Kumamoto T, Chishaki A, Tochihara Y. Effects of vertical air temperature gradients on physiological and psychological responses in the elderly. J Hum Environ Syst. 2011;14(1):9–17.
6. Kenney WL, Havenith G. Thermal physiology of the elderly and handicapped. Heat stress and age: skin blood flow and body temperature. J Therm Biol. 1993;18:341–4.
7. Ellis FP, Exton-Smith AN, Foster KG, Weiner JS. Eccrine sweating and mortality during heat waves in very young and very old person. Israel J Med Sci. 1976;12(8):815–7.
8. Natsume K, Ogawa T, Sugenoya J, Ohnishi N, Imai K. Preferred ambient temperature for old and young men in summer and winter. Int J Biometeorol. 1992;36(1):1–4.
9. Uchida Y, Tamura T, Iwasaki F. Changes in cold/warm thresholds with advancing age -the case of adult women aged from the 20s to the 80s. Sen'i Gakkaishi. 2009;65(5):132–8.
10. Guergova S, André D. Thermal sensitivity in the elderly: a review. Ageing Res Rev. 2011;10(1):80–92.
11. Takeda R, Imai D, Suzuki A, Ota A, Naghavi N, Ymashina Y, et al. Lower thermal sensation in normothermic and mildly hyperthermic older adults. Eur J Appl Physiol. 2016;116(5):975–84.
12. Inoue Y, Gerrett N, Ichinose-Kuwabara T, Umino Y, Kiuchi S, Amano T, et al. Sex differences in age-related changes on peripheral warm and cold innocuous thermal sensitivity. Physiol Behav. 2016;164(Pt A):86–92.
13. Shibata Y, Kitamura E, Matsubara N. A study of the actual indoor thermal environments and the consciousness and behavior for the prevention to heat disorders in the elderly. Jpn J Biometeor. 2018;55(1):33–50.
14. Fujitani Y, Otani S, Majbauddin A, Amano H, Masumoto T, Kurozawa Y. Impact of maximum air temperature on ambulance transports owing to heat stroke during spring and summer in Tottori prefecture, Japan: a time-stratified case-crossover analysis. Yonago Acta Med. 2019;62:47–52.
15. Tokairin T, Kiritoshi K, Ihara T. Investigation of the number of indoor heatstroke cases by perception time and quantitative evaluation of the effect of air conditioner use on reducing the severity of indoor heatstroke. lpn J Biomet. 2022;59(3–4):115–22.
16. Rohles FH Jr, Johnson MA. Thermal comfort in the elderly. ASHRAE Tans. 1972;78 part I:131–7.
17. Fanger PO. Thermal comfort. New York: Mcgraw-Hill Book Company; 1970. p. 75–86.
18. Iwashige H, Igarashi Y, Miyazawa M, Sakakibara N, Mizuno Y, Kubo H, et al. A survey of residential thermal environments on the aged in the daily living (part 2) aspects of ambient temperature on the aged. Home Econ Jpn. 1994;46(5):455–64.
19. Kawashima Y, Tochihara Y, Gotoh S, Uryu Y, Ohmori M, Masuda J, et al. A study on body temperature regulation and residential thermal environments of the elderly (1) study methods and residential thermal environment in Japan. Japanese Society of Human-Environment System. In: International conference on human-environment system (Tokyo 1991); 1991. p. 185–8.
20. Collins KJ, Exton-Smith AN, Doré C. Urban hypothermia: preferred temperature and thermal perception in old age. Br Med J. 1981;282(6259):175–7.
21. Ohnaka T, Tochihara Y, Tsuzuki K, Nagai Y, Tokuda T, Kawashima Y. Preferred temperature of the elderly after cold and heat exposures determined by individual self-selection of air temperature. J Therm Biol. 1993;18(5–6):349–53.
22. Rohles FH. Humidity, human factors, and the energy shortage. ASHRAE J. 1975;17:38–40.
23. Bornehag CG, Blomquist G, Gyntelberg F, Järvholm B, Malmberg P, Nordvall L, et al. Dampness in buildings and health. Nordic interdisciplinary review of the scientific evidence on associations between exposure to "dampness" in buildings and health effects (NORDDAMP). Indoor Air. 2001;11(2):72–86.
24. Bornehag CG, Sundell J, Bonini S, Custovic A, Malmberg P, Skerfving S, et al. Dampness in buildings as a risk factor for health effects,

EUROEXPO: a multidisciplinary review of the literature (1998–2000) on dampness and mite exposure in buildings and health effects. Indoor Air. 2004;14(4):243–57.
25. Goad N, Gawkrodger DJ. Ambient humidity and the skin: the impact of air humidity in healthy and diseased states. J Eur Acad Dermatol Venereol. 2016;30(8):1285–94.
26. Azuma K, Ikeda K, Kagi N, Yanagi U, Osawa H. Physicochemical risk factors for building-related symptoms in air-conditioned office buildings: ambient particles and combined exposure to indoor air pollutants. Sci Total Environ. 2018;616-617:1649–55.
27. Wolkoff P. Indoor air humidity, air quality, and health—an overview. Int J Hyg Environ Health. 2018;221(3):376–90.
28. Moriyama M, Hugentobler WJ, Iwasaki A. Seasonality of respiratory viral infections. Annu Rev Virol. 2020;7(1):83–101.
29. MHWL (Ministry of Health, Labour and Welfare in Japan). Law for maintenance of sanitation in buildings. 1970.
30. Tanaka M. Heat stress standard for hot work environments in Japan. Ind Health. 2007;45:85–90.
31. Saito H, Sawada S, Yasuda A, Oka T, Hagiwara M, Tai T, et al. Survey of temperature and humidity and health effects in power-saving office environments (in Japanese). Special research report of National Institute of Occupational Safety and Health, Japan, vol. 43; 2013. p. 157–63.
32. Hashiguchi N, Hirakawa M, Tochihara Y, Kaji Y, Karaki C. Effects of setting up of humidifiers on thermal conditions and subjective responses of patients and staff in a hospital during winter. Appl Ergon. 2008;39(2):158–65.
33. Enomoto-Koishimizu H, Tochihara Y, Ikeda K. Changes in thermal comfort and subjective symptoms of factory workers by the use of humidifiers in low-humidity rooms in winter. J Architect Plann (Trans AIJ). 2002;67:9–14.
34. Arundel AV, Sterling EM, Biggin JH, Sterling TD. Indirect health effects of relative humidity in indoor environments. Environ Health Perspect. 1986;65:351–61.
35. Saijo Y, Kishi R, Sata F, Katakura Y, Urashima Y, Hatakeyama A, et al. Symptoms in relation to chemicals and dampness in newly built dwellings. Int Arch Occup Environ Health. 2004;77(7):461–70.
36. Bornehag CG, Blomquist G, Gyntelberg F, Jarvholm B, Malmberg P, Nordvall L, et al. Dampness in buildings and health. Nordic interdisciplinary review of the scientific evidence on associations between exposure to "dampness" in buildings and health effects (NORDDAMP). Indoor Air. 2001;11(2):72–86.
37. Chung NK, Pin CH. Obesity and the occurrence of heat disorders. Mil Med. 1996;161(12):739–42.
38. Azuma K, Ikeda K, Kagi N, Yanagi U, Osawa H. Prevalence and risk factors associated with nonspecific building-related symptoms in office employees in Japan: relationships between work environment, Indoor Air Quality, and occupational stress. Indoor Air. 2015;25:499–511.
39. Azuma K, Ikeda K, Kagi N, Yanagi U, Osawa H. Evaluating prevalence and risk factors of building-related symptoms among office workers: seasonal characteristics of symptoms and psychosocial and physical environmental factors. Environ Health Prev Med. 2017;22:38.
40. Zhang H, Yoshino H, Murakami S, Bogaki K, Tanaka T, Akabayashi S-i, et al. Analyses of indoor humidity environment in nationally residential houses of Japan. AIJ J Technol Design. 2009;15(30):453–7.
41. Hasegawa K, Yoshino H, Mitamura T. Investigation of feeling dryness and its adverse health effect in residential buildings of Japan. J Environ Eng (Trans AIJ). 2019;84(760):587–96.
42. Sato M, Fukayo S, Yano E. Adverse environmental health effects of ultra-low relative humidity indoor air. J Occup Health. 2003;45:133–6.
43. Reinikainen LM, Jaakkola JJ, Seppanen O. The effect of air humidification on symptoms and perception of indoor air quality in office workers: a six-period cross-over trial. Arch Environ Health. 1992;47(1):8–15.
44. Reinikainen LM, Jaakkola JJ. Effects of temperature and humidification in the office environment. Arch Environ Health. 2001;56(4):365–8.
45. Reinikainen LM, Jaakkola JJ. Significance of humidity and temperature on skin and upper airway symptoms. Indoor Air. 2003;13(4):344–52.
46. Takata S. Study on the lower limit of indoor humidity subject experiments on psychological responses. Indoor Air. 2021;31(1):250–63.
47. Egawa M, Oguri M, Kuwahara T, Takahashi M. Effect of exposure of human skin to a dry environment. Skin Res Technol. 2002;8:212–8.
48. Egawa M, Koizumi K, Hirao T. Changes in facial moisture distribution and feelings of moisture/dryness among various environmental temperatures and humidities in summer and winter. Skin Res Technol. 2020;26:937–48.
49. Engebretsen KA, Johansen JD, Kezic S, Linneberg A, Thyssen JP. The effect of environmental humidity and temperature on skin barrier function and dermatitis. J Eur Acad Dermatol Venereol. 2016;30:223–49.
50. Kawakami R, Yamaguchi M, Otsuka T. Study on relationship between hydration level in stratum corneum, sebum level and skin dryness sensation. J Environ Eng (Trans AIJ). 2020;85(778):913–21.
51. Abusharha AA, Pearce EI. The effect of low humidity on the human tear film. Cornea. 2013;32(4):429–34.
52. Andersen I, Lundqvist GR, Jensen PL, Proctor DF. Human response to 78-hour exposure to dry air. Arch Environ Health. 1974;29(6):319–24.
53. Sunwoo Y, Chou C, Takeshita J, Murakami M, Tochihara Y. Physiological and subjective responses

to low relative humidity. J Physiol Anthropol. 2006;25(1):7–14.
54. Sunwoo Y, Chou C, Takeshita J, Murakami M, Tochihara Y. Physiological and subjective responses to low relative humidity in young and elderly men. J Physiol Anthropol. 2006;25(3):229–38.
55. Wyon D, Fang L, Lagercrantz L, Fanger P. Experimental determination of the limiting criteria for human exposure to low winter humidity indoors (RP1160). HVAC&R Res. 2006;12(2):201–13.
56. Hashiguchi N, Takeda A, Yasuyama Y, Chishaki A, Tochihara Y. Effects of 6-h exposure to low relative humidity and low air pressure on body fluid loss and blood viscosity. Indoor Air. 2013;23:430–6.
57. Hemmes JH, Winkler KC, Kool SM. Virus survival as a seasonal factor in influenza and polimyelitis. Nature. 1960;188:430–1.
58. Sale CS. Humidification to reduce respiratory illnesses in nursery school children. South Med J. 1972;65(7):882–5.
59. Mäkinen TM, Juvonen R, Jokelainen J, Harju TH, Peitso A, Bloigu A, et al. Cold temperature and low humidity are associated with increased occurrence of respiratory tract infections. Respir Med. 2009;103(3):456–62.
60. Nordström K, Norbäck D, Wieslander G. Subjective indoor air quality in geriatric hospitals. Indoor Built Envion. 1999;8(1):49–57.
61. Katagiri C, Sato J, Nomura J, Denda M. Changes in environmental humidity affect the water-holding property of the stratum corneum and its free amino acid content, and the expression of filaggrin in the epidermis of hairless mice. J Dermatol Sci. 2003;31(1):29–35.
62. Denda M, Sato J, Masuda Y, Tsuchiya T, Koyama J, Kuramoto M, et al. Exposure to a dry environment enhances epidermal permeability barrier function. J Invest Dermatol. 1998;111(5):858–63.
63. Yoshikawa N, Imokawa G, Kimoto K, Jin K, Higaki Y, Kawashima M. Regional analysis of ceramides within the stratum corneum in relation to seasonal changes. Dermatology. 1994;188(3):207–14.
64. Statistics Bureau, Ministry of Internal Affairs and Communications, 2018. https://www.stat.go.jp/data/topics/topi1135.html. 14 Nov 2023.
65. Wilhelm KP, Cua AB, Maibach HI, Skin aging. Effect on transepidermal water loss, stratum corneum hydration, skin surface pH, and casual sebum content. Arch Dermatol. 1991;127(12):1806–9.
66. Hirokawa K, Utsuyama M, Kasai M, Kurashima C. Aging and immunity. Acta Pathol Jpn. 1992;42(8):537–48.
67. Tochihara Y, Yamashita K, Fujii K, Kaji Y, Wakabayashi H, Kitahara H. Thermoregulatory and cardiovascular responses in the elderly towards a broad range of gradual air temperature changes. J Therm Biol. 2021;99:103007.
68. Inoue Y, Shibasaki M. Regional differences in age-related decrements of the cutaneous vascular and sweating responses to passive heating. Eur J Appl Physiol Occup Physiol. 1996;74(1–2):78–84.
69. Hashiguchi N, Tochihara Y, Takeda A, Yasuyama Y. Effects of indoor summer dehumidification and winter humidification on the physiological and subjective responses of the elderly. J Therm Biol 2023;111:1–10.
70. Mollayeva T, Thurairajah P, Burton K, Mollayeva S, Shapiro CM, Colantonio A. The Pittsburgh sleep quality index as a screening tool for sleep dysfunction in clinical and non-clinical samples: a systematic review and meta-analysis. Sleep Med Rev. 2016;25:52–73.
71. Itani O, Jike M, Watanabe N, Kaneita Y. Short sleep duration and health outcomes: a systematic review, meta-analysis, and meta-regression. Sleep Med. 2017;32:246–56.
72. Hasler G, Buysse DJ, Klaghofer R, Gamma A, Ajdacic V, Eich D, et al. The association between short sleep duration and obesity in young adults: a 13-year prospective study. Sleep. 2004;27(4):661–6.
73. Smiley A, King D, Bidulescu A. The association between sleep duration and metabolic syndrome: the NHANES 2013/2014. Nutrients. 2019;11(11):2582.
74. Jike M, Itani O, Watanabe N, Buysse DJ, Kaneita Y. Long sleep duration and health outcomes: a systematic review, meta-analysis and meta-regression. Sleep Med Rev. 2018;39:25–36.
75. Vogel M, Braungardt T, Meyer W, Schneider W. The effects of shift work on physical and mental health. J Neural Transm (Vienna). 2012;119(10):1121–32.
76. Faraut B, Virginie B, Damien L. Neuroendocrine, immune and oxidative stress in shift workers. Sleep Med Rev. 2013;17(6):433–44.
77. Sunderajan P, Gaynes BN, Wisniewski SR, Miyahara S, Fava M, Akingbala F, et al. Insomnia in patients with depression: a STAR*D report. CNS Spectr. 2010;15(6):394–404.
78. Baglioni C, Battagliese G, Feige B, Spiegelhalder K, Nissen C, Voderholzer U, et al. Insomnia as a predictor of depression: a meta-analytic evaluation of longitudinal epidemiological studies. J Affect Disord. 2011;135(1–3):10–9.
79. Gulia KK, Kumar VM. Sleep disorders in the elderly: a growing challenge. Psychogeriatrics. 2018;18:155–65.
80. Nakazaki C, Noda A, Koike Y, Yamada S, Murohara T, Ozaki N. Association of insomnia and short sleep duration with atherosclerosis risk in the elderly. Am J Hypertens. 2012;25(11):1149–55.
81. Ensrud KE, Blackwell TL, Redline S, Ancoli-Israel S, Paudel ML, Cawthon PM, et al. Sleep disturbances and frailty status in older community-dwelling men. J Am Geriatr Soc. 2009;57(11):2085–93.
82. Endeshaw YW, Unruh ML, Kutner ME, Newman AB, Bliwise DL. Sleep-disordered breathing and frailty in the cardiovascular health study cohort. Am J Epidemiol. 2009;170(2):193–202.

83. Ensrud KE, Blackwell TL, Ancoli-Israel S, Redline S, Cawthon PM, Paudel ML, et al. Sleep disturbances and risk of frailty and mortality in older men. Sleep Med. 2012;13(10):1217–25.
84. Nakakubo S, Doi T, Makizako H, Tsutsumimoto K, Hotta R, Ono R, et al. Sleep duration and excessive daytime sleepiness are associated with incidence of disability in community-dwelling older adults. J Am Med Dir Assoc. 2016;17(8):768.e1–5.
85. Pourmotabbed A, Ghaedi E, Babaei A, Mohammadi H, Khazaie H, Jalili C, et al. Sleep duration and sarcopenia risk: a systematic review and dose-response meta-analysis. Sleep Breath. 2020;24(4):1267–78.
86. Rubio-Arias JÁ, Rodríguez-Fernández R, Andreu L, Martínez-Aranda LM, Martínez-Rodriguez A, Ramos-Campo DJ. Effect of sleep quality on the prevalence of sarcopenia in older adults: a systematic review with meta-analysis. J Clin Med. 2019;8(12):2156.
87. Foley D, Monjan A, Masaki K, Ross W, Havlik R, White L, et al. Daytime sleepiness is associated with 3-year incident dementia and cognitive decline in older Japanese-American men. J Am Geriatr Soc. 2005;49(12):1628–32.
88. Tsapanou A, Gu Y, Manly J, Schupf N, Tang MX, Zimmerman M, et al. Daytime sleepiness and sleep inadequacy as risk factors for dementia. Dement Geriatr Cogn Dis Extra. 2015;5(2):286–95.
89. Manabe K, Matsui T, Yamaya M, Sato-Nakagawa T, Okamura N, Arai H, et al. Sleep patterns and mortality among elderly patients in a geriatric hospital. Gerontology. 2000;46(6):318–22.
90. Dew MA, Hoch CC, Buysse DJ, Monk TH, Begley AE, Houck PR, et al. Healthy older adults' sleep predicts all-cause mortality at 4 to 19 years of follow-up. Psychosom Med. 2003;65(1):63–73.
91. Kräuchi K, Cajochen C, Werth E, Wirz-Justice A. Warm feet promote the rapid onset of sleep. Nature. 1999;401:36–7.
92. Van den Heuvel CJ, Noone JT, Lushington K, Dawson D. Changes in sleepiness and body temperature precede nocturnal sleep onset: evidence from a polysomnographic study in young men. J Sleep Res. 1998;7:159–66.
93. Van Someren EJ. Mechanisms and functions of coupling between sleep and temperature rhythms. Prog Brain Res. 2006;153:309–24.
94. Gilbert SS, Van den Heuvel CJ, Ferguson SA, Dawson D. Thermoregulation as a sleep signaling system. Sleep Med Rev. 2004;8(2):81–93.
95. Lan L, Pan L, Lian Z, Huang H, Lin Y. Experimental study on thermal comfort of sleeping people at different air temperatures. Build Environ. 2014;73:24–31.
96. Haskell EH, Palca JW, Walker JM, Berger RJ, Heller HC. The effects of high and low ambient temperatures on human sleep stages. Electroencephalogr Clin Neurophysiol. 1981;51:494–501.
97. Strøm-Tejsen P, Mathiasen S, Bach M, Petersen S. The effects of increased bedroom air temperature on sleep and next-day mental performance. In: The 14th international conference of indoor air quality and climate-indoor air; 2016. p. 640.
98. Okamoto-Mizuno K, Mizuno K, Michie S, Maeda A, Iizuka S. Effects of humid heat exposure on human sleep stages and body temperature. Sleep. 1999;15(22):767–73.
99. Tsuzuki K, Okamoto-Mizuno K, Mizuno K. Effects of humid heat exposure on sleep, thermoregulation, melatonin, and microclimate. J Therm Biol. 2004;29:31–6.
100. Tsuzuki K, Okamoto-Mizuno K, Mizuno K, Iwaki T. Effects of airflow on body temperatures and sleep stages in a warm humid climate. Int J Biometeorol. 2008;52:261–70.
101. Okamoto-Mizuno K, Tsuzuki K. Effects of season on sleep and skin temperature in the elderly. Int J Biometeorol. 2010;54:401–9.
102. Okamoto-Mizuno K, Mizuno K. Effects of thermal environment on sleep and circadian rhythm. J Physiol Anthropol. 2012;31:14. http://www.jphysiolanthropol.com/content/31/1/14
103. Lan L, Xia L, Tang J, Zhang X, Lin Y, Wang Z. Elevated airflow can maintain sleep quality and thermal comfort of the elderly in a hot environment. Indoor Air. 2019;29:1040–9.
104. Yan Y, Zhang H, Kang M, Lan L, Wang Z, Lin Y. Experimental study of the negative effects of raised bedroom temperature and reduced ventilation on the sleep quality of elderly subjects. Indoor Air. 2022;32:e13159.
105. Ohnaka T, Tochihara Y, Kanda K. Body movements of the elderly during sleep and thermal conditions in bedrooms in summer. Appl Hum Sci. 1995;14(2):89–93.
106. Yamasaki S, Maeda T, Horiuchi T. Association between habitual hot spring bathing and depression in Japanese older adults: a retrospective study in Beppu. Complement Ther Med. 2022;72:102909. https://doi.org/10.1016/j.ctim.2022.102909. Epub 2022 Dec 13.
107. Hayashi E, Aoyama M, Fukano F, Takano J, Shimizu Y, Miyashita M. Effects of bathing in a tub on physical and psychological symptoms of end-of-life cancer patients an observational, controlled study. J Hosp Palliat Nurs. 2022;24:30–9.
108. Naumann J, Grebe J, Kaifel S, Weinert T, Sadaghiani C, Huber R. Effects of hyperthermic baths on depression, sleep and heart rate variability in patients with depressive disorder: a randomized clinical pilot trial. BMC Complement Altern Med. 2017;17:172.
109. Ukai T, Iso H, Yamagishi K, Saito I, Kokubo Y, Yatsuya H, et al. Habitual tub bathing and risks of incident coronary heart disease and stroke. Heart. 2020;106:732–7.
110. Katsuyama H, Hakoshima M, Adachi H, Masui Y, Sako A, Inokuma S, et al. Habitual hot-tub bathing and cardiovascular risk factors in patients with type

2 diabetes mellitus: a cross-sectional study. Cardiol Res. 2022;13:144–53.
111. Yagi A, Hayasaka S, Ojima T. The association between tub-bathing frequency and the incidence of depression symptoms among older people in Japan: JAGES longitudinal study (in Japanese). Jap J Heal Res. 2019;40:67–73.
112. Liao WC. Effects of passive body heating on body temperature and sleep regulation in the elderly: a systematic review. Int J Nurs Stud. 2002;39:803–10.
113. Haghayegh S, Khoshnevis S, Smolensky MH, Diller KR, Castriotta RJ. Before-bedtime passive body heating by warm shower or bath to improve sleep: a systematic review and meta-analysis. Sleep Med Rev. 2019;46:124–35.
114. Sung EJ, Tochihara Y. Effects of bathing and hot footbath on sleep in winter. J Physiol Anthropol Appl Hum Sci. 2000;19:21–7.
115. Ohtsuka Y, Inokuma S, Sugimoto H. Analysis of bathing related accidents in Japan -collaboration with Japanese association for acute medicine. J Jpn Soc Balneol Clim Phys Med. 2011;74:239–45.
116. Takahashi R, Sakamoto Y, Tsuzuki K, Asakawa Y. Actual conditions of cardiopulmonary arrest (CPA) during bathing in Japan -analysis of 9,360 cases of emergency transport in 47 prefectures-(in Japanese). Tokyo Metrop Geriatr Hosp Inst Gerontol. 2014:99–106.
117. Suzuki M, Ikaga T, Hori S. Relationship between bath-related deaths and lower air temperature. Intern Med. 2017;56:3173–7.
118. Suzuki M, Shimbo T, Ikaga T, Hori S. Sudden death phenomenon while bathing in Japan—mortality data. Jpn Circ Soc. 2017;81:1144–9.
119. Suzuki M, Shimbo T, Ikaga T, Hori S. Incidence and characteristics of bath-related accidents. Intern Med. 2019;58:53–62.
120. Chiba T, Yamauchi M, Nishida N, Kaneko T, Yoshizaki K, Yoshioka N. Risk factors of sudden death in the Japanese hot bath in the senior population. Forensic Sci Int. 2005;149:151–8.
121. Hashiguchi N, Ni F, Tochihara Y. Effects of room temperature on physiological and subjective responses during whole-body bathing, half-body bathing and showering. J Physiol Anthropol Appl Hum Sci. 2002;21:277–83.
122. Miwa C, Shimasaki H, Mizutani M, Mori Y, Maeda K, Deguchi A. Effect of aging on thermoregulatory and subjective responses during a 15-minute bath at 41 °C. J Hum Environ Syst. 2020;23:45–53.
123. Tochihara Y. A review of Japanese-style bathing: its demerits and merits. J Physiol Anthropol. 2022;41:5.
124. Kanda K, Tsuchiya J, Seto M, Ohnaka T, Tochihara Y. Thermal conditions in the bathroom in winter and summer, and physiological responses of the elderly during bathing. Jpn J Hyg. 1995;50:595–603.
125. Kataoka Y, Yoshida F. The change of hemodynamics and heart rate variability on bathing by the gap of water temperature. Biomed Pharmacother. 2005;59:S92–9.
126. Allison TG, Maresh CM, Armstrong LE. Cardiovascular responses in a whirlpool bath at 40 degrees C versus user-controlled water temperatures. Mayo Clin Proc. 1998;73:210–5.
127. Ono J, Hashiguchi N, Sawatari H, Ohkusa T, Miyazono M, Son SY, et al. Effect of water bath temperature on physiological parameters and subjective sensation in older people. Geriatr Gerontol Int. 2017;17:2164–70.
128. Nagasawa Y, Komori S, Sato M, Tsuboi Y, Umetani K, Watanabe Y, et al. Effects of hot bath immersion on autonomic activity and hemodynamics—comparison of the elderly patient and the healthyyoung—. Jpn Circ J. 2001;65:587–92.
129. Tai Y, Obayashi K, Yamagami Y, Yoshimoto K, Kurumatani N, Nishio K, et al. Hot-water bathing before bedtime and shorter sleep onset latency are accompanied by a higher distal-proximal skin temperature gradient in older adults. J Clin Sleep Med. 2021;17:1257–2166.
130. Takasaki Y, Nagai Y, Inoue K, Maki M, Ohnaka T, Tochihara Y. Bathing habits of the elderly in winter and factors affecting regional differences in bathing date rate (in Japanese). J Hum Living Environ. 2011;18:99–106.
131. Kanda K, Ohnaka T, Tochihara Y, Tsuzuki K, Shodai Y, Nakamura K. Effects of the thermal conditions of the dressing room and bathroom on physiological responses during bathing. Appl Hum Sci. 1996;15:19–24.
132. Tochihara Y, Hashiguchi N, Yadoguchi I, Kaji Y, Shoyama S. Effects of room temperature on physiological and subjective responses to bathing in the elderly. J Hum Environ Syst. 2012;15:13–9.
133. Masuda Y, Marui S, Kato I, Fujiki M, Nakada M, Nagashima K. Thermal and cardiovascular responses and thermal sensation during hot-water bathing and the influence of room temperature. J Therm Biol. 2019;82:83–9.
134. Taylor NA, Allsopp NK, Parkes DG. Preferred room temperature of young vs aged males: the influence of thermal sensation, thermal comfort, and affect. J Gerontol A Biol Sci Med Sci. 1995;50:M216–21.

International Standards for the Thermal Environment

10

Hajime Oi

10.1 Overview

The International Organization for Standardization (ISO), based in Geneva, Switzerland, is the world's largest standards development organization. ISO develops and publishes international standards in all areas of industry, except electrical/electronic engineering, which is the responsibility of the International Electrotechnical Commission (IEC), and telecommunications, which is the responsibility of the International Telecommunication Union (ITU).

As of 2025, 173 nations participate in ISO, and 25,768 standards have been published [1]. ISO develops standards in Technical Committees (TCs). The first Technical Committee, TC1, which dealt with screw threads, was established in 1947. Currently 274 TCs [2] exist, working in each industrial sector. The actual standardization work is carried out in Sub Committees (SCs) and Working Groups (WGs), both organized under each TC.

International standards related to the thermal environment are developed by ISO/TC159/SC5/WG1. TC159 has responsibility for "ergonomics." SC5, one of the four active SCs in TC159, is responsible for "Ergonomics of the physical environment." WG1, one of the four active WGs in SC5, is responsible for "thermal environments." The scope of TC159/SC5/WG1 is to develop standards on human responses to the thermal environment, effects of the thermal environment on humans (health, safety, comfort, and performance), and methods of evaluation (indices, surveys, and theoretical and analytical methods).

ISO/TC159/SC5/WG1 has developed and published 21 standards up to 2025 February. The document numbers, latest edition numbers, years of publication, and titles are listed in Table 10.1. Each document can be purchased from the ISO web store [3].

Some standards published by TC159/SC5/WG1 are entitled "ISO/TS," instead of "ISO," as shown in Table 10.1. TS indicates Technical Specifications. TS is published when the subject in question is still under development or when for any other reason, a future but not immediate possibility exists of an agreement to publish an International Standard [4]. The TS is regularly reviewed by TC159/SC5, and when possible to achieve the agreement necessary for publication as a full International Standard (IS) to replace the TS, it is converted to an IS.

In addition to the ISO standards, some other standards for thermal environments are used worldwide, including ASHRAE Standard 55. The American Society of Heating, Refrigerating and Air-Conditioning Engineers (ASHRAE) publishes standards for Heating, Ventilation, and

H. Oi (✉)
Integrated CAE Department, Nissan Motor Co., Ltd., Atsugi, Japan
e-mail: h-oi@mail.nissan.co.jp

Y. Tochihara (ed.), *The Thermal Environment*, https://doi.org/10.1007/978-981-96-3009-7_10

Table 10.1 List of ISO standards developed by TC159/SC5/WG1

Document No.	Edition	Year	Title
ISO 7243	ed.3	2017	Ergonomics of the thermal environment—Assessment of heat stress using the WBGT (wet bulb globe temperature) index
ISO 7726	ed.2	1998	Ergonomics of the thermal environment—Instruments for measuring physical quantities
ISO 7730	ed.3	2005	Ergonomics of the thermal environment—Analytical determination and interpretation of thermal comfort using calculation of the PMV and PPD indices and local thermal comfort criteria
ISO 7933	ed.3	2023	Ergonomics of the thermal environment—Analytical determination and interpretation of heat stress using calculation of the predicted heat strain
ISO 8996	ed.3	2021	Ergonomics of the thermal environment—Determination of metabolic rate
ISO 9886	ed.2	2004	Ergonomics—Evaluation of thermal strain by physiological measurements
ISO 9920	ed.2	2007	Ergonomics of the thermal environment—Estimation of thermal insulation and water vapour resistance of a clothing ensemble
ISO 10551	ed.2	2019	Ergonomics of the physical environment—Subjective judgement scales for assessing physical environments
ISO 11079	ed.1	2007	Ergonomics of the thermal environment—Determination and interpretation of cold stress when using required clothing insulation (IREQ) and local cooling effects
ISO 11399	ed.1	1995	Ergonomics of the thermal environment—Principles and application of relevant International Standards
ISO 12894	ed.1	2001	Ergonomics of the thermal environment—Medical supervision of individuals exposed to extreme hot or cold environments
ISO 13731	ed.1	2001	Ergonomics of the thermal environment—Vocabulary and symbols
ISO 13732-1	ed.1	2006	Ergonomics of the thermal environment—Methods for the assessment of human responses to contact with surfaces—Part 1: Hot surfaces
ISO/TS 13732-2	ed.1	2001	Ergonomics of the thermal environment—Methods for the assessment of human responses to contact with surfaces—Part 2: Human contact with surfaces at moderate temperature
ISO 13732-3	ed.1	2005	Ergonomics of the thermal environment—Methods for the assessment of human responses to contact with surfaces—Part 3: Cold surfaces
ISO/TS 14505-1	ed.1	2007	Ergonomics of the thermal environment—Evaluation of thermal environments in vehicles—Part 1: Principles and methods for assessment of thermal stress
ISO 14505-2	ed.1	2006	Ergonomics of the thermal environment—Evaluation of thermal environments in vehicles—Part 2: Determination of equivalent temperature
ISO 14505-3	ed.1	2006	Ergonomics of the thermal environment—Evaluation of thermal environments in vehicles—Part 3: Evaluation of thermal comfort using human subjects
ISO 14505-4	ed.1	2021	Ergonomics of the thermal environment—Evaluation of thermal environments in vehicles—Part 4: Determination of the equivalent temperature by means of a numerical manikin
ISO 15265	ed.1	2004	Ergonomics of the thermal environment—Risk assessment strategy for the prevention of stress or discomfort in thermal working conditions
ISO 15743	ed.1	2008	Ergonomics of the thermal environment—Cold workplaces—Risk assessment and management

Air Conditioning (HVAC) systems that are often referenced internationally. ASHRAE Standard 55—Thermal Environmental Conditions for Human Occupancy—specifies conditions for acceptable thermal environments and comfort. The American Conference of Governmental Industrial Hygienists (ACGIH) publishes Threshold Limit Values (TLVs®). TLVs for heat stress and strain, and cold stress are described in the section on TLVs for Physical Agents (TLV-PA). These are also widely referenced internationally.

These standards can be divided into five groups: hot environment, moderate environment, cold environment, application, and supporting standards (Table 10.2). The following sections

Table 10.2 Grouping of international standards

Group	ISO		OTHERS
Hot environments	7243 (WBGT) 7933 (PHS/SWreq) 13732-1 (surface)	9886 (physiology)	ACGIH (heat stress and strain)
Moderate environments	7730 (PMV/PPD) 13732-2 (surface)	9886 (physiology)	ASHRAE Standard 55
Cold environments	11079 (IREQ/WCI) 13732-3 (surface)	9886 (physiology)	ACGIH (cold stress)
Applications	[Vehicle] 14505-1 (principles), 14505-2 (teq), 14505-3 (human subjects), 14505-4 (numerical manikin)		
	[working conditions] 15265 (risk assessment), 15743 (cold)		
Supporting standards	11399 (principles), 7726 (instruments), 8996 (metabolic rate), 9920 (clothing), 10551 (subjective scales), 12894 (subject screening), 13731 (vocabulary and units)		

introduce the ISO standards, ASHRAE Standard 55, and ACGIH TLVs by group.

10.2 Hot Environments

10.2.1 ISO 7243:2017 Ergonomics of the Thermal Environment: Assessment of Heat Stress Using the WBGT (Wet Bulb Globe Temperature) Index

ISO 7243 [5] provides a screening method for the assessment of heat stress for a person exposed to a hot environment. It applies to the assessment of the effect of heat on a person during total exposure over a working day (up to 8 h). It does not apply to very short exposures to heat. It can be applied to the assessment of both indoor and outdoor environments and to male and female adults who are fit for work.

ISO 7243 uses the Wet Bulb Globe Temperature (WBGT) index to assess heat stress. The WBGT is a simple index of heat stress that is considered together with metabolic rate. In an environment without solar load, the WBGT can be calculated from the natural wet bulb temperature (t_{nw}) and the black globe temperature (t_g). Where the sensors are influenced by solar load, the weighting of the black globe temperature is reduced by including the air temperature (t_a). WBGT can assess heat stress by considering the effects of air temperature, humidity, air velocity, and radiation, which are the four basic thermal factors.

WGBT (°C) is calculated by the following equations:

– Without solar load:

$$\mathrm{WBGT} = 0.7t_{nw} + 0.3t_g \text{ and} \quad (10.1)$$

– With solar load:

$$\mathrm{WBGT} = 0.7t_{nw} + 0.2t_g + 0.1t_a, \quad (10.2)$$

where

t_{nw}: natural wet-bulb temperature (°C)
t_g: black globe temperature (°C) and
t_a: air temperature (°C)

The WBGT index was originally developed using cotton clothes (clothing insulation, 0.6 clo; permeability index, 0.38) as the reference clothing. Wearing different clothing is likely to have a different effect on heat stress levels. Clothing adjustment values (CAVs) are provided to correct for differences in clothing. For example, if nonwoven polyolefin Coveralls as a single layer are worn, CAV = 2. The CAV is added to the measured WBGT to give an effective WBGT ($\mathrm{WBGT_{eff}}$), which represents an estimated heat stress with the actual clothing worn.

$$\mathrm{WBGT_{eff}} = \mathrm{WBGT} + \mathrm{CAV}, \quad (10.3)$$

where

$\mathrm{WBGT_{eff}}$: effective WBGT (°C) and
CAV: clothing adjustment values (°C).

Table 10.3 WBG_{Teff} reference values

Metabolic rate	Metabolic rate (W)	WBGT reference limit for persons acclimatized to heat (°C)	WBGT reference limit for persons unacclimatized to heat (°C)
Class 0: Resting metabolic rate	115	33	32
Class 1: Low metabolic rate	180	30	29
Class 2: Moderate metabolic rate	300	28	26
Class 3: High metabolic rate	415	26	23
Class 4: Very high metabolic rate	520	25	20

The reference values (exposure limits) are given in Table 10.3.

WBGT values should normally be measured at the abdomen (ISO 7726), which is 1.1 m high when standing and 0.6 m high when sitting. If the parameters in the environment are not spatially uniform, the measurement should be taken at the position where the heat stress is the greatest. If the parameters are not constant over time, a representative mean value shall be determined. A method for calculating the time-weighted mean is described in this standard.

The annexes of this standard describe the required specifications for the globe temperature sensor and the natural wet-bulb temperature sensor. For the globe temperature sensor, that with a 150-mm diameter should be used. However, globe temperature sensors of other sizes can be used if a valid correction is made. The correction method is also given in an annex. In addition, a method for estimating the natural wet-bulb temperature using a relative humidity sensor instead of the natural wet bulb sensor is provided; thus, the WBGT can be evaluated by electrical sensors.

10.2.2 ISO 7933:2023 Ergonomics of the Thermal Environment: Analytical Determination and Interpretation of Heat Stress Using Calculation of the Predicted Heat Strain

ISO 7933 [6] provides a method for the analytical evaluation and interpretation of the thermal stress experienced by a subject in a hot environment. It provides a method based on the predicted heat stress (PHS) model. This method is based on an analysis of the heat exchange between a person and the environment. In hot environments, ISO proposes the use of the WBGT heat stress index, defined in ISO 7243, as a screening method to assess the presence or absence of heat stress. If the WBGT exceeds the reference values, or if a more detailed analysis is required, ISO 7933 provides methods to make a more accurate assessment of stress, to determine the allowable duration of work in the condition, and to optimize protective methods.

The required evaporative heat flow, E_{req}, is the evaporative heat flow required to maintain the body heat balance. It is given by the following equation:

$$E_{req} = M - W - C_{res} - E_{res} - C - R, \quad (10.4)$$

where

E_{req}: required evaporative heat flow at the skin surface (W/m^2)

M: metabolic rate (W/m^2)

W: effective mechanical power (W/m^2)

C_{res}: respiratory convective heat flow (W/m^2)

E_{res}: respiratory evaporative heat flow (W/m^2)

C: convective heat flow (W/m^2) and

R: radiative heat flow (W/m^2)

Each parameter in the equation can be calculated from thermal factors, such as air temperature, radiant temperature, air velocity and humidity, metabolic rate, and clothing insulation.

The required skin wettedness, W_{req}, is the ratio between the required evaporative heat flow and the maximum evaporative heat flow at the skin surface.

$$W_{req} = E_{req} / E_{max}, \quad (10.5)$$

where

W_{req}: required skin wettedness (N.D.) and

E_{max}: maximum evaporative heat flow at the skin surface (W/m^2)

The calculation of the required sweat rate is based on the required evaporative heat flow, taking into account the evaporative efficiency of sweating, r_{req}, as follows:

$$S_{wreq} = E_{req} / r_{req}, \tag{10.6}$$

where

S_{wreq}: required sweat rate (W/m^2) and

r_{req}: required evaporative efficiency of sweating (N.D.)

The interpretation of the calculated values is as follows. The required sweat rate, S_{wreq}, cannot exceed the maximum sweat rate, S_{wmax}. The required skin wettedness, w_{req}, cannot exceed the maximum skin wettedness, w_{max}. In the case of nonequilibrium of the thermal balance, the rectal temperature increase should be limited at a maximum value, $t_{re,max.}$ Finally, whatever the thermal balance, the water loss should be restricted to a value, D_{max}. Annex B includes reference values for the S_{wmax}, w_{max}, $t_{re,max}$, and D_{max}. Different values are presented for acclimatized and nonacclimatized subjects.

10.2.3 ISO 9886:2004 Ergonomics: Evaluation of Thermal Strain by Physiological Measurements

ISO 9886 [7] provides methods for measuring and interpreting the following physiological parameters: body core temperature, skin temperature, heart rate, and body-mass loss. When the WBGT (ISO 7243) exceeds the reference value or when the heat stress and strain evaluated by ISO 7933 are extremely high, the individual's thermal strain should be monitored directly. In these cases or for other reasons such as research, this standard may be useful. Please note that this standard can be applied not only to hot environments but also to other environments, including moderate and cold environments.

The body core temperature can be approximated by measuring the temperature at different points on the body. The methods of measurement of the following temperatures are explained in this standard: esophageal temperature, rectal temperature, intra-abdominal temperature, oral temperature, tympanic temperature, auditory canal temperature, and urine temperature.

This standard describes the local skin temperature measured at a specific point on the body surface and the mean skin temperature over the entire body surface. The mean skin temperature can be estimated by weighting an ensemble of local skin temperatures. The 4-point, 8-point, and 14-point methods of estimating mean skin temperature are described.

Heart rate over a time interval t (min) is defined as $HR = n/t$, where n is the number of heartbeats observed during the time interval. At any given time, the heart rate, HR, can be considered as the sum of several components that are not independent of each other. In the context of this standard, only the thermal component, ΔHR_T, is considered.

$$HR = HR_0 + \Delta HR_M + \Delta HR_s + \Delta HR_T + \Delta HR_N + \Delta HR_\varepsilon, \tag{10.7}$$

where

HR: heart rate (beats/min)

HR_0: average heart rate (beats/min) of the subject at rest while sitting under neutral conditions

ΔHR_M: increase in heart rate (beats/min) linked with work metabolism

ΔHR_s: increase in heart rate (beats/min) linked with static exertion

ΔHR_T: increase in heart rate (beats/min) connected with the thermal strain experienced by the subject

ΔHR_N: increase in heart rate (beats/min) due to psychological factors and

ΔHR_ε: residual component in heart rate (beats/min) connected with rhythm of breathing circadian rhythm, etc.

The body-mass loss of a person during a given time interval is the difference between the body mass measured at the beginning and at the end of the interval. The measurement should consider the sweat remaining in the clothing. In the context of this standard, the sweat loss and the net water balance of the body are considered.

10.2.4 ISO 13732-1:2006 Ergonomics of the Thermal Environment: Methods for the Assessment of Human Responses to Contact with Surfaces—Part 1: Hot Surfaces

ISO 13732-1 [8] provides temperature threshold values for burns that occur when human skin is in contact with a hot solid surface. It also provides a method for assessing the risk of burning, protective measures, and guidance for setting surface temperature limit values. This standard addresses the risk when a part of the skin is in contact with the surface for ≥0.5 s. It is not applicable when a large area of the skin (approximately >10% of the whole body or >10% of the head) can be in contact with the hot surface.

In the case of short contact (contact periods between 0.5 and 10 s), the burn thresholds are expressed in graphs as a function of the contact period. The relationship between contact period and risk is different across materials. The graphs for uncoated metal, coated metal, ceramic/glass/stone, plastics, and wood are given in this standard.

Where the contact period is ≥1 min, burn thresholds are defined in temperature. The data for 1 min, 10 min, and ≥8 h are given in this standard, as shown in Table 10.4. When the contact period is 1 min, the threshold temperature is still different depending on the material (uncoated metal, coated metal, ceramic/glass/stone, plastics, or wood). However, when the contact period is 10 min and ≥8 h, the threshold temperatures are independent of the contact material. Threshold temperatures are 48 °C for a contact period of 10 min and 43 °C for a contact period of ≥8 h.

Table 10.4 Burn thresholds for contact periods of 1 min and longer

	Burn thresholds (°C) for contact periods of		
	1 min	10 min	>= 8 h
Uncoated metal	51	48	43
Coated metal	51	48	43
Ceramics, glass, and stone material	56	48	43
Plastics	60	48	43
Wood	60	48	43

10.2.5 ACGIH-TLV 2022 (Heat Stress and Strain) 2022 TLVs and BEIs: Based on the Documentation of the Threshold Limit Values for Chemical Substances and Physical Agents and Biological Exposure Indices

ACGIH-TLV [9] is published by the American Conference of Governmental Industrial Hygienist (ACGIH). TLVs and BEIs indicate "Threshold Limit Values and Biological Exposure Indices." TLV® for heat stress and strain is described in the Threshold Limit Values for Physical Agents (TLV-PA) section. It is intended for use in risk assessment for the safety and health of workers and is used worldwide.

In this standard, WBGT is used to screen the risk of heat stress exposure. The definition of WBGT and the equation for calculating WBGT are the same as those in ISO 7243:2017. The heat stress exposure criteria are shown in Table 10.5.

These criteria were developed for a typical work uniform of a long-sleeve shirt and pants. If more clothing is required, the clothing adjustment values given in this standard should be added when calculating the WBGT.

The TLV represents conditions to which almost all heat-acclimatized, adequately hydrated, unmedicated, healthy workers may be repeatedly exposed without adverse health effects. The action limit (AL) is similarly protective for unacclimatized workers and represents

Table 10.5 Criteria of WBGT (°C) for TLV and action limit for heat stress exposure at each activity level (W) and allocation of work in a cycle of working and recovery (%)

Working ratio in a cycle	WBGT criteria							
	TLV				Action limit			
	Light	Moderate	Heavy	Very heavy	Light	Moderate	Heavy	Very heavy
	180 W	300 W	415 W	520 W	180 W	300 W	415 W	520 W
75–100%	31.0	28.0	–	–	28.0	25.0	–	–
50–70%	31.0	29.0	27.5	–	28.5	26.0	24.0	–
25–50%	32.0	30.0	29.0	28.0	29.5	27.0	25.5	24.5
0–25%	32.5	31.5	30.5	30.0	30.0	29.0	28.0	27.0

conditions for which a heat stress management program should be considered.

If the heat stress cannot be assessed by WBGT or if WBGT exceeds TLV or action limit, detailed analysis, such as PHS (see ISO 7933) or heat strain monitoring, can be used. This standard provides guidance on limit of heat strain. The following status may indicate excessive heat strain, and an individual's exposure to heat stress should be discontinued if any of the following occur:

– Peak heart rate:	greater than (180-age) bpm for several min
– Recovery heart rate:	greater than 120 bpm at 1 min after a peak work effort
– Body core temperature:	greater than 38 °C (38.5 °C for selected and acclimatized personnel)
– Symptoms:	sudden and severe fatigue, nausea, dizziness, or light-headedness
– Sweating:	profuse sweating over hours
– Weight loss:	greater than 1.5% of body weight
– Urinary sodium excretion:	less than 50 mmol/24 h

10.3 Moderate Environments

10.3.1 ISO 7730:2005 Ergonomics of the Thermal Environment: Analytical Determination and Interpretation of Thermal Comfort Using Calculation of the PMV and PPD Indices and Local Thermal Comfort Criteria

ISO 7730 [10] presents methods for predicting the overall thermal sensation and the level of discomfort (thermal dissatisfaction) experienced by people exposed to moderate thermal environments. It enables the analytical determination and interpretation of thermal comfort by calculating the predicted mean vote (PMV) and predicted percentage of dissatisfied (PPD).

The PMV is an index that predicts the mean of the votes of a large group of people on the 7-point thermal sensation scale (+3: hot, +2: warm, +1: slightly warm, 0: neutral, −1: slightly cool, −2: cool, −3: cold) based on the thermal balance of the human body. The PMV is calculated from air temperature, mean radiant temperature, humidity, air velocity, metabolic rate, and thermal insulation of clothing. The equations for calculating the PMV are described in detail in clause 8 of this standard. The advantage of PMV is that the result of the calculation directly indicates the thermal sensation on the 7-point scale shown above. PMV should only be used in moderate environments where the value of PMV is between −2 and +2 and the thermal factors are not extreme.

The PPD is an index that predicts the percentage of thermally dissatisfied people who feel too cool or too warm (voting other than −1, 0, 1). The PPD is calculated from the PMV. Therefore, once the PMV has been determined, the PPD can be calculated immediately using the following equation, which has been established based on experiments with 1300 subjects:

$$PPD = 100 - 95 \cdot \exp\begin{pmatrix} -0.03353 \cdot PMV^4 \\ -0.2179 \cdot PMV^2 \end{pmatrix}. \quad (10.8)$$

Figure 10.1 shows the relationship between PMV and PPD. This result shows that even when PMP = 0, a proportion of people (5%) report discomfort. This means that no thermal environment can satisfy everyone.

PMV and PPD express hot and cold discomfort for the whole body. However, thermal discomfort can also be caused by unwanted local cooling or heating. This standard also provides methods for assessing local thermal discomfort.

The most common cause of local discomfort is draughts. The effect is expressed as the percentage of people predicted to be bothered by draughts. The method of calculating the draught rate (DR) is given in this standard.

Local discomfort can also be caused by an abnormally high vertical temperature difference between the head and ankles. The calculation method of the percentage of dissatisfied (PD) as a function of the vertical air temperature difference between the head and the ankles is provided in this standard.

If the floor is too warm or too cool, the occupants may feel uncomfortable due to the thermal sensation of their feet. This standard includes a graph showing the percentage dissatisfied as a function of the floor temperature, based on studies with standing and/or sedentary people. In addition, similar values can be used for people sitting or lying on the floor. The methods for calculating the PD for these postures are also provided in this standard.

Radiant asymmetry can also cause discomfort. People are most sensitive to radiant asymmetry caused by warm ceilings or cool walls (windows). The methods for calculating the PD as a function of the radiant asymmetry for the warm ceiling, cool wall, cool ceiling, and warm ceiling are provided in this standard.

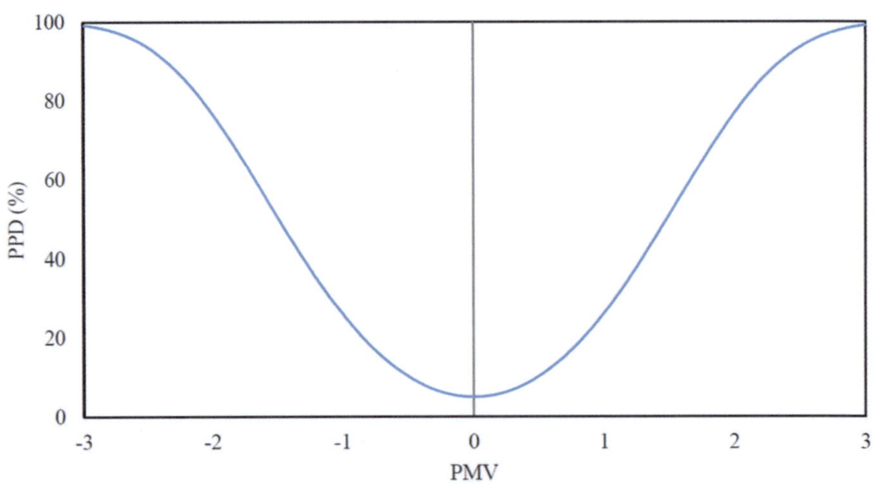

Fig. 10.1 Relationship between PMV and PPD

10.3.2 ISO/TS 13732-2:2001 Ergonomics of the Thermal Environment: Methods for the Assessment of Human Responses to Contact with Surfaces—Part 2: Human Contact with Surfaces at Moderate Temperature

ISO/TS 13732-2 [11] provides principles and methods for predicting the thermal sensation and level of discomfort experienced by humans when parts of the body are in contact with solid surfaces at moderate surface temperatures (approximately 10–40 °C). Skin contact with solid surfaces can cause thermal discomfort depending on the part of the body in contact, the temperature of the material, and the type of material. Bare skin in contact with metal at room temperature may feel cold, whereas contact with wood may feel comfortable. This difference should be taken into account when assessing thermal comfort or designing hand tools.

This standard includes a graph showing the relationship between the temperature of the material and the thermal sensation of the hand on initial contact with the different types of materials (wood, plastic, steel, and aluminium). At a surface temperature of 36 °C, the contact sensation is the same for all materials, and the thermal sensation of the hand is neutral because the temperature of the materials is the same as that of the skin. In contact with wood/plastic, the difference in thermal sensation is small according to the temperature of the material. However, in contact with steel/aluminium, the difference is large. This contrast is caused by the difference in thermal diffusivity. This standard also includes a diagram showing the relationship between the thermal sensation of the bare hand, the thermal diffusivity of the material, and the temperature of the material. Using this diagram, the thermal sensation can be predicted from the thermal diffusivity and temperature of the material.

This standard also includes a graph showing the relationship between floor temperatures and local discomfort as a percentage of dissatisfied people standing with normal shoes (sole thickness between 5 and 15 mm). At a surface temperature of around 25 °C, most people (>95%) report no discomfort. Seated people prefer a floor temperature about 1 °C higher, while standing people prefer 1 °C lower. This standard additionally provides design guidance for seated and standing people with shoes and for people seated on floors.

10.3.3 ASHRAE Standard 55-2020 Thermal Environmental Conditions for Human Occupancy

ASHRAE Standard 55 [12] is an American National Standard published by the American Society of Heating, Refrigerating and Air-Conditioning Engineers (ASHRAE). However, this standard is used worldwide, not just in America, to specify conditions for acceptable thermal environments for the majority occupants.

This standard basically assesses thermal environments by PMV, like ISO 7730. However, it differs from ISO 7730 in two important ways. First, the "Elevated air speed comfort zone method" is used for assessing the high air speed environment. This method uses the adjusted PMV, corrected by standard effective temperature (SET). Second, an adaptive model can be applied when assessing buildings with occupant-controlled naturally conditioned space. The following sections explain the two methods.

10.3.3.1 Elevated Air Speed Comfort Zone Method

In the low air speed environment (air speed <0.20 m/s), the environment can be assessed by the PMV (see ISO 7730). If the PMV is between −0.5 and 0.5, the thermal environment is acceptable. If the air speed is >0.20 m/s, the "Elevated air

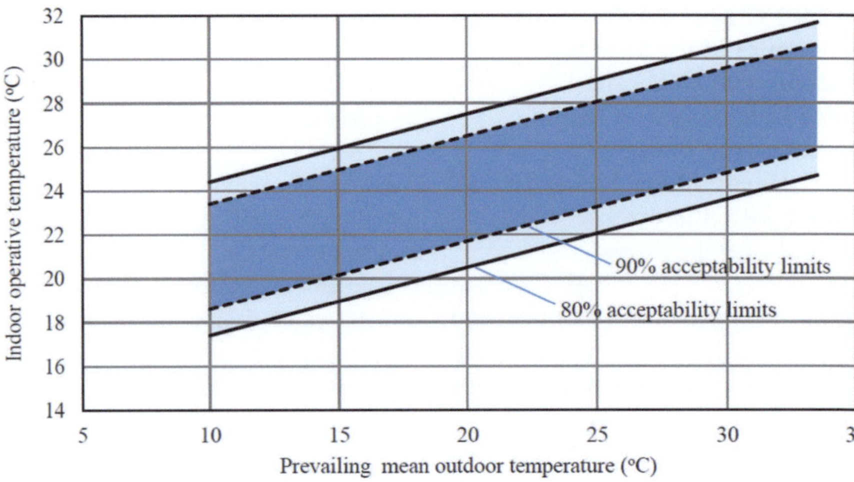

Fig. 10.2 Acceptable indoor operative temperature ranges by the adaptive model

speed comfort zone method" should be used. In this method, the cooling effect (CE) is first calculated from the SET to calculate the adjusted PMV. SET is a kind of thermal index to evaluate perceived temperature. The computer program for calculating SET is shown in the appendix of this standard. CE is defined by the following equation:

$$\mathrm{SET}\left(t_{\mathrm{a}}, t_{\mathrm{r}}, v_{\mathrm{elev}}, *\right) = \mathrm{SET}\begin{pmatrix} t_{\mathrm{a}} - \mathrm{CE}, t_{\mathrm{r}} \\ - \mathrm{CE}, v_{\mathrm{still}}, * \end{pmatrix}, \quad (10.9)$$

where

t_a: average air temperature (°C)

t_r: mean radiant temperature (°C)

v_{elev}: elevated air speed (average air speed in the real environment) (m/s)

v_{still}: air speed in the still air environment = 0.1 m/s

CE: cooling effect (°C) and

: Other parameters. The important variables are listed explicitly, while the invariant parameters are indicated by ""

CE is the corrected amount of air temperature and radiant temperature at which SET is equivalent in a still air environment. The CE is used to calculate the adjusted PMV. The adjusted PMV (PMV_{adj}) is calculated using the following equation:

$$\mathrm{PMV}_{\mathrm{adj}} = \mathrm{PMV}\left(t_{\mathrm{a}} - \mathrm{CE}, t_{\mathrm{r}} - \mathrm{CE}, v_{\mathrm{still}}, *\right). \quad (10.10)$$

10.3.3.2 Adaptive Model

The adaptive model can be used to assess buildings with occupant-controlled naturally conditioned space. The idea of the adaptive model is based on the adaptive hypothesis that the outdoor climate influences the indoor comfort because humans can adapt to different temperatures during different seasons. Figure 10.2 shows the relationship between the acceptable indoor operative temperature ranges (80% and 90%) and the prevailing outdoor temperature based on this standard. If the outdoor temperature is higher, a higher indoor temperature is acceptable.

10.4 Cold Environments

10.4.1 ISO 11079:2007 Ergonomics of the Thermal Environment: Determination and Interpretation of Cold Stress when Using Required Clothing Insulation (IREQ) and Local Cooling Effects

ISO 11079 [13] specifies methods and strategies for assessing thermal stress associated with exposure to cold environments. This standard assesses thermal stress using the required clothing insulation index (IREQ), which is the clothing insulation required to maintain body heat balance at defined levels of physiological strain. In addition to whole body assessment by IREQ, wind chill temperature (t_{WC}) is proposed to assess local cooling effects.

The calculation of the IREQ is based on the human heat exchange with the environment. The IREQ is calculated using the following equation:

$$\begin{aligned} \text{IREQ} &= (t_{sk} - t_{cl}) / (R + C) \\ &= (t_{sk} - t_{cl}) / \begin{pmatrix} M - W - E_{res} \\ -C_{res} - E - S \end{pmatrix}, \end{aligned} \quad (10.11)$$

where

t_{sk}: mean skin temperature (°C)

t_{cl}: clothing surface temperature (°C)

R: radiative heat flow (exchange) (W/m^2)

C: convective heat flow (exchange) (W/m^2)

M: metabolic rate (W/m^2)

W: effective mechanical power (W/m^2)

E_{res}: respiratory evaporative heat flow (loss) (W/m^2)

C_{res}: respiratory convective heat flow (loss) (W/m^2)

E: evaporative heat flow (exchange) at the skin (W/m^2) and

S: body heat storage rate (W/m^2)

The equation expresses that the value of IREQ increases in cold environments where the body's cooling power is high. When the air temperature is lower, or the wind velocity is higher, IREQ is higher. In addition, under the same conditions, the IREQ value decreases with increased heat production due to high metabolic rates. IREQ is defined at the following two levels of physiological strain:

(a) IREQ_{min} defines a minimal thermal insulation required to maintain body thermal equilibrium at a subnormal level of mean body temperature; and.

(b) $\text{IREQ}_{neutral}$ is defined as the thermal insulation required to provide conditions of thermal neutrality, i.e., thermal equilibrium maintained at a normal level of mean body temperature.

The relevant physiological criteria are given in Annex B.

If the value of clothing insulation, corrected for the effects of wind penetration and activity, is between $\text{IREQ}_{neutral}$ and IREQ_{min}, no action is required. If the value of clothing insulation is less than IREQ_{min}, a risk of progressive cooling of the body is present. In this case, the clothing insulation must be increased, or the exposure time must be limited to less than duration limited exposure (DLE). The method of calculation of DLE is also described in this standard.

In addition to the whole-body assessment by the IREQ, wind chill temperature (t_{WC}) is proposed to assess the effect of local skin cooling by cold wind. The t_{WC} defines the ambient temperature at which a wind speed of 4.2 km/h produces the same cooling power (sensation) as the actual environmental conditions. The wind chill temperature (°C) is determined by the following formula:

$$\begin{aligned} t_{WC} &= 13.12 + 0.6215 \cdot t_a - 11.37 \cdot v_{10}^{0.16} \\ &+ 0.3965 \cdot t_a \cdot v_{10}^{0.16}, \end{aligned} \quad (10.12)$$

where

t_a: air temperature (°C) and

v_{10}: wind speed measured 10 m above ground level (m/s).

Wind velocity (v_{10}) is defined as the standard meteorological value measured at 10 m above the ground. This value is obtained from weather stations and weather forecasts. Therefore, the wind chill temperature can be estimated without measuring on site. Table 10.6 shows the relationship between wind chill temperature and freezing time of exposed skin.

Table 10.6 Wind chill temperature (t_{WC}) and freezing time of exposed skin

Classification of risk	t_{WC} (°C)	Effect
1	−10 to −24	Uncomfortably cold
2	−25 to −34	Very cold, risk of skin freezing
3	−35 to −59	Bitterly cold, exposed skin may freeze in 10 min
4	−60 and colder	Extremely cold, exposed skin may freeze within 2 min

10.4.2 ISO 13732-3:2005 Ergonomics of the Thermal Environment: Methods for the Assessment of Human Responses to Contact with Surfaces—Part 3: Cold Surfaces

ISO 13732-3 [14] provides methods and data for assessing the risk of cold injury and other adverse effects from bare hand−/finger-skin contact with a cold surface. It can be used to set temperature limits for cold solid surfaces. The data presented in this standard are not limited to the hands but can be applied generally to healthy human skin of male and female adults.

When bare skin comes into contact with cold surfaces, the skin temperature drops, causing pain, numbness, and frostbite. When the contact skin temperature drops to around 15 °C, a subjective sensation of pain is experienced on the contacting skin. When the contact skin temperature drops to around 7 °C, the sensory receptors of the contacting skin are blocked, and numbness occurs. When the contact skin temperature falls below 0 °C, the contacting skin tissue freezes, and frostbite occurs.

This standard provides the surface temperature data for cooling thresholds from the ergonomics database. The thresholds are shown by the relationship between contact period and surface temperature. The relationship varies with material and contact conditions. This standard provides pain, numbness, and frostbite thresholds for aluminium, stone, steel, nylon, and wood by finger touch. In addition, pain thresholds by hand gripping are provided.

10.4.3 ACGIH-TLV 2022 (Cold Stress) 2022 TLVs and BEIs: Based on the Documentation of the Threshold Limit Values for Chemical Substances and Physical Agents and Biological Exposure Indices

ACGIH-TLV [9] also describes Threshold Limit Values (TLVs®) for cold stress in the Threshold Limit Values for Physical Agents (TLV-PA) section, as it does for heat stress. The Cold Stress TLVs are designed to protect workers from cold stress and describe exposures to cold working conditions to which it is believed that almost all workers can be repeatedly exposed without adverse health effects. The main aim of the TLV is to prevent hypothermia and frostbite.

Hypothermia is defined as a core body temperature of <35 °C. If the core body temperature is <35 °C, blood pressure may be elevated. If it is <28 °C, which causes severe hypothermia, more severe symptoms are observed, and the risk of death increases as the core temperature drops. In an occupational setting, workers should be protected from exposure to cold so that the core body temperature does not fall below 36 °C. Cold, wet, and windy weather poses the greater risk of hypothermia. Whole-body protection must be provided in these conditions. This standard provides a graph showing the approximate amount of clothing insulation required in cold environments.

Frostbite occurs when the tissue temperature drops below 0 °C. ACGIH uses the Wind Chill Temperature (WCT) index to assess the risk of frostbite. The WCT integrates wind speed and air temperature to provide an estimate of the cooling power of the cold environment. The definition and calculation method of the WCT is the same

Table 10.7 Wind chill temperature index and frostbite time for exposed facial skin

		Air temperature (°C)											
		5	0	-5	-10	-15	-20	-25	-30	-35	-40	-45	-50
Wind speed (km/h)	5	4	-2	-7	-13	-19	-24	-30	-36	-41	-47	-53	-58
	10	3	-3	-9	-15	-21	-27	-33	-39	-45	-51	-57	-63
	15	2	-4	-11	-17	-23	-29	-35	-41	-48	-54	-60	-66
	20	1	-5	-12	-18	-24	-30	-37	-43	-49	-56	-62	-68
	25	1	-6	-12	-19	-25	-32	-38	-44	-51	-57	-64	-70
	30	0	-6	-13	-20	-26	-33	-39	-46	-52	-59	-65	-72
	35	0	-7	-14	-20	-27	-33	-40	-47	-53	-60	-66	-73
	40	-1	-7	-14	-21	-27	-34	-41	-48	-54	-61	-68	-74
	45	-1	-8	-15	-21	-28	-35	-42	-48	-55	-62	-69	-75
	50	-1	-8	-15	-22	-29	-35	-42	-49	-56	-63	-69	-76
	55	-2	-8	-15	-22	-29	-36	-43	-50	-57	-63	-70	-77
	60	-2	-9	-16	-23	-30	-36	-43	-50	-57	-64	-71	-78
	65	-2	-9	-16	-23	-30	-37	-44	-51	-58	-65	-72	-79
	70	-2	-9	-16	-23	-30	-37	-44	-51	-58	-65	-72	-80
	75	-3	-10	-17	-24	-31	-38	-45	-52	-59	-66	-73	-80
	80	-3	-10	-17	-24	-31	-38	-45	-52	-60	-67	-74	-81

- Low risk of frostbite for most people
- Increasing risk of frostbite for most people in 10 to 30 minutes of exposure
- High risk for most people in 5 to 10 minutes of exposure
- High risk for most people in 2 to 5 minutes of exposure
- High risk for most people in 2 minutes of exposure or less

as that of the wind chill temperature (t_{WC}) defined in ISO 11079. In this standard, the relative risk of frostbite and the predicted times to freezing of exposed facial skin are presented in Table 10.7, while the classification and criteria are slightly different from those of the ISO 11079. The facial skin was chosen because this area of the body is not normally protected by clothing. Frostbite cannot occur if the air temperature is >0 °C. Wet skin exposed to the wind will cool even faster, and if the skin is wet and exposed to the wind, the air temperature used for the WCT calculation should be 10 °C lower than the actual air temperature.

10.5 Applications

10.5.1 ISO/TS 14505-1:2007 Ergonomics of the Thermal Environment: Evaluation of Thermal Environments in Vehicles—Part 1: Principles and Methods for Assessment of Thermal Stress

The ISO 14505 series defines the evaluation methods of the thermal environment in vehicles. The ISO 14505 series is different from other conventional evaluation standards for unspecified spaces, as it limits the evaluation space to the vehicle cabin (including ships, aircraft, etc.).

ISO/TS 14505-1 [15] is one of the standards in the ISO 14505 series. This part of ISO 14505 provides guidelines for the assessment of thermal stress in vehicles. It provides information on the assessment of hot, cold, and moderate thermal stress.

ISO 14505-1 describes that the conventional ISO 7243 (WBGT), 7933 (PHS), and 11079 (IREQ) should be used in hot and cold environments, and ISO 7730 (PMV) should be used in moderate environments in which the environment is uniform. However, vehicle cabin rarely has uniform thermal environments. For example, because vehicle air conditioning systems are powerful, the influence on each part of the occupant's body is different depending on the direction of the air outlet. In addition, the thermal environments in vehicle cabins are more likely to be nonuniform due to the influence of solar radiation.

For the detailed assessment of nonuniform thermal environments in vehicle cabins, the use of the equivalent temperature defined in ISO 14505-2 is recommended. The equivalent temperature is a real-feeling temperature based on heat loss from the human body. It can take into account the effects of air temperature, radiation, and air velocity to evaluate each part of the body. ISO 14505-2 provides the measurement methods of equivalent temperature, and ISO 14505-4 provides the calculation methods of equivalent temperature by computer simulation. The other method to evaluate the nonuniform thermal environment is the evaluation by subjects. Methods for subjective assessment of the thermal environment are described in ISO 14505-3. Figure 10.3

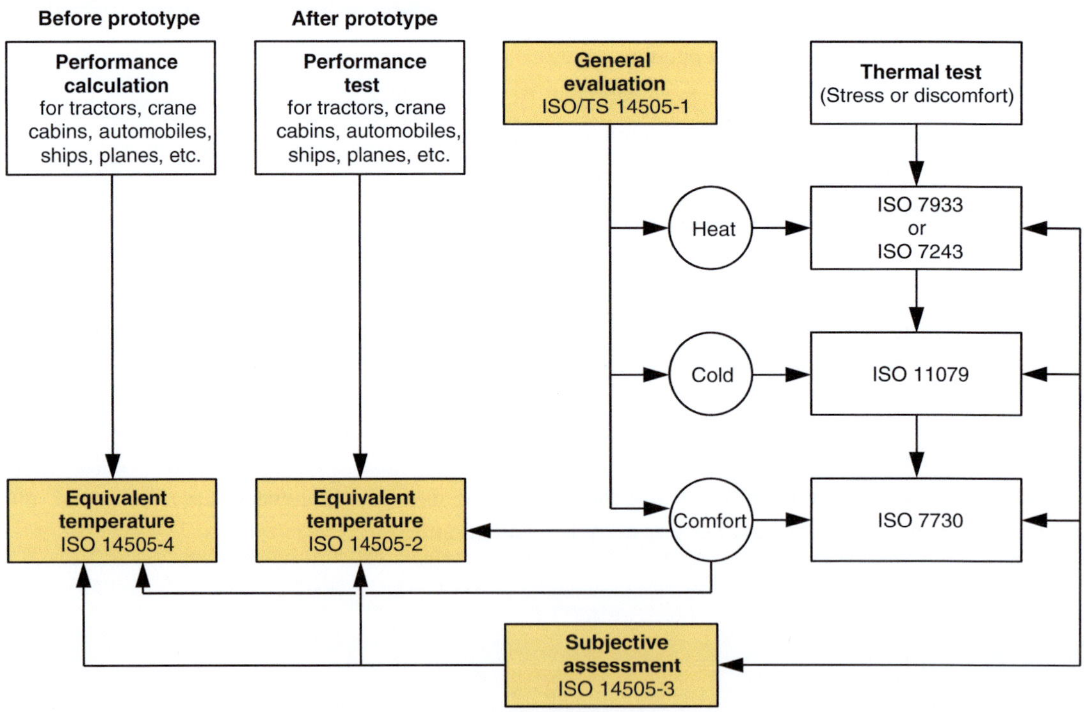

Fig. 10.3 Structure of the ISO 14505 series and related standards

shows the structure of the ISO 14505 series and related standards.

10.5.2 ISO 14505-2:2006 Ergonomics of the Thermal Environment: Evaluation of Thermal Environments in Vehicles—Part 2: Determination of Equivalent Temperature

ISO 14505-2 [16] is one of the standards in the ISO 14505 series. This part of ISO 14505 provides guidelines for the assessment of thermal conditions in a vehicle cabin by measurement. It is primarily intended for the assessment of thermal conditions in which deviations from thermal neutrality are relatively small.

ISO 14505-2 assesses the thermal environment using the equivalent temperature. ISO 14505-2 provides its measurement methods and interpretation of equivalent temperature in terms of thermal sensations. The equivalent temperature is defined as "temperature of a homogenous space, with mean radiant temperature equal to air temperature and zero air velocity, in which a person exchanges the same heat loss by convection and radiation as in the actual conditions under assessment." The equivalent temperature is determined by the following equation:

$$t_{eq} = t_{sk} - Q / h_{cal}, \quad (10.13)$$

where

t_{eq}: equivalent temperature (°C)

t_{sk}: skin surface temperature (°C)

Q: total heat loss from skin surface (W/m^2) and

h_{cal}: total heat transfer coefficient between skin surface and environment under the standard environment (W/m^2 °C)

Each parameter required to determine the equivalent temperature can be measured using heated measuring devices such as thermal manikins. The technical specifications required for the measuring devices are given in Annexes A and B. Examples of comfort zones at 16 body segments and for the whole body are given in Annex D. For example, in summer, wearing light clothing of 0.6 clo, the comfort zones of equivalent temperature at the face, chest, and foot are approximately 19–27, 22–28, and 21–25 °C, respectively.

10.5.3 ISO 14505-3:2006 Ergonomics of the Thermal Environment: Evaluation of Thermal Environments in Vehicles—Part 3: Evaluation of Thermal Comfort Using Human Subjects

ISO 14505-3 [17] is one of the standards in the ISO 14505 series. This part of ISO 14505 provides guidelines and a standard test method for the evaluation of thermal comfort in vehicles using subjective judgements. In vehicles, the method can be applied to both passengers and drivers if its application does not interfere with the safe operation of the vehicle.

The selection of human subjects is important for this assessment, and a random sampling of human subjects from the population of interest should be used. In a repeated measures design, all subjects are exposed to all conditions. A number of eight or more is often considered to be an acceptable sample size.

The responses of the subjects to be collected are selected according to the objectives of the test. Typically, subjective responses are taken to quantify thermal comfort. The following subjective scales are provided in this standard: thermal sensation scale, uncomfortable scale, stickiness scale, and preference scale. Objective measures, such as mean skin temperature, are sometimes used to complement the subjective measures.

This evaluation method can be applied to both climatic chamber tests and field trials. If consistent conditions are required, conditions such as vehicle HVAC settings, starting conditions, time of day, direction of sunlight, and subject clothing should be controlled. This may only be possible in a climate chamber. Because control of experimental variables is difficult in field tests, identification of the influencing factors and control of factors, when possible, are important.

The tests should be carried out in accordance with Annex A. Annex A provides a specific test method based on the principles and guidelines described in the standard and an interpretation of the results. A standard questionnaire form for collecting subjective votes is also provided in Annex B.

10.5.4 ISO 14505-4:2021 Ergonomics of the Thermal Environment: Evaluation of Thermal Environments in Vehicles—Part 4: Determination of the Equivalent Temperature by Means of a Numerical Manikin

ISO 14505-4 [18] is one of the standards in the ISO 14505 series. This part of ISO 14505 provides guidelines for the calculation of the equivalent temperature, defined in ISO 14505-2, by computer simulation. As described in ISO 14505-2, the equivalent temperature assessment can be applied to vehicle cabins and other similar spaces with nonuniform environments. ISO 14505-4 is applicable to various situations, such as unprepared test facilities or prototype vehicles.

ISO 14505-4 provides two numerical methods for calculating the equivalent temperature, the numerical thermal manikin method, and the thermal factors method.

The numerical thermal manikin method reproduces the thermal manikin in three-dimensional (3D) simulations such as computational fluid dynamics (CFD) and calculates the equivalent temperature from the heat loss simulated by the numerical thermal manikin. This standard specifies the structure of the numerical thermal manikin, the calculation method of the heat exchange between the manikin and the environment through the clothing, and the calculation method of the equivalent temperature. For the control logic of the thermal manikin, ISO 14505-2 allows the use of three methods: constant temperature mode, constant heat flux mode, and comfort control mode. ISO 14505-4 provides calculation methods corresponding to the three control methods.

The numerical thermal manikin method requires a geometric model of a thermal manikin. The heat transfer model is shown in Fig. 10.4 and assumes that the inside of the numerical thermal manikin is adiabatic and that its skin surface is heated under the clothing. In the simulation, the surface of the manikin model is assumed to be the surface of the clothing, regardless of its shape. Once the skin surface temperature (t_{sk}) and the

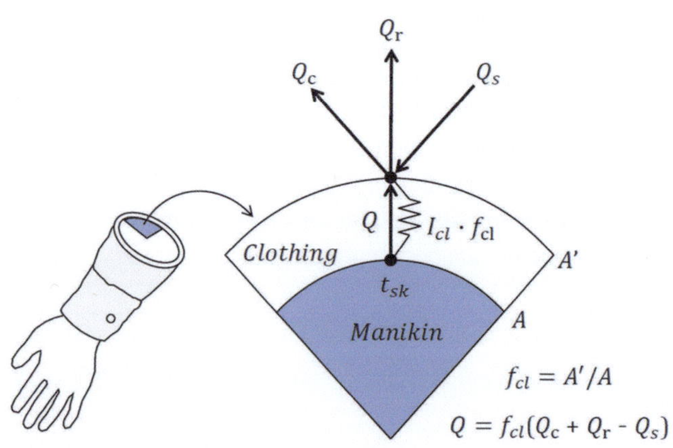

Fig. 10.4 Cross section of a part of the thermal manikin model and its heat transfer model

total heat loss from the skin surface (Q) have been determined in the simulation, the equivalent temperature (t_{eq}) in each body segment can be calculated using the following equation, as defined in ISO 14505-2:

$$t_{eq} = t_{sk} - Q / h_{cal}, \tag{10.14}$$

where

t_{eq}: equivalent temperature (°C)

t_{sk}: skin surface temperature (°C)

Q: total heat loss from skin surface (W/m²) and

h_{cal}: total heat transfer coefficient between skin surface and environment under the standard environment (W/m² °C)

Although h_{cal} can be measured using the thermal manikin, it can also be estimated using the following equation:

$$h_{cal} = \frac{1}{I_{cl} + \frac{1}{f_{cl} \cdot (h_{cs} + h_{rs})}}, \tag{10.15}$$

where

I_{cl}: thermal insulation of clothing (m² °C/W)

f_{cl}: area factor (N.D.), which can be estimated by $f_{cl} = 1.0 + 0.3\, I_{cl} / 0.155$

h_{cs}: convective heat transfer coefficient under the standard environment (W/m² °C) and

h_{rs}: radiant heat transfer coefficient under the standard environment (W/m² °C)

The following section explains the other calculation method, the thermal factors method. The thermal factors method uses air temperature, radiant temperature, air velocity, solar radiation, and clothing insulation. This standard gives equations to calculate the equivalent temperature from these thermal factors. Figure 10.5 shows the heat transfer model between a thermal manikin and the environment used in this method.

ISO 14505-4 provides calculation methods corresponding to the three control methods described before. For example, the equation for calculating the equivalent temperature from thermal factors in the constant temperature mode is expressed as follows:

$$t_{eq} = t_{skset} - \frac{t_{skset} - \frac{h_c \cdot t_a + h_r \cdot t_r + \alpha \cdot S}{h_c + h_r}}{I_{cl} + \frac{1}{f_{cl} \cdot (h_c + h_r)}} \cdot \left(I_{cl} + \frac{1}{f_{cl} \cdot (h_{cs} + h_{rs})} \right), \tag{10.16}$$

where

t_{skset}: set value of skin temperature at constant temperature mode (°C)

h_c: convective heat transfer coefficient (W/m² °C), calculated from air velocity

t_a: air temperature (°C)

h_r: radiant heat transfer coefficient (W/m² °C)

t_r: radiant temperature (°C)

α: solar absorptivity on clothing surface (skin surface on unclothed area) (N.D.) and

S: solar radiation reached on clothing surface (skin surface on unclothed area) (W/m²)

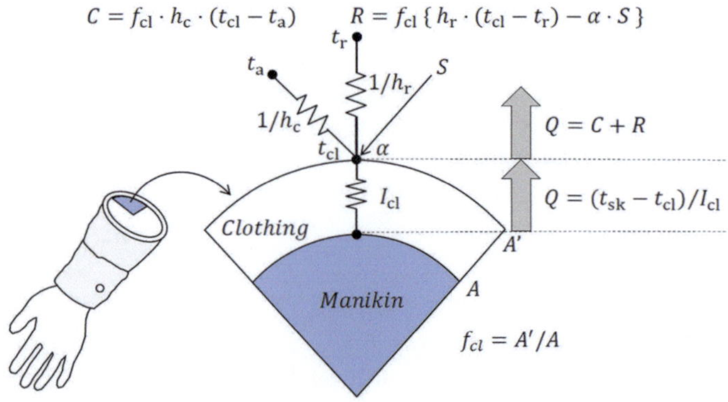

Fig. 10.5 Cross section of a part of the thermal manikin showing heat transfer between the manikin surface and the environment

10.5.5 ISO 15265:2004 Ergonomics of the Thermal Environment: Risk Assessment Strategy for the Prevention of Stress or Discomfort in Thermal Working Conditions

ISO 15265 [19] describes a strategy for assessing and interpreting the risk of physiological constraints or discomfort while working in a given climatic environment. It is applicable to any working situation including hot, moderate, and cold environments. This standard provides a strategy in three stages that can be used in sequence to gain a deeper insight into the working conditions.

10.5.5.1 Stage 1: Observation

Stage 1 should be carried out by people from the company with full knowledge of the working conditions but who are not necessarily trained in ergonomics. The aim is to characterize the working situation in all circumstances, throughout the day and throughout the year, rather than at a specific time.

At this stage, air temperature, humidity, thermal radiation, air movements, physical workload, clothing, and opinions of the workers are evaluated by scoring scales. The scales range from −3 to +3 points. A point of 0 indicates ideal conditions. If the point value is large (positive value), the risk of heat strain is increased. If the point value is small (negative value), the risk of cold strain is increased.

From this "Observation," the technical measures that can be directly implemented to prevent/control the risk are defined. If a more detailed "Analysis" is required to quantify and solve the problem, proceed to the next stage.

10.5.5.2 Stage 2: Analysis

Stage 2 is carried out by the same persons with the assistance of specialists with a specific training in ergonomics of the thermal environment. It deals with the working situation in particular circumstances (summer, night, etc.) identified during the first stage and requires common measurements.

The aims of this stage are to quantify the risk of thermal discomfort or constraint and to determine whether an "Expertise" (Stage 3) is required. This stage involves measuring or estimating the maximum/minimum values of the thermal factor and then calculating the following thermal indices: IREQ (ISO 11079); PMV; PPD (ISO 7730); WBGT (ISO 7243); Predicted Heat Strain (PHS); and DLE (ISO 7933).

From this "Analysis," the remaining risk after implementation of prevention/control measures is assessed, and the need for Stage 3 ("Expertise") is determined.

10.5.5.3 Stage 3: Expertise

Stage 3 is carried out by the same persons, with the additional assistance of highly specialized experts. It deals with highly complex thermal working circumstances and requires sophisticated or special measurements.

The aims at this stage are to better characterize some heat or cold sources and/or some thermal phenomena in the working environment by means of specific measurements and to characterize the overall exposure of the workers and look for special prevention/control measures to be implemented through a more refined analysis of the activities and the climatic parameters. In this stage, the risk per activity and global risk is assessed more precisely using the thermal indices described in Stage 2.

10.5.6 ISO 15743:2008 Ergonomics of the Thermal Environment: Cold Workplaces—Risk Assessment and Management

ISO 15743 [20] provides a strategy and practical tools for assessing and managing cold risks in the workplace. This standard is applicable to both indoor and outdoor work situations. Cold risk assessment in the workplace follows the principles presented in ISO 15265. This standard focuses on the cold workplace and provides the following information and tools: models and methods for cold risk assessment and management; a checklist for identifying cold-related problems at work; a model, method, and questionnaire intended for use by occupational healthcare professionals in identifying those individuals with symptoms that increase their cold sensitivity and, with the aid of such identification, offering optimal guidance and instructions for individual cold protection; guidelines on how to apply thermal standards and other validated scientific methods when assessing cold-related risks; and a practical example from cold work.

10.6 Supporting Standards

10.6.1 ISO 11399:1995 Ergonomics of the Thermal Environment: Principles and Application of Relevant International Standards

ISO 11399 [21] provides information to enable the correct, effective, and practical use of international standards relating to the ergonomics of the thermal environment.

For the assessment of hot environments, ISO 7243 provides a simple, rapid method based on the WBGT index. If a more detailed analysis is required, the ISO 7933 method, based on the human thermal balance, is applicable. In addition, if monitoring of individuals is required, a physiological measurement based on ISO 9886 should be used.

For the assessment of moderate environments, ISO 7730 provides methods for calculating the PMV and PPD. These indices can be used to estimate the average thermal sensation and the percentage of people who are thermally dissatisfied. If individual responses are required, ISO 10551, subjective judgement scales, can be used to measure them.

For the assessment of cold environments, ISO 11079 can be used. It provides the IREQ. Once the appropriate clothing insulation for a cold environment has been estimated by IREQ, IOS 9920, estimation of thermal insulation of a clothing ensemble can be applied. In addition, ISO 9986 provides guidance on physiological response, and ISO 10551 provides guidance on subjective measurement.

This standard was published in 1995. Since then, several international standards have been published or revised in the field of ergonomics of the thermal environment. Please note that the information in this standard may not always be up to date. Therefore, if detailed information is needed, it is recommended to refer to the latest ISO documents rather than to the explanations in this standard.

10.6.2 ISO 7726:1998 Ergonomics of the Thermal Environment: Instruments for Measuring Physical Quantities

ISO 7726 [22] provides minimum specifications for instruments used to measure thermal parameters such as air temperature, mean radiant temperature, humidity, and air velocity. Measurements of these basic parameters are very important for assessing the thermal environment.

This standard defines two classes (specifications) required for instruments. Class C (comfort) specifications relate to measurements made in moderate environments close to comfort conditions. Class S (thermal stress) specifications relate to measurements made in environments with greater thermal stress or extreme thermal stress.

This standard specifies the measurement ranges, accuracies, and 90% response times of the sensors for each thermal parameter. In class C (comfort), the measurement range can be narrower, but more accuracy is required. Class S (thermal stress) usually requires a wider range than class C, but a lower accuracy may be acceptable. Please note that the specifications given in this standard are minimum requirements. Depending on the purpose and needs, more precise specifications can be adopted. For example, very accurate thermal stress measurements for certain quantities may require the use of instruments with Class S ranges and Class C accuracy.

In addition, the definition of each thermal parameter, measurement principle, measurement method, and type of measuring instrument are explained in the Annex.

10.6.3 ISO 8996:2021 Ergonomics of the Thermal Environment: Determination of Metabolic Rate

ISO 8996 [23] specifies different methods for determining the metabolic rate in the context of the ergonomics of the thermal working environment. Metabolic rate is an important human factor in assessing the thermal environment together with clothing. It is also used to calculate PMV (ISO 7730) and IREQ (ISO 11079).

Methods for determining metabolic rate are classified into four levels of increasing accuracy: Level 1, Screening; Level 2, Observation; Level 3, Analysis; and Level 4, Expertise.

10.6.3.1 Level 1: Screening

Metabolic rate can be estimated approximately using the table given in Annex A. The table defines five classes of metabolic rate: resting, low, moderate, high, and very high. For each class, a range of metabolic rate values is given with examples. For example, the resting metabolic rate (sitting at ease) is 100–125 W. As this method provides only a rough estimate with considerable error, it should only be used for classification purposes.

10.6.3.2 Level 2: Observation

Metabolic rate is estimated by a time and motion study. Mean values or formulae for estimating the metabolic rates for some activities, including those at rest, with displacements, and without displacement, are given in Annex B. For example, the formula for walking (<6 km/h) is as follows:

$$M = \begin{pmatrix} 0.5 + 0.37 \cdot v_{\mathrm{w}} \\ + 0.2 \cdot v_{\mathrm{w}} \cdot G \end{pmatrix} \cdot \left(W_{\mathrm{b}} + L\right), \tag{10.17}$$

where
M: metabolic rate (W)
v_{w}: walking speed (km/h)
G: ground slope (%)
W_{b}: body mass (kg) and
L: load carried by the person (kg)

10.6.3.3 Level 3: Analysis

Metabolic rate can be estimated from heart rate or accelerometer measurements. The heart rate method is described in detail in this standard. In the case of purely dynamic work involving the major muscle groups, with no static muscular, thermal, or mental load, the metabolic rate can be estimated by measuring the heart rate during the work. Under these conditions, a linear relationship exists between metabolic rate and heart rate. Considering the above limitations, this method can be more accurate than the Level 1 and Level 2 methods and is considerably less complex than the Level 4 method. In this case, the relationship between heart rate and metabolic rate is given by the formula:

$$M = a + b \cdot HR, \tag{10.18}$$

where
M: metabolic rate (W)
HR: measured heart rate (beats/min) and
a and b: coefficients

The *HR-M* relationship can be determined by several methods. Details are given in clause 8 and Annex C.

10.6.3.4 Level 4: Expertise

Three methods of assessing the metabolic rate are explained by measuring the oxygen consumption rate, by the doubly labelled water method, and by direct calorimetry. The details of the first method and a calculation example are explained in clause 9 and Annex D.

While muscles can work without oxygen for short periods, oxidative metabolism is the main source of energy for longer periods of work. Therefore, the metabolic rate can be assessed by measuring the oxygen consumption rate. Metabolic rate can be calculated using the following formulae:

$$RQ = \dot{V}_{co2} / \dot{V}_{o2}, \qquad (10.19)$$

$$EE = (0.23 \cdot RQ + 0.77) \cdot 5.88, \text{and} \qquad (10.20)$$

$$M = EE \cdot \dot{V}_{o2}, \qquad (10.21)$$

where
RQ: respiratory quotient
V_{O2}: oxygen consumption rate (L O_2/h)
V_{CO2}: carbon dioxide production rate (L CO_2/h)
EE: energetic equivalent, in watt hours per litre of oxygen (Wh/L O_2) and
M: metabolic rate (W)

The *EE* of oxygen is used to convert the oxygen consumption rate to the metabolic rate. The *EE* depends on the type of metabolism, which is indicated by the *RQ*. When assessing the metabolic rate, it is often sufficient to use an average *RQ* of 0.85 and an *EE* of 5.68. In this case, measurement of the carbon dioxide production rate is unnecessary.

10.6.4 ISO 9920:2007 Ergonomics of the Thermal Environment: Estimation of Thermal Insulation and Water Vapor Resistance of a Clothing Ensemble

ISO 9920 [24] specifies methods for estimating thermal parameters such as thermal resistance to dry heat loss and evaporative heat loss for a clothing ensemble. These clothing parameters, together with metabolic rate, are important human factors for assessing the thermal environment. The clothing parameters estimated by this standard can be used for calculation of PMV (ISO 7730) and PHS model (ISO 7933) and for assessment of IREQ (ISO 11079).

The following parameters and their relationships are defined in this standard as representative fundamental parameters for clothing (Fig. 10.6).

Total Insulation I_T (m^2 K/W)

Thermal insulation from the body surface to the environment (including all clothing, enclosed air layers and boundary air layer) under reference conditions (static).

Basic Insulation (Intrinsic Insulation) I_{cl} (m^2 K/W or clo)

Thermal insulation from the skin surface to the outer clothing surface (including enclosed air layers) under reference conditions (static). The unit "clo" is often used for basic insulation, as is the SI unit "m^2 K/W." These units can be converted to each other. 1 clo = 0.155 m^2 K/W.

Air Insulation I_a (m^2 K/W)

Thermal insulation of the boundary (surface) air layer around the outer clothing or, when nude, around the skin surface.

Total Water Vapor Resistance $R_{e,T}$ (m^2 kPa/W)

Vapor resistance from the body surface to the environment (including all clothing, enclosed air layers, and boundary air layers) under reference conditions (static).

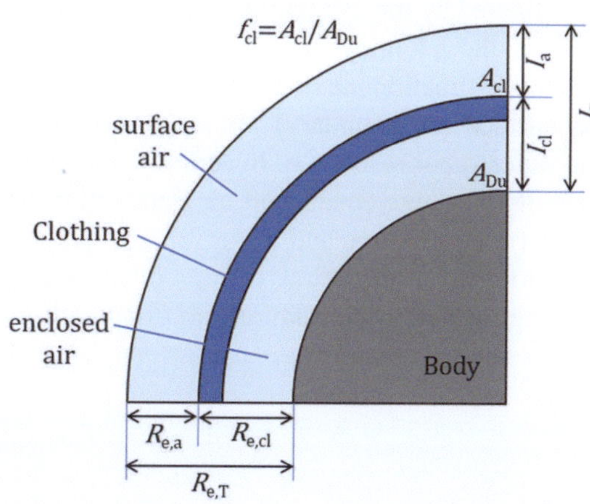

Fig. 10.6 Fundamental parameters for clothing

Basic Water Vapor Resistance $R_{e,cl}$ (m^2 kPa/W)
Vapor resistance from the body surface to the outer clothing surface (including enclosed air layers) under reference conditions (static).

Air Water Vapor Resistance $R_{e,a}$ (m^2 kPa/W)
Vapor resistance of the boundary (surface) air layer around the outer clothing or when nude, around the skin surface.

Clothing Area Factor f_{cl} (−)
Ratio of the outer surface area of the clothed body (A_{cl}) to the surface area of the nude body (A_{Du}).

The main part of this standard is the large database of clothing insulation measured by thermal manikins. In the Annex, clothing insulation values are listed for a selection of clothing ensembles. Instead of using the ensembles list, the clothing insulation for an ensemble may also be estimated based on a summation of the insulation of individual garments using the following empirical equation:

$$I_{cl} = 0.161 + 0.835 \sum I_{clu}, \qquad (10.22)$$

where
I_{clu}: effective thermal insulation of the individual garments making up the ensemble (clo).

The I_{clu} values of the individual garments are listed in the Annex. For example, the basic insulation of an ensemble consisting of an underwear shirt (I_{clu} = 0.06), long-sleeved shirt (I_{clu} = 0.33), briefs (I_{clu} = 0.04), and work pants (I_{clu} = 0.24) is calculated as 0.72 clo.

Followed by the estimation of thermal insulation of the clothing ensemble, this standard explains estimation methods of clothing area factor, surface (or boundary) air layer insulation, and water vapor resistance. In addition, the influence of body movement and air penetration on the thermal insulation and water vapor resistance of a clothing ensemble, as well as other factors influencing clothing insulation, are discussed.

10.6.5 ISO 10551:2019 Ergonomics of the Physical Environment: Subjective Judgment Scales for Assessing Physical Environments

ISO 10551 [25] provides principles and examples of subjective scales for assessing physical environments, including thermal, visual, air quality, acoustic, and vibration environments. This standard does not intend to standardize specific scales. However, some scales, based on the principles, are presented in the Annex as examples. For thermal environments, thermal sensation, thermal comfort, thermal preference, personal acceptability, and personal tolerance scales are presented. Table 10.8 shows thermal sensation and thermal comfort scales. Although this standard does not present principles of surveys or a questionnaire, design, scales developed using this standard can be used in surveys or questionnaires.

10.6.6 ISO 12894:2001 Ergonomics of the Thermal Environment: Medical Supervision of Individuals Exposed to Extreme Hot or Cold Environments

ISO 12894 [26] provides advice on the safety of human exposure to extreme hot or cold thermal environments. Definition of such extreme conditions is difficult, as they depend on clothing and activity, as well as the parameters of the climatic environment. However, as a guide, they can be considered as follows: for hot environments, WBGT of ≥25 °C; for cold environments, air temperature of ≤0 °C. This guidance is applicable to laboratory and field studies in extreme environments.

A change of body core temperature can be used as an index of heat storage. When core tem-

Table 10.8 Examples of thermal sensation and thermal comfort scales

Degrees	9-degree thermal sensation scale	7-degree thermal sensation scale	5-degree thermal comfort scale
+5	Extremely hot		
+4	Very hot		Extremely uncomfortable
+3	Hot	Hot	Very uncomfortable
+2	Warm	Warm	Uncomfortable
+1	Slightly warm	Slightly warm	Slightly uncomfortable
0	Neutral	Neutral	Comfortable
−1	Slightly cool	Slightly cool	
−2	Cool	Cool	
−3	Cold	Cold	
−4	Very cold		
−5	Extremely cold		

perature is maintained between 36.0 and 38.0 °C, serious general health effects, due to changes in body heat storage, are unlikely. The desirable aim for exposures in hot or cold conditions is to arrange that core temperature does not exceed a value in this range. However, in laboratory experiments, more severe environments can sometimes be accepted with appropriate measures. Where core temperature is expected to exceed the range, core temperatures are typically monitored. Core temperature is defined in ISO 13731 and described in ISO 9886. If experiments are planned in which the core temperature is expected to be >38.5 or <35.0 °C, a medical officer or other trained personnel, such as an emergency medical technician/paramedic, should present in the laboratory and assist with first aid, if required. The detailed guidance and recommendations on the following information can be found in the Annexes in this standard.

- General principles relating to the ethics of experimentation on human subjects: Example of a consent form for a volunteer subject in an ergonomic investigation involving exposure to hot or cold is provided in Annex A.
- Medical effects of exposure to hot and cold environments: The main illnesses which can arise from changes in body heat storage in hot or cold environments is provided in Annex B.
- Medical supervision of individuals exposed to hot or cold environments in the laboratory or field studies: The details of the recommendations (a minimum level of medical supervision) and the pre-exposure questionnaire to assess medical fitness are given in Annex C-F.

10.6.7 ISO 13731:2001 Ergonomics of the Thermal Environment: Vocabulary and Symbols

ISO 13731 [27] provides terms, definitions, symbols, and units of quantity used in the standards published by ISO in the field of ergonomics of the thermal environment. The purpose of this standard is to provide terms and symbols for the quantities used in those standards and to provide a reference of terms and symbols to be used in the preparation of future International Standards or other publications in the field of ergonomics of the thermal environment.

Approximately 120 terms used in published standards are listed in clause 2, together with their definitions. The definitions are mainly based on the "Glossary of terms for thermal physiology" [28]. Clause 3 provides approximately 170 symbols and units associated with the terms listed in clause 2 and other parameters related to the ergonomics of the thermal environment. The units used in this International Standard are based on the International System of Units (SI). However, where a unit outside the SI is commonly used for practical purposes, both units are given (e.g., clo and m^2K/W for clothing insulation and met and W/m^2 for metabolic rate).

References

1. https://www.iso.org/about-us.html. Browsed on 26 Feb 2025.
2. https://www.iso.org/technical-committees.html. Browsed on 26 Feb 2025.
3. https://www.iso.org/store.html. Browsed on 26 Feb 2025.
4. ISO/ICE Directives, Part 1. Consolidated ISO Supplement—procedures for the technical work. 2022 International Organization for Standardization, Geneva.
5. ISO 7243:2017, Ergonomics of the thermal environment—assessment of heat stress using the WBGT (wet bulb globe temperature) index. 2017 International Organization for Standardization, Geneva.
6. ISO 7933:2023, Ergonomics of the thermal environment—analytical determination and interpretation of heat stress using calculation of the predicted heat strain. 2023 International Organization for Standardization, Geneva.
7. ISO 9886:2004, Ergonomics—evaluation of thermal strain by physiological measurements. 2004 International Organization for Standardization, Geneva.
8. ISO 13732-1:2006, Ergonomics of the thermal environment—methods for the assessment of human responses to contact with surfaces—Part 1: Hot surfaces. 2006 International Organization for Standardization, Geneva.
9. ACGIH-TLV 2022, TLVs and BEIs: based on the documentation of the threshold limit values for chemical substances and physical agents & biological exposure indices. In: 2022 American Conference of Governmental Industrial Hygienist, Cincinnati.
10. ISO 7730:2005, Ergonomics of the thermal environment—analytical determination and interpretation of thermal comfort using calculation of the PMV and PPD indices and local thermal comfort criteria. 2005 International Organization for Standardization, Geneva.
11. ISO/TS 13732-2:2001, Ergonomics of the thermal environment—methods for the assessment of human responses to contact with surfaces—Part 2: Human contact with surfaces at moderate temperature. 2001 International Organization for Standardization, Geneva.
12. ASHRAE Standard 55-2020, Thermal environmental conditions for human occupancy. 2020 American National Standard published by the American Society of Heating, Refrigerating and Air-Conditioning Engineers, Norcross.
13. ISO 11079:2007, Ergonomics of the thermal environment—determination and interpretation of cold stress when using required clothing insulation (IREQ) and local cooling effects. 2007 International Organization for Standardization, Geneva.
14. ISO 13732-3:2005, Ergonomics of the thermal environment—methods for the assessment of human responses to contact with surfaces—Part 3: Cold surfaces. 2005 International Organization for Standardization, Geneva.
15. ISO/TS 14505-1:2007, Ergonomics of the thermal environment—evaluation of thermal environments in vehicles—Part 1: Principles and methods for assessment of thermal stress. 2007 International Organization for Standardization, Geneva.
16. ISO 14505-2:2006, Ergonomics of the thermal environment—evaluation of thermal environments in vehicles—Part 2: Determination of equivalent temperature. 2006 International Organization for Standardization, Geneva.
17. ISO 14505-3:2006, Ergonomics of the thermal environment—evaluation of thermal environments in vehicles—Part 3: Evaluation of thermal comfort using human subjects. 2006 International Organization for Standardization, Geneva.
18. ISO 14505-4:2021, Ergonomics of the thermal environment—evaluation of thermal environments in vehicles—Part 4: Determination of the equivalent temperature by means of a numerical manikin. 2021 International Organization for Standardization, Geneva.
19. ISO 15265:2004, Ergonomics of the thermal environment—risk assessment strategy for the prevention of stress or discomfort in thermal working conditions. 2004 International Organization for Standardization, Geneva.
20. ISO 15743:2008, Ergonomics of the thermal environment—cold workplaces—risk assessment and management. 2008 International Organization for Standardization, Geneva.
21. ISO 11399:1995, Ergonomics of the thermal environment—principles and application of relevant International Standards. 1995 International Organization for Standardization, Geneva.
22. ISO 7726:1998, Ergonomics of the thermal environment—instruments for measuring physical quantities. 1998 International Organization for Standardization, Geneva.

23. ISO 8996:2021, Ergonomics of the thermal environment—determination of metabolic rate. 2021 International Organization for Standardization, Geneva.
24. ISO 9920:2007, Ergonomics of the thermal environment—estimation of thermal insulation and water vapour resistance of a clothing ensemble. 2007 International Organization for Standardization, Geneva.
25. ISO 10551:2019, Ergonomics of the physical environment—subjective judgement scales for assessing physical environments. 2019 International Organization for Standardization, Geneva.
26. ISO 12894:2001, Ergonomics of the thermal environment—medical supervision of individuals exposed to extreme hot or cold environments. 2001 International Organization for Standardization, Geneva.
27. ISO 13731:2001, Ergonomics of the thermal environment—vocabulary and symbols. 2001 International Organization for Standardization, Geneva.
28. The Commission for Thermal Physiology of the International Union of Physiological Sciences (IUPS Thermal Commission). Glossary of terms for thermal physiology. Pflugers Arch. 1987;410:567–87. https://doi.org/10.1007/BF00586542.

Index

Y. Tochihara (ed.), *The Thermal Environment*, https://doi.org/10.1007/978-981-96-3009-7